Lecture Notes in Mathematics

2102

For further volumes:
http://www.springer.com/series/304

Peter E. Kloeden • Christian Pötzsche
Editors

Nonautonomous Dynamical Systems in the Life Sciences

Editors
Peter E. Kloeden
Institut für Mathematik
Goethe-Universität Frankfurt
Frankfurt am Main, Germany

Christian Pötzsche
Institut für Mathematik
Alpen-Adria Universität Klagenfurt
Klagenfurt am Wörthersee, Austria

ISBN 978-3-319-03079-1 ISBN 978-3-319-03080-7 (eBook)
DOI 10.1007/978-3-319-03080-7
Springer Cham Heidelberg New York Dordrecht London

Lecture Notes in Mathematics ISSN print edition: 0075-8434
ISSN electronic edition: 1617-9692

Library of Congress Control Number: 2013956227

Mathematics Subject Classification (2010): 37B55, 92XX, 34C23, 34C45, 37HXX

Printed on acid-free paper

Springer is part of Springer Science+Business Media (www.springer.com)

Preface

The theory of dynamical systems is a well-developed and successful mathematical framework to describe time-varying phenomena. Its applications in the life sciences range from simple predator–prey models to complicated signal transduction pathways in biological cells, in physics from the motion of a pendulum to complex climate models, and beyond that to further fields as diverse as chemistry (reaction kinetics), economics, engineering, sociology, demography, and biosciences. Indeed, Systems Biology relies heavily on methods from Dynamical Systems. Moreover, these diverse applications have provided a significant impact on the theory of dynamical systems itself and is one of the main reasons for its popularity over the last decades.

As a general principle, before abstract mathematical tools can be applied to real-world phenomena from the above areas, one needs corresponding models in terms of some kind of evolutionary difference or differential equation. Their goal is to provide a realistic and tractable picture for the actual behavior of, e.g., a biodynamical system. A thorough understanding helps to optimize time-consuming and costly experiments, like the development of harvesting or dosing strategies and might even enable field studies to be avoided.

From a conceptional level, in developing such models one distinguishes an actual dynamical *system* from its surrounding *environment*. The system is given in terms of physical or internal feedback laws that yield an evolutionary equation. The parameters in this equation describe the current state of the environment. The latter may or may not vary in time, but is assumed to be unaffected by the system.

For *autonomous dynamical systems* the basic law of evolution is static in the sense that the environment does not change with time. However, in many applications such a static approach is too restrictive and a temporally fluctuating environment is required:

- Parameters in real-world situations and particularly in the life sciences are rarely constant over time. This has various reasons, like absence of lab conditions, adaption processes, seasonal effects on different time scales, changes in nutrient supply, or an intrinsic "background noise."

- On the other hand, sometimes it is desirable to include regulation or control strategies into a model (e.g., harvesting and fishing, dosing of drugs or radiation, stimulating chemicals or catalytic submissions), as well as extrinsic noise, and to study their effects. In particular, biochemical signaling within and into cells is a nonautonomous process.
- Several problems can be decomposed into coupled subsystems. Provided the influence of some of them is understood, without the requirement to precisely know their explicit form, they can be seen as a non-constant time-varying input.

These temporal fluctuations might be deterministic or random and in the first case often more than just periodic in time. The evidence of time-dependent parameters can also be verified statistically, when it comes to the problem of fitting parameters to actual measured data or in phenomena like cardiovascular ageing.

Consequently, in reasonable models that are adapted to and well suited for problems in temporally fluctuating environments, the resulting evolutionary equations have to depend explicitly on time. In order to study such realistic problems, the classical theory of dynamical systems has to be extended. The field of *nonautonomous and random dynamical systems* has thus received a wide attraction over the recent 10–15 years and is expected to develop to further maturity. Both the fields of nonautonomous and of random dynamical systems are parallel theories featuring very similar concepts.

In the area of biomathematics, for example, the corresponding contributions deal with nonautonomous equations and have provided major progress in our understanding of classical boundedness, global stability, persistence, permanence, or positivity aspects. Nevertheless, often more subtle questions are crucial. For instance, it is of utmost importance to identify "key players" in biodynamical processes, i.e., the variables or parameters crucially affecting the long-term behavior of a system. Knowing these quantities enables researchers to reduce the dimension of a system significantly and thus makes it amenable for analytical tools, as opposed to sometimes problematic numerical methods and simulations. Such questions clearly fit into the framework of *bifurcation theory* describing qualitative changes. However, when dealing with nonautonomous and random equations new ideas and concepts are required: For instance, the classical notions of invariance, attraction, hyperbolicity, and invariant manifolds have had to be extended.

Despite being well motivated, one rarely finds biological processes modeled using nonautonomous equations. A cause seems to be the problem that classical methods from autonomous dynamical systems theory do not apply to them, while more recent tools tailor-made for time-dependent problems still need to be popularized.

For these reasons, the contemporary fields of nonautonomous and random dynamical systems on the one side, and biodynamics on the other side, strongly benefit from each other. Actually, it is essential to

- illustrate such a modern theory using successful and convincing real-world applications,

- gain input from real-life application inducing a further development of the theory into directions of a broader interest.

On the other hand, researchers interested in mathematical approaches to life sciences will

- find suitable mathematical methods promising and fruitful in various applications,
- obtain a solid toolbox for an understanding and iterative refinement of models in fluctuating environments,
- lower an inhibition threshold to use nonautonomous models from the beginning.

In conclusion, the main motivation for this book is to bring readers' attention to various recent developments and methods from the field of nonautonomous and random dynamical systems and promising applications originating in life sciences. For this reason we collected three articles (Chaps. 1–3) illustrating theoretical aspects, as well as six further papers (Chaps. 4–8) focussing on more concrete applications, where these new ideas and tools could be used:

1. The introductory contribution of the editors shares the title with this book and surveys several key concepts from the mathematical theory of deterministic nonautonomous dynamical systems. They are exemplified using various simple models from the life sciences.
2. The chapter on *Random Dynamical Systems with Inputs* by Michael Marcondes de Freitas and Eduardo D. Sontag describes the concept of a random dynamical system and extends it to problems with inputs and outputs. Applications to feedback connections are given.
3. Multiple time scales are an important feature of physiological systems and functions. Martin Wechselberger, John Mitry, and John Rinzel give a modern introduction to the basic geometrical singular perturbation theory and use it in their chapter *Canard Theory and Excitability* to tackle related problems and their transient behavior.
4. *Stimulus-Response Reliability of Biological Networks* by Kevin K. Lin reviews some basic concepts and results from the ergodic theory of random dynamical systems and explains how these ideas can be used (partly in combination with numerical simulations) to study the reliability of networks, i.e., the reproducibility of a network's response when repeatedly presented with a given stimulus.
5. The "Lancaster group" Philip Clemson, Spase Petkoski, Tomislav Stankovski, and Aneta Stefanovska explain how networks of nonautonomous self-sustained oscillators can model a virtual physiological human. Their chapter *Coupled Nonautonomous Oscillators* includes novel methods suitable to reconstruct nonautonomous dynamics using data from a real living system by studying time-dependent coupling between cardiac and respiratory rhythms.
6. Germán Enciso's contribution *Multisite Mechanisms for Ultrasensitivity in Signal Transduction* gives a mathematical review of the most important molecular models featuring ultrasensitive behavior.

7. The chapter *Mathematical Concepts in Pharmacokinetics and Pharmacodynamics with Application to Tumor Growth* by Gilbert Koch and Johannes Schropp describes corresponding models and their applications in the pharmaceutical industry. Moreover, a model for tumor growth and anticancer effects is developed and discussed.
8. In *Viral Kinetic Modeling of Chronic Hepatitis C and B Infection*, Eva Herrmann and Yusuke Asai demonstrate the interplay between mathematical and statistical analysis of compartment ODE models for hepatitis B and C. They give an account of clinical use of models in treatment. Moreover, the most relevant models for such infections are surveyed.
9. Finally, Christina and Nicolae Surulescu study *Some Classes of Stochastic Differential Equations as an Alternative Modeling Approach to Biomedical Problems*. In detail, models for an intracellular signaling pathway, a radio-oncological treatment, and cell dispersal are presented and studied.

Finally, we cordially thank all the contributors to this volume and hope to have contributed to building a bridge between nonautonomous/random dynamics and the life sciences.

Frankfurt am Main, Germany
Klagenfurt, Austria
September 2013

Peter E. Kloeden
Christian Pötzsche

Acknowledgments

The idea of this book arose during discussions at a workshop

Nonautonomous and Random Dynamical Systems in Life Sciences

held in Inzell, Germany from August 1 to 5, 2011 and organized by the editors in collaboration with Prof. Dr. Rupert Lasser (Helmholtz Centre Munich and Munich University of Technology). The workshop was funded by the Volkswagen Stiftung and the Institute for Biomathematics and Biometrics at the Helmholtz Centre Munich. We are grateful for their support. We also thank Prof. Dr. R. Lasser for his assistance in organizing the workshop.

We are grateful to the anonymous referees—their suggestions and constructive criticism led to a more coherent and accessible presentation.

Contents

Part I Theoretical Basics

List of Contributors

Yusuke Asai Department of Medicine, Institute of Biostatistics and Mathematical Modeling, Goethe University Frankfurt, Deutschland

Philip Clemson Physics Department, Lancaster University, Lancaster, UK

Germán A. Enciso Department of Mathematics, University of California, Irvine, Irvine, CA, USA

Eva Herrmann Department of Medicine, Institute of Biostatistics and Mathematical Modeling, Goethe University Frankfurt, Deutschland

Peter E. Kloeden Institut für Mathematik, Goethe-Universität Frankfurt, Postfach, Frankfurt am Main, Deutschland

Gilbert Koch Department of Pharmaceutical Sciences, School of Pharmacy and Pharmaceutical Sciences, State University of New York at Buffalo, Buffalo, NY, USA

Kevin K. Lin Department of Mathematics, University of Arizona, Tucson, AZ, USA

Michael Marcondes de Freitas Department of Mathematics, Rutgers University, Piscataway, NJ, USA

John Mitry School of Mathematics and Statistics, University of Sydney, NSW, Australia

Spase Petkoski Physics Department, Lancaster University, Lancaster, UK

Christian Pötzsche Institut für Mathematik, Alpen-Adria Universität Klagenfurt, Universitätsstr., Klagenfurt am Wörthersee, Österreich

John Rinzel Courant Institute, New York University, New York, NY, USA

Johannes Schropp FB Mathematik und Statistik, Universität Konstanz, Postfach, Konstanz, Deutschland

Eduardo D. Sontag Department of Mathematics, Rutgers University, Piscataway, NJ, USA

Tomislav Stankovski Physics Department, Lancaster University, Lancaster, UK

Aneta Stefanovska Physics Department, Lancaster University, Lancaster, UK

Christina Surulescu Felix Klein Zentrum für Mathematik, Universität Kaiserslautern, Kaiserslautern, Deutschland

Nicolae Surulescu Mathematisches Institut, Universität Münster, Münster, Deutschland

Martin Wechselberger School of Mathematics and Statistics, University of Sydney, NSW, Australia

Part I
Theoretical Basics

Chapter 1
Nonautonomous Dynamical Systems in the Life Sciences

Peter E. Kloeden and Christian Pötzsche

Abstract Nonautonomous dynamics describes the qualitative behavior of evolutionary differential and difference equations, whose right-hand side is explicitly time-dependent. Over recent years, the theory of such systems has developed into a highly active field related to, yet recognizably distinct from that of classical autonomous dynamical systems. This development was motivated by problems of applied mathematics, in particular in the life sciences where genuinely nonautonomous systems abound.

In this survey, we introduce basic concepts and tools for appropriate nonautonomous dynamical systems and apply them to various representative biological models.

Keywords Nonautonomous dynamical system • Exponential dichotomy • Dichotomy spectrum • Bohl exponent • Pullback attractor • Bifurcation • Integral manifold • Life sciences

1.1 Motivation

The theory of dynamical systems is a well-developed and successful mathematical framework to describe time-varying phenomena in various applied sciences, especially in mathematical and theoretical biology. Its areas of applications range

P.E. Kloeden (✉)
Institut für Mathematik, Goethe-Universität Frankfurt, Postfach 11 19 32, 60054 Frankfurt am Main, Deutschland
e-mail: kloeden@math.uni-frankfurt.de

C. Pötzsche (✉)
Institut für Mathematik, Alpen-Adria Universität Klagenfurt, Universitätsstraße 65–67, 9020 Klagenfurt am Wörthersee, Österreich
e-mail: christian.poetzsche@aau.at

P.E. Kloeden and C. Pötzsche (eds.), *Nonautonomous Dynamical Systems in the Life Sciences*, Lecture Notes in Mathematics 2102, DOI 10.1007/978-3-319-03080-7_1,

from simple predator-prey models to complicated signal transduction pathways in biological cells, from epidemiological models to tumor growth and beyond that to further fields as pattern formation or wound healing, see e.g., [39, 62]. In particular, this broad range of its applications has provided a significant impact on the development of the theory of dynamical systems itself and is one of the main reasons for its popularity over recent decades.

As a general principle, appropriate models have to be developed before abstract mathematical tools can be applied to real-world phenomena in the above areas. From a conceptional level, in developing such models, one must distinguish the actual dynamical *system* from its surrounding *environment*. The system is given in terms of physical or internal feedback laws involving evolutionary equations, which are typically difference, ordinary differential or delay equations or an integro-difference or reaction diffusion equation when spatial effects are relevant. The parameters in this equations describe the current state of the environment that may or may not be variable in time, but is assumed to be unaffected by the system.

For *autonomous dynamical systems* the basic law of evolution is static in the sense that the environment does not change with time. In many applications, however, such a static approach is too restrictive and a temporally fluctuating environment must be taken into account.

- Parameters in real-world situations are rarely constant over time. This has various reasons, like absence of lab conditions, adaption processes, seasonal effects on different time scales, changes in nutrient supply, or an intrinsic "background noise".
- Sometimes it is desirable to include regulation or control strategies into a model (e.g. harvesting, dosing of drugs or radiation, stimulating chemicals or catalytic submissions) and to study their influence.

Evidence of time-dependent parameters can often be verified statistically, especially when parameters have to be determined from actual measured data (cf., e.g., [1, 7]).

Consequently, in reasonable models adapted to and well-studied for problems in temporally fluctuating environments, the evolutionary equations have to depend explicitly on time through time-dependent parameters or external inputs. Then the classical theory of dynamical systems is no longer applicable and has to be extended. Alone in the area of biomathematics, there has been significant progress in treating boundedness, global stability, persistence, permanence or positivity issues in nonautonomous systems [11, 28, 30, 55, 59, 87, 88, 91–93]. Although well-motivated, one rarely finds biological processes being modeled directly in terms of nonautonomous equations. The reason for this seems to be the problem that classical methods from autonomous dynamical systems theory do not apply to them. There are, of course, many papers in the literature with nonautonomous modifications of existing models, but their analysis has been somewhat ad hoc and problem specific.

In recent years the theory of *nonautonomous dynamical systems* has undergone extensive development and is providing tools, concepts and results to describe the behavior of nonautonomous systems in a more systematic way. The main motivation for this survey and tutorial is to show how these developments in the theory of nonautonomous dynamical systems can be applied to problems in biomathematics and the life sciences. Here, we restrict to analytical methods for continuous time dynamical systems. It is not our goal to develop, motivate and explain the various models presented here, nor do we intend to be exhaustive in the topics chosen. Instead we take representative models from the literature and illustrate how new ideas about nonautonomous dynamical systems might be of use in understanding them.

We point out that biological applications have a stimulating and symbiotic influence on related mathematical fields like control theory (see, e.g., [61, 85, 86]). This chapter restricts to deterministic problems since they, compared to a stochastic approach, have the advantage that their behavior is easier to interpret, in particular for non-mathematicians (cf. [89]). On the other hand, there are many publications of biomathematical models with noise in one form or another, involving either random (see, e.g., Chap. 2 by de Freitas and Sontag in this volume) or stochastic differential equations (see Chap. 9 by Surulescu and Surulescu in this volume). These are also intrinsically nonautonomous and can be formulated as random dynamical systems with analogous concepts of nonautonomous attractors, we refer to e.g., [14, 49, 50]. The latter possess a measure theoretic skew-product structure with the noise as a driving system.

We organize this introductory chapter as follows:

- Our initial Sect. 1.2 consists of a biased list of problems from the life science, where nonautonomous models are well-motivated.
- In Sect. 1.3 we present the basic ingredients and geometrical intuition behind our nonautonomous theory for parameter-dependent ordinary differential equations (ODEs for short). This includes an appropriate invariance notion and the helpful concept of the equation of perturbed motion.
- Due to its robustness properties, we advocate uniform asymptotic stability as the appropriate stability notion in a nonautonomous context. The related hyperbolicity concept is given in terms of an exponential dichotomy. Thus, when dealing with time-varying equations it turns out that eigenvalue real parts become spectral intervals (in the sense of Sacker and Sell), whose location w.r.t. 0 indicates stability or hyperbolicity.
- Equilibria of autonomous equations generically persist as entire bounded solutions under time-varying perturbations and we describe a procedure how to approximate them in terms of a Taylor series in the parameters.
- Section 1.6 illustrates that nonautonomous attractors are whole families of sets rather than single sets and discusses two convergence concepts, namely forward and pullback convergence.
- Nonautonomous bifurcation theory is briefly sketched in Sect. 1.7, basically using two examples.
- The theoretical part of this survey is concluded by explanations on nonautonomous invariant manifolds, so-called integral manifolds. We explain how to obtain

a Taylor approximation of them. This enables a corresponding center manifold reduction and applications in bifurcation theory.

- The final Sect. 1.9 briefly indicates parallels between the theory of random dynamical systems, control systems and nonautonomous systems described by means of skew-product flows—a concept of fundamental theoretical importance.

Throughout, we illuminate our results and techniques using corresponding continuous time nonautonomous models from the life sciences. Furthermore, a survey of related results in discrete time can be found in [54] or, with a focus on bifurcation theory, in [71].

1.2 Examples of Nonautonomous Models from Life Sciences

There exists a large number of models describing biological phenomena, where an aperiodic temporally fluctuating, and deterministic environment is well-motivated. The models presented here have been chosen to give a hint at the wide range of applications and to illustrate different aspects of nonautonomous behavior. We use a combined numerical and analytical framework to set up a local bifurcation analysis. Indeed, the above methods shall be exemplified using such applications w.r.t. the following aspects:

- Identification of the bounded entire solutions and their hyperbolicity properties.
- Stability and bifurcation analysis to detect sensitive parameters. Under which parameter changes is the behavior robust or leads to significant qualitative changes?
- Suggest and theoretically verify immunization, dosing and treatment strategies.

The models presented are low-dimensional and might be considered as caricatures. Nevertheless, we think a thorough application of methods from Sect. 1.3 etc. to such models is challenging and will provide an insight into more complex phenomena.

1.2.1 Bacterial Growth

In [8] the growth of bacterial cultures is modeled using the ODE

$$\dot{x} = \alpha(t)\mu(x)x, \tag{1.1}$$

where x denotes the cell concentration (i.e. the number of individual bacteria per unit). The function μ describes the specific growth rate and is assumed to continuously differentiable. A lag phase in form of an adjustment period is given using the time-dependent function α.

1.2.2 Epidemiology

The realistic assumption of a time-varying total population size N, sometimes given as solution of an independent ODE (cf., e.g., [55]), naturally leads to nonautonomous problems. In particular, as a model for the spread of infectious childhood diseases, [87, 88] study the following time-heterogeneous, i.e., nonautonomous *SIR model*

$$\begin{cases} \dot{S} = \mu(t)I + (\mu(t) + \xi(t))R - \alpha(t)SI, \\ \dot{I} = -\mu(t)I + \alpha(t)SI - \gamma(t)I, \\ \dot{R} = -\mu(t)R + \gamma(t)I - \xi(t)R, \end{cases}$$

where the constant population size $N = S + I + R$ splits into susceptible S, infective I and recovered R individuals. As motivation, for childhood diseases the time-heterogeneity in the per capita/capita infection rate $\alpha(t)$ is induced by the school system, because the chain of infections is interrupted or at least weakened by the vacations and new individuals are recruited into a scene with higher infection risk at the beginning of each school year (cf. also [87, 88] for references). A further stability and bifurcation analysis of nonautonomous SIRS models with, as well as without constant population size can be found in [48, 51].

Similar models with time-varying external forcing and hence a time-varying population were investigated by Kloedenm and Kozyakin [48] in the SI case, where explicit entire solutions are given, and in [15], where the system behaves chaotically. Corresponding four-dimensional SEIRS models were considered in [30, 88, 91], while [11] investigate a three-dimensional model leading to Tuberculosis elimination in the USA, which also incorporates the effect of HIV/AIDS after 1983.

SEIR models of microparasitic infections featuring time-periodic, hence nonautonomous nonlinearities were investigated in [78]. Similar contact rates were used in [24] in order to understand resonance phenomena in influenza epidemics.

Optimal control problems for a chemotherapy in the interaction of the immune system with the human immunodeficiency virus (HIV) have been studied in [40–43]. See also [60] for a linear nonautonomous system.

1.2.3 Tumor Drug Treatment Models

Logistic autonomous models for tumor growth have been analyzed in [36, 58], and an extension in form of

$$\begin{cases} \dot{x} = r_1 x \left(1 - \frac{x}{K_1} - c_1 y\right) - (p + d_0(t))x - d_1(t)y, \\ \dot{y} = r_2 y \left(1 - \frac{y}{K_2} - c_2 x\right) + (p + d_0(t))x - d_2(t)y \end{cases}$$

has been investigated in [28]. Roughly speaking, x and y describe populations of differentiated and undifferentiated tumor cells, respectively. The constant mutation rate p is influenced by the time-dependent induction rate $d_0(t)$ of a cytotoxic drug, and $d_1(t), d_2(t)$ denote death rates due to cytotoxicity.

The temporal change of a tumor mass V was considered in [29, 76] using

$$\begin{cases} \dot{V} = -\lambda_1 V \log\left(\frac{V}{K}\right), \\ \dot{K} = -\lambda_2 K + bS(V,K) - dI(VK) + eg(t)K, \end{cases}$$

where the carrying capacity K is also time-dependent and given by the coupled second equation. The concentration $g(t)$ represents the effect of a chemotherapeutic treatment. We refer to [21, 38] for further time-dependent tumor models as control problem. Further reference are [10], the tridiagonal model from [20] or the scalar tumor growth model from [76]. A survey of non-spatial models describing the interaction between cancer and the immune system is [25].

1.2.4 Pharmacodynamics

For a general introduction into pharmacodynamic models we refer to the Chap. 7 by Koch and Schropp in this book (in particular, see Sect. 7.3).

Moreover, models describing the effects of antibiotics dosing on a bacterial population whose growth is checked by nutrient-limitation and possibly host defenses, are studied in [34, 35]. They are of the form

$$\begin{cases} \dot{S} = D(S_0 - S) - \gamma f(S)u, \\ \dot{u} = (f(S) - D - g(S, a(t)))u, \end{cases}$$

where S denotes the concentration of a nutrient sustaining microbial growth, u is the density of bacteria and a the concentration of an antibiotic. Here, $a(t)$ is time-dependent and given as solution of a scalar ODE

$$\dot{a} = D(i(t) - a) - up(a)$$

and $i(t)$ is the antibiotic input and p a nondecreasing function with $p(0) = 0$.

References to various forms of the so-called *pharmacodynamic functions* f, g have been given in [34]. For related investigations we refer to [16].

1.2.5 Cardiovascular System

Oscillations are a very basic phenomenon in the life sciences ranging from biochemical reactions, though circadian rhythms to the respiratory and cardiovascular system [39, 62].

Chapter 5 in this book by Stefanovska and her coworkers investigates the intrinsically nonautonomous nature of the coupled respiratory and cardiovascular system, which they model as autonomous oscillators with time-dependent coupling. Much of their work involves the analysis of time-series of data obtained from real systems to identify such effects and to determine appropriate parameters in their models.

1.3 Nonautonomous Differential Equations

Here we briefly review of some basic concepts for nonautonomous systems. In order to determine the temporal horizon of interest, let us abbreviate $\mathbb{R}_+ := [0, \infty)$ and $\mathbb{R}_- := (-\infty, 0]$. For simplicity we restrict attention to ODEs

$$\boxed{\dot{x} = f_\lambda(t, x)} \qquad (\mathfrak{D}_\lambda)$$

in a d-dimensional Euclidean state space $\mathbb{R}^d$. We suppose $(\mathfrak{D}_\lambda)$ has a right-hand side $f_\lambda : \mathbb{R} \times \Omega \to \mathbb{R}^d$ such that $(x, \lambda) \mapsto f_\lambda(t, x)$ is of class C^m with continuous partial derivatives in (t, x, λ), although in some applications such as in control theory the time-dependence may only be measurable. Typically, the nonempty set $\Omega \subseteq \mathbb{R}^d$ is open or the nonnegative cone $\mathbb{R}^d_+$, while the parameter λ is assumed to be a real number or a vector in $\mathbb{R}^p$.

Since $(\mathfrak{D}_\lambda)$ is nonautonomous, the asymptotic behavior of its solutions will depend crucially on the initial time and not just on the elapsed time as in autonomous systems. Thus, we denote the solution of the ODE $(\mathfrak{D}_\lambda)$ with the initial value problem $x(t_0) = x_0$ by $\varphi_\lambda(t, t_0, x_0)$.

Such nonautonomous problems typically occur in two forms:

- investigate the *behavior near fixed reference solutions* ϕ^*_λ of $(\mathfrak{D}_\lambda)$, which may be periodic or almost periodic or even aperiodic such as a heteroclinic trajectory joining two steady state solutions; even in a purely autonomous setting, where f_λ is independent of the time variable, this yields a nonautonomous problem.
- replace constant parameters λ by time-dependent functions $\lambda(t)$, which can be an external stimulus or the solution of an independent equation with known behavior; then one speaks of *parametric perturbations*.

As a general principle, the appropriate geometrical setting to describe the dynamical behavior of $(\mathfrak{D}_\lambda)$ is the *extended state space* $\mathbb{R} \times \Omega$. Accordingly, invariant subspaces, manifolds or attractors will be subsets $\mathscr{A}$ of $\mathbb{R} \times \Omega$ rather than of Ω, or equivalently families of subsets $(\mathscr{A}(t))_{t \in \mathbb{R}}$ of Ω parametrized over the real numbers, which represents time. Here the sets $\mathscr{A}(t) \subseteq \Omega$ are called *fibers* of $\mathscr{A}$ and

$$\mathscr{A} = \bigcup_{t \in \mathbb{R}} \{t\} \times \mathscr{A}(t) = \{(t, x) \in \mathbb{R} \times \Omega : x \in \mathscr{A}(t)\}.$$

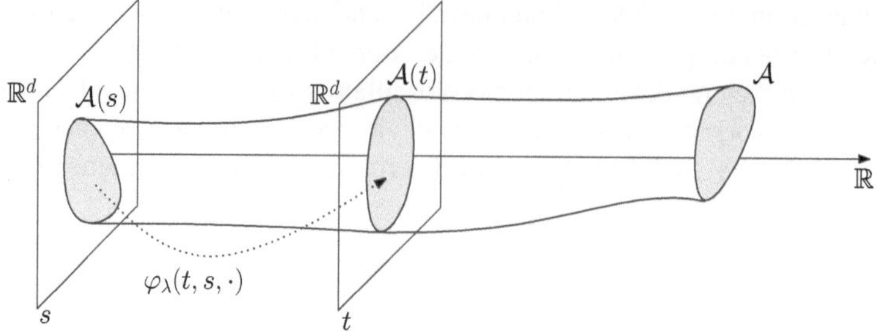

Fig. 1.1 Extended state space $\mathbb{R} \times \Omega$ and an invariant set $\mathscr{A} \subseteq \mathbb{R} \times \Omega$

We say that $\mathscr{A}$ is *invariant* if its fibers fulfill

$$\varphi_\lambda(t,s,\mathscr{A}(s)) = \mathscr{A}(t) \quad \text{for all } s \le t. \tag{1.2}$$

See the textbook [52] for an elementary but detailed exposition and Fig. 1.1 for an illustration of (1.2).

The behavior of $(\mathfrak{D}_\lambda)$ close to ϕ_λ^* is studied often using the associated *equation of perturbed motion*

$$\boxed{\dot{x} = A_\lambda(t)x + F_\lambda(t,x)} \tag{1.3}$$

with linear part $A_\lambda(t) := D_2 f_\lambda(t, \phi_\lambda^*(t))$ and a nonlinearity F_λ satisfying the limit relation $F_\lambda(t,x) = o(x)$ as $x \to 0$ uniformly in t given by

$$F_\lambda(t,x) := f_\lambda(t, x + \phi_\lambda^*(t)) - f_\lambda(t, \phi_\lambda^*(t)) - D_2 f_\lambda(t, \phi_\lambda^*(t))x.$$

Clearly, the behavior of (1.3) near the trivial solution is the same as the behavior of $(\mathfrak{D}_\lambda)$ close to ϕ_λ^*.

1.4 Linear Stability Theory

An important tool for investigating the dynamics of $(\mathfrak{D}_\lambda)$ near ϕ_λ^* is provided by linear stability theory. For instance, an attractive zero solution $\phi_\lambda^* = 0$ is a necessary condition for the extinction of all populations (or tumor cells, cf. Example 1.6) with small initial size in population dynamics (tumor models, respectively). Similarly, a solution ϕ_λ^* on the coordinate hyperplanes $x_i = 0$ for certain $1 \le i \le d$ indicates asymptotically vanishing species (or for example, HIV populations like in Example 1.10). Asymptotic stability in linear systems implies that all solutions share the same long term behavior.

Now classical examples (cf. [18, p. 3]) show that the time-dependent eigenvalues of $A_\lambda(t)$ are of little use for stability investigations in a nonautonomous framework, unless the time-dependence is periodic (replace eigenvalues by Floquet multipliers, cf. [3]) or "slow" (see, e.g., [18, 66]). Moreover, Lyapunov exponents merely yield asymptotic stability and therefore do not give a stability theory that is robust to nonlinear perturbations (see, e.g., [2]). We advocate that the appropriate stability notion for nonautonomous linear problems

$$\boxed{\dot{x} = A_\lambda(t)x = D_2 f_\lambda(t, \phi_\lambda^*(t))x} \tag{$\mathfrak{L}_\lambda$}$$

in a d-dimensional space $\mathbb{R}^d$ is *uniform exponential stability* on a subinterval $I \subseteq \mathbb{R}$, while the corresponding hyperbolicity concept is as follows:

Definition 1.1 (cf. [18]). A linear ODE $(\mathfrak{L}_\lambda)$ is said to possess an *exponential dichotomy* (ED for short) on I, if there exists a projection $P_\lambda \in \mathbb{R}^{d\times d}$ and real numbers $K \geq 1$, $\alpha > 0$ such that the transition or fundamental matrix of $(\mathfrak{L}_\lambda)$ denoted by $\Phi_\lambda(t,s) \in \mathbb{R}^{d\times d}$ fulfills

$$\|\Phi_\lambda(t,0)P_\lambda\Phi_\lambda(0,s)\| \leq Ke^{-\alpha(t-s)}, \quad \|\Phi_\lambda(s,0)[\mathrm{id}-P_\lambda]\Phi_\lambda(0,t)\| \leq Ke^{\alpha(t-s)}$$

for all $s, t \in I$ with $s \leq t$.

The linear equation $(\mathfrak{L}_\lambda)$ is also known as the *variational equation* associated with a reference solution ϕ_λ^* and in case $(\mathfrak{L}_\lambda)$ has an ED, ϕ_λ^* is called *hyperbolic*. The corresponding spectral notion for $(\mathfrak{L}_\lambda)$ is given in terms of the *dichotomy* (also called the *Sacker-Sell* or *dynamical*) *spectrum* (cf. [77, 81])

$$\Sigma_I(A_\lambda) := \{\gamma \in \mathbb{R} : \dot{x} = [A_\lambda(t) - \gamma\,\mathrm{id}]x \text{ does not have an ED on } I\} \subseteq \mathbb{R}.$$

Theorem 1.1 (Spectral Theorem, cf. [77, 81]). *For unbounded intervals $I \subseteq \mathbb{R}$ the dichotomy spectrum $\Sigma_I(A_\lambda)$ of $(\mathfrak{L}_\lambda)$ is the disjoint union of $0 \leq n \leq d$ closed* spectral intervals, *i.e., $\Sigma_I(A_\lambda) = \emptyset$, $\Sigma_I(A_\lambda) = \mathbb{R}$ or one of the four cases*

$$\Sigma_I(A_\lambda) = \left\{\begin{array}{c} [a_1, b_1] \\ or \\ (-\infty, b_1] \end{array}\right\} \cup [a_2, b_2] \cup \ldots \cup [a_{n-1}, b_{n-1}] \cup \left\{\begin{array}{c} [a_n, b_n] \\ or \\ [a_n, \infty) \end{array}\right\}$$

applies, where $a_i \leq b_i < a_{i+1}$.

Remark 1.1. (a) The dichotomy spectrum $\Sigma_I(A_\lambda)$ depends crucially on the time interval, where it is usually $I = \mathbb{R}$, $I = \mathbb{R}_+$ or $I = \mathbb{R}_-$. We often use the abbreviations

$$\Sigma(A_\lambda) := \Sigma_{\mathbb{R}}(A_\lambda), \quad \Sigma^+(A_\lambda) := \Sigma_{\mathbb{R}_+}(A_\lambda), \quad \Sigma^-(A_\lambda) := \Sigma_{\mathbb{R}_-}(A_\lambda).$$

(b) For bounded $A_\lambda : I \to \mathbb{R}^{d\times d}$ in $(\mathfrak{L}_\lambda)$ the dichotomy spectrum $\Sigma_I(A_\lambda)$ is compact and depends upper-semicontinuously on perturbations in A_λ.

For asymptotically autonomous or periodic equations, the spectral intervals in $\Sigma^\pm(A_\lambda)$ are singleton sets given by real parts for the eigenvalues (or Floquet multipliers) of the limit system. In contrast, the linearization along a heteroclinic solution yields spectral intervals in $\Sigma(A_\lambda)$ of positive length. Using [19] we obtain the following examples:

Example 1.1 (Autonomous Equations). For autonomous problems $\dot{x} = A_\lambda x$ with a coefficient matrix $A_\lambda \in \mathbb{R}^{d\times d}$ the dichotomy spectrum consists of singletons

$$\Sigma(A_\lambda) = \Sigma^+(A_\lambda) = \Sigma^-(A_\lambda) = \{\mathrm{Re}\mu \in \mathbb{R} : \mu \in \sigma(A_\lambda)\}.$$

Example 1.2 (Periodic Equations). For periodic problems $\dot{x} = A_\lambda(t)x$, i.e., with coefficient matrices $A_\lambda(t) = A_\lambda(t+T)$ for some $T > 0$, the dichotomy spectrum consists of singleton sets

$$\begin{aligned}\Sigma(A_\lambda) &= \Sigma^+(A_\lambda) = \Sigma^-(A_\lambda)\\ &= \left\{\gamma \in \mathbb{R} : \Phi_\lambda(T,0) \text{ has an eigenvalue with modulus } e^{\gamma T}\right\};\end{aligned}$$

here, the Floquet multipliers for $(\mathfrak{L}_\lambda)$ are the eigenvalues of $\Phi_\lambda(T,0)$.

Example 1.3 (Bohl Exponents). For scalar equations $\dot{x} = a(t)x$ with a coefficient function $a \in L^\infty(I)$ the dichotomy spectrum reads as

$$\Sigma(a) = [\underline{\beta}(a), \overline{\beta}(a)],$$

with the lower and upper *Bohl exponents*

$$\underline{\beta}(a) := \liminf_{\substack{t-s\to\infty\\ s,t\in I}} \frac{1}{t-s}\int_s^t a(\tau)\,d\tau, \qquad \overline{\beta}(a) := \limsup_{\substack{t-s\to\infty\\ s,t\in I}} \frac{1}{t-s}\int_s^t a(\tau)\,d\tau.$$

This example illustrates that the dichotomy spectrum actually depends on the temporal interval I under consideration. For a linear equation $(\mathfrak{L}_\lambda)$ with triangular matrix A_λ the dichotomy spectrum $\Sigma^\pm(A_\lambda)$ is given as union of the spectra for the diagonal elements.

Concrete examples on the dichotomy spectrum and its computation will be given below. Once the dichotomy spectrum is known, it has the following consequences on the stability of the reference solution ϕ_λ^*:

Proposition 1.1 (cf. [69]). *Let $\lambda \in \Lambda$ and the interval I be unbounded above.*

(a) If $\max \Sigma_I(A_\lambda) < 0$, *then ϕ_λ^* is uniformly asymptotically stable on I,*
(b) if there exists a spectral interval σ with $\min\sigma > 0$, *then ϕ_λ^* is unstable.*

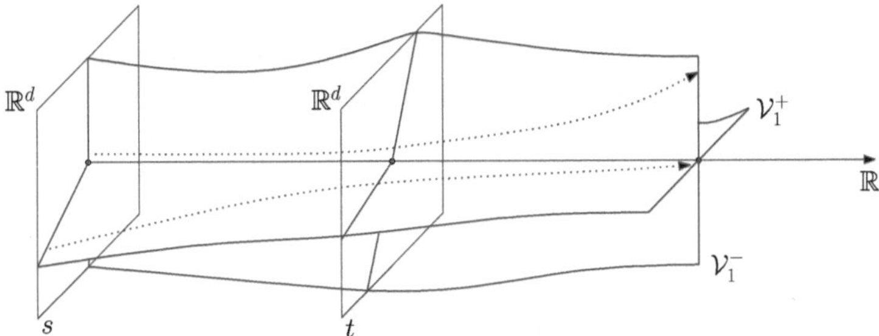

Fig. 1.2 Spectral manifolds $\mathcal{V}_1^+, \mathcal{V}_1^- \subseteq \mathbb{R} \times \mathbb{R}^d$ for $(\mathfrak{L}_\lambda)$ associated to a hyperbolic situation $\Sigma(A_\lambda) = [a_1, b_1] \cup [a_2, b_2]$ with $b_1 < 0 < a_2$. The *stable spectral manifold* $\mathcal{V}_1^+$ consists of all solutions decaying exponentially to 0 in forward time, while the *unstable spectral manifold* $\mathcal{V}_1^-$ is formed of solutions with the corresponding asymptotics in backward time. Solution curves are indicated by *dotted lines*

Each gap in the dichotomy spectrum induces two associated *spectral manifolds* $\mathcal{V}_i^+, \mathcal{V}_i^- \subseteq \mathbb{R} \times \mathbb{R}^d$, $1 \leq i < n$, consisting of solutions for $(\mathfrak{L}_\lambda)$ with a specific growth behavior (see Fig. 1.2). The spectral manifolds $\mathcal{V}_i^\pm$ extend the generalized eigenspaces known from the autonomous case (see, e.g., [31]).

Example 1.4 (Gene Transcription, cf. [9]). The transcription concentration $x_j(t)$ of a gene j is given by a system of decoupled linear nonautonomous differential equation

$$\dot{x}_j = -D_j x_j + B_j + S_j f(t) \quad \text{for all } j = 1, \ldots, n, \tag{1.4}$$

where $D_j > 0$ is the degradation rate. The production term $B_j + S_j f(t)$ comprises of a basal transcription rate $B_j > 0$, a sensitivity S_j and a transcription factor activity f, which is assumed to be a bounded function. This system has the dichotomy spectra

$$\Sigma = \Sigma^\pm = \bigcup_{j=1}^{n} \{-D_j\}$$

and is therefore uniformly asymptotically stable.

Example 1.5 (Insulin Absorption, cf. [65]). According to Palumbo et al. [65, Eq. (1)] the insulin absorption from a subcutaneous injection is described by the linear inhomogeneous ODE

$$\begin{cases} \dot{x} = -K_1(t)x + u_f(t), \\ \dot{y} = -K_2(t)y + u_s(t), \\ \dot{z} = V_I^{-1} K_1(t)x + V_I^{-1} K_2(t)y - K_3(t)z, \end{cases} \tag{1.5}$$

where x and y are the respective fast and slow insulin masses, z is the plasma insulin concentration. The absorption rates of the fast and slow masses of insulin are K_1 and K_2, while K_3 denotes the rate of plasma insulin disappearance and V_I stands for the distribution volume. The fast and slow insulin injections u_f and u_s, as well as the other above time-dependent parameters, are assumed to be L^∞-functions.

The dichotomy spectrum of (1.5) can be computed as

$$\Sigma^\pm = \bigcup_{i=1}^{3} \left[\underline{\beta}(-K_i), \overline{\beta}(-K_i)\right]$$

and under the assumption $\Sigma^+ \subseteq (-\infty, 0)$ the system (1.5) is uniformly asymptotically stable.

Example 1.6 (Tumor Growth, cf. [82]). We investigate a model for tumor growth under the effect of anticancer treatment of the form

$$\begin{cases} \dot{x}_1 = \dfrac{\lambda_0 x_1}{1 + \dfrac{\lambda_0}{\lambda_1} \displaystyle\sum_{j=1}^{4} x_j} - k_{pot} c(t) x_1, \\ \dot{x}_2 = k_{pot} c(t) x_1 - k x_2, \\ \dot{x}_3 = k(x_2 - x_3), \\ \dot{x}_4 = k(x_3 - x_4). \end{cases} \tag{$\mathfrak{T}$}$$

The tumor cells can be subdivided into four classes, where only the class x_1 is actually proliferating and x_2, x_3, x_4 show different stages of degeneracy until cell death. Concerning the parameters, k is the first-order rate constant of transit, k_{pot} measures the drug potency and $c(t)$ gives the plasma concentration of the anticancer agent.

System ($\mathfrak{T}$) has the trivial solution with corresponding variational equation

$$\dot{x} = \begin{pmatrix} \lambda_0 - k_{pot} c(t) & 0 & 0 & 0 \\ k_{pot} c(t) & -k & 0 & 0 \\ 0 & k & -k & 0 \\ 0 & 0 & k & -k \end{pmatrix} x \tag{1.6}$$

and the dichotomy spectrum

$$\Sigma^\pm = \{-k\} \cup \left[\underline{\beta}(\lambda_0 - k_{pot} c), \overline{\beta}(\lambda_0 - k_{pot} c)\right].$$

Hence, an effective dosing strategy for the anticancer treatment in form of the function c must satisfy the Bohl exponent condition $\overline{\beta}(\lambda_0 - k_{pot}c) < 0$.

A model related to $(\mathfrak{T})$ allowing n classes of tumor cells can be found in Chap. 7 by Koch and Schropp (see the PKPD model in Sect. 7.4).

How to Verify an Exponential Dichotomy?

Although the approach via the dichotomy spectrum provides a natural extension of the classical autonomous theory, an explicit expression for $\Sigma_I(A_\lambda)$ is hard to obtain in general. Effective numerical approximation methods are quite recent (see [22,23]) and robust for so-called integrally separated systems. These algorithms are based on triangulation techniques via QR or singular value decompositions of $A_\lambda(t)$ over I.

One might argue that the computation of the dichotomy spectrum requires a priori information on an infinite interval I. However, under certain recurrence assumptions (e.g. almost periodicity) it is possible to extend an ED from a finite interval to the whole real axis (cf. [64]).

1.5 Entire Solutions

A nonautonomous problem usually does not share equilibria or periodic solutions with its autonomous counterpart, so an appropriate substitute for equilibria in a time-variant framework is needed. Since the equilibria points of autonomous systems generically persist as bounded, globally defined solutions $\phi(\lambda)$ under parametric perturbations (see [33, 67, 70]), such entire solutions are the appropriate and adequate concept in a nonautonomous setting. They are called *nonautonomous equilibria* in [14].

Theorem 1.2 (Hyperbolic Solutions on $\mathbb{R}$, cf. [70]). *Let a parameter $\lambda^* \in \Lambda$ be fixed. If $\phi^* \in BC(\mathbb{R}, \Omega)$ is an entire solution of $(\mathfrak{D}_{\lambda^*})$ staying away from the boundary of Ω and satisfying*

$$0 \notin \Sigma(A_{\lambda^*}), \tag{1.7}$$

then there exist a C^m-function $\phi : B_\rho(\lambda^*) \subseteq \mathbb{R}^p \to B_\epsilon(\phi_{\lambda^*}) \subseteq BC^1(\mathbb{R}, \Omega)$, $\rho, \epsilon > 0$, such that:

(a) $\phi(\lambda^*) = \phi^*$,
(b) $\phi(\lambda)$ is the unique bounded entire solution of $(\mathfrak{D}_\lambda)$ in $B_\epsilon(\phi^*) \times B_\rho(\lambda^*)$,
(c) $\phi(\lambda)$ is hyperbolic with the same Morse index, *i.e., the same dimension, as the kernel $N(P_\lambda)$.*

Here, BC or BC^1 abbreviates the bounded continuous respectively bounded continuously-differentiable functions.

Remark 1.2 (Autonomous Case). Suppose that $(\mathfrak{D}_{\lambda^*})$ is autonomous and ϕ^* is a nontrivial periodic solution to $\dot{x} = f(x, \lambda^*)$. Since the derivative $\dot{\phi}^*$ is a nontrivial periodic solution to the variational equation $\dot{x} = D_1 f(\phi^*(t), \lambda^*)x$ corresponding to the Floquet multiplier equal to 1, the hyperbolicity assumption (1.7) cannot hold in this situation. Adequate continuation results for periodic solutions can be found in [3].

In the following, we illustrate the nonautonomous alternative to a steady state solution in some typical models in the life sciences. The first examples are linear, where Theorem 1.2 becomes global:

Example 1.7 (Gene Transcription, cf. [9]). In Example 1.4 we established that the gene transcription model (1.4) is uniformly asymptotically stable. Now we tackle its limit behavior. Indeed, every solution converges to the unique bounded solution

$$\begin{aligned} x_j^*(t) &= \int_{-\infty}^{t} e^{-D_j(t-s)} \big(B_j + S_j f(s)\big)\, ds \\ &= \frac{B_j}{D_j} + S_j \int_{-\infty}^{t} e^{-D_j(t-s)} f(s)\, ds \quad \text{for all } j = 1, \ldots, n. \end{aligned}$$

This solution exists for all $t \in \mathbb{R}$, i.e., is an *entire solution*. It is obtained by taking the *pullback limit*, i.e., $t_0 \to -\infty$ with t fixed, of the explicit solution

$$x_j(t) = x_0 e^{-D_j(t-t_0)} + \int_{t_0}^{t} e^{-D_j(t-s)} \big(B_j + S_j f(s)\big)\, ds \quad \text{for all } j = 1, \ldots, n.$$

Example 1.8 (Insulin Absorption, cf. [65]). The Insulin absorption model (1.5) from Example 1.5 features the following asymptotics: There exists a unique globally bounded entire solution that is given explicitly by

$$\begin{aligned} x^*(t) &:= \int_{-\infty}^{t} e^{-\int_s^t K_1(\tau)d\tau} u_f(s)\, ds, \\ y^*(t) &:= \int_{-\infty}^{t} e^{-\int_s^t K_2(\tau)d\tau} u_s(s)\, ds, \\ z^*(t) &:= \frac{1}{V_I} \int_{-\infty}^{t} e^{-\int_s^t K_3(\tau)d\tau} \int_{-\infty}^{s} \Big(K_1(\sigma) e^{-\int_\sigma^t K_1(\tau)d\tau} u_f(\sigma) \\ &\qquad + K_2(\sigma) e^{-\int_\sigma^t K_2(\tau)d\tau} u_s(\sigma) \Big)\, d\sigma\, ds. \end{aligned}$$

We point out that intrinsically nonautonomous systems often occur as models for glucose-insulin regulatory system; see [57, 90] for surveys.

The proof of Theorem 1.2 is based on the implicit mapping theorem. Thus, there exist constructive methods to obtain the perturbed solution $\phi(\lambda)$ for parameters λ near λ^*. Due to the C^m-smoothness of ϕ, a finite Taylor approximation is possible. This highlights a typical phenomenon in the nonautonomous theory (see also [72]): Instead of solving an algebraic equation to obtain the Taylor coefficients, one has to determine the bounded entire solutions of nonautonomous linear differential equations, i.e., an algebraic problem in the autonomous case has become a dynamical one in the nonautonomous case. The corresponding differential equation for the unknowns is called *homological equation* (cf. $(\mathfrak{I}_n)$ below).

More precisely, to deduce a formal scheme we begin with the Taylor ansatz

$$\phi(\lambda) = \phi^* + \sum_{n=1}^{m} \frac{1}{n!} D^n \phi(\lambda^*)(\lambda - \lambda^*)^n + R_m(\lambda) \tag{1.8}$$

for coefficients $D^n\phi(\lambda^*) \in L_n(\mathbb{R}^p, \mathbb{R}^d)$ and a remainder R_m satisfying

$$\lim_{\lambda \to 0} \frac{R_m(\lambda)}{|\lambda|^m} = 0.$$

For $1 \leq n \leq m$ we apply the higher order chain rule (see [72] for a reference in our notation) to the solution identity

$$\dot{\phi}(t, \lambda) \equiv f_\lambda(t, \phi(t, \lambda)) \quad \text{on } B_\rho(\lambda^*) \quad \text{on } \mathbb{R}.$$

For all $y_1, \ldots, y_n \in \mathbb{R}^p$ this yields the relation

$$\begin{aligned} D_2^n \dot{\phi}(t, \lambda) y_1 \cdot \ldots \cdot y_n &= D_2 f_\lambda(t, \phi(t, \lambda)) D_2^n \phi(t, \lambda) y_1 \cdot \ldots \cdot y_n \\ &+ \sum_{j=2}^{n} \sum_{(N_1, \ldots, N_j) \in P_j^<(l)} D_2^j f_\lambda(t, \phi(t, \lambda)) g^{\#N_1}(t, \lambda) y_{N_1} \cdots g_k^{\#N_j}(t, \lambda) y_{N_j}, \end{aligned}$$

where $P_j^<(l)$ is the totality of ordered partitions $\{N_1, \cdots, N_j\}$ of the finite set $\{0, 1, \ldots, l\}$ into disjoint subsets (see [70] for details) and the abbreviation

$$g_k^{\#N_1}(t, \lambda) := \frac{d^{\#N_1}(\phi(t, \lambda), \lambda)}{d\lambda^{\#N_1}}$$

has been used. Setting $\lambda = \lambda^*$ in this relation it follows that the Taylor coefficients $D^n\phi(\lambda^*) \in L_n(\mathbb{R}^p, BC) \cong BC(L_n(\mathbb{R}^p, \mathbb{R}^d))$ in demand fulfill the linearly inhomogeneous differential equation

$$\dot{X} = D_2 f_{\lambda^*}(t, \phi^*(t)) X + H_n(t) \tag{$\mathfrak{I}_n$}$$

in $L_n(\mathbb{R}^p, \mathbb{R}^d)$, where the inhomogeneity $H_n : \mathbb{R} \to L_n(\mathbb{R}^p, \mathbb{R}^d)$ reads as

$$H_n(t) y_1 \cdot \ldots \cdot y_n := \sum_{j=2}^{n} \sum_{(N_1,\ldots,N_j)\in P_j^<(l)} D_2^j f_{\lambda^*}(t, \phi^*(t)) \times g^{\# N_1}(t, \lambda^*) y_{N_1} \cdots g^{\# N_j}(t, \lambda^*) y_{N_j}.$$

In particular, $H_1(t) = D_2 f_{\lambda^*}(t, \phi^*(t))$. We denote $(\mathfrak{I}_n)$ as *homological equations.*

Corollary 1.1 (Taylor Approximation of Hyperbolic Solutions, cf. [70]). *The coefficients* $D^n\phi(\lambda^*) : \mathbb{R} \to L_n(\mathbb{R}^p, \mathbb{R}^d)$, $1 \leq n \leq m$, *in the Taylor expansion* (1.8) *are determined recursively by the* Lyapunov-Perron integrals

$$D_2^n \phi(t, \lambda^*) = \int_{\mathbb{R}} \Gamma_{\lambda^*}(t, s) H_n(s)\, ds \quad \text{for all } 1 \leq n \leq m,$$

where Γ_λ *is the* Green's function *associated to* $(\mathfrak{L}_\lambda)$*, which is defined by*

$$\Gamma_\lambda(t, s) := \begin{cases} \Phi_\lambda(t, 0) P_\lambda \Phi_\lambda(0, s), & s \leq t, \\ -\Phi_\lambda(t, 0)[\mathrm{id} - P_\lambda]\Phi_\lambda(0, s), & t < s. \end{cases}$$

The following example illustrates the formal procedure described above:

Example 1.9 (Neural Networks of Hopfield-type, cf. [59]). The dynamics of an artificial and isolated neuron under a temporally changing stimulus $c(t)$ is given by the scalar ODE

$$\dot{x} = f_\lambda(t, x) := -a(t)x + b(t)\tanh x + \lambda c(t), \tag{1.9}$$

where x is the membrane potential. The function $a : \mathbb{R} \to (0, \infty)$ yields a dissipative or negative feedback term and the bounded $b : \mathbb{R} \to (0, \infty)$ describes the neuron gain; both are assumed to be continuous.

For $\lambda = \lambda^* = 0$ it is clear that (1.9) has the trivial solution $\phi^*(t) \equiv 0$, whose continuation for $\lambda \neq 0$ we like to approximate. The corresponding variational equation reads as

$$\dot{x} = (b(t) - a(t))x$$

and has the dichotomy spectrum $\Sigma = [\underline{\beta}(b - a), \overline{\beta}(b - a)]$. Hence, ϕ^* is a hyperbolic solution, if the parameter functions have Bohl exponents satisfying $\Sigma \cap \{0\} = \emptyset$. It persists due to Theorem 1.2 for small values of λ. If we abbreviate $\phi_i(t) := D_2^i \phi(t, \lambda^*)$, then the corresponding linear equations $(\mathfrak{I}_n)$ become

$$\dot{x} = [b(t) - a(t)]x + c(t),$$

$$\dot{x} = [b(t) - a(t)]x,$$

$$\dot{x} = [b(t) - a(t)]x - 2b(t)\phi_1^3(t),$$

$$\dot{x} = [b(t) - a(t)]x - 12b(t)\phi_2(t)\phi_1^2(t),$$

$$\dot{x} = [b(t) - a(t)]x + 2b(t)\phi_1(t)(8\phi_1^4(t) - 10\phi_3(t)\phi_1(t) - 15\phi_2^2(t)),$$

$$\dot{x} = [b(t) - a(t)]x + 30b(t)(8\phi_2(t)\phi_1^4(t) - \phi_4(t)\phi_1^2(t) - 4\phi_2(t)\phi_3(t)\phi_1(t) - \phi_2^3(t))$$

and we can apply Corollary 1.1 in order to obtain its bounded entire solutions. Above all, one sees that $\phi_2(t) \equiv \phi_4(t) \equiv \phi_6(t) \equiv 0$ on $\mathbb{R}$ and we obtain successively

$$\phi_1(t) = \begin{cases} \int_{-\infty}^t c(s)e^{\int_s^t (b(r)-a(r))\,dr}\,ds, & \overline{\beta}(b-a) < 0, \\ \int_t^\infty c(s)e^{\int_s^t (b(r)-a(r))\,dr}\,ds, & \underline{\beta}(b-a) > 0, \end{cases}$$

$$\phi_3(t) = -2\begin{cases} \int_{-\infty}^t b(s)\phi_1(s)^3 e^{\int_s^t (b(r)-a(r))\,dr}\,ds, & \overline{\beta}(b-a) < 0, \\ \int_t^\infty b(s)\phi_1(s)^3 e^{\int_s^t (b(r)-a(r))\,dr}\,ds, & \underline{\beta}(b-a) > 0, \end{cases}$$

$$\phi_5(t) = 4\begin{cases} \int_{-\infty}^t b(s)\phi_1(s)^2(4\phi_1(t)^3 - 5\phi_3(t))e^{\int_s^t (b(r)-a(r))\,dr}\,ds, & \overline{\beta}(b-a) < 0, \\ \int_t^\infty b(s)\phi_1(s)^2(4\phi_1(t)^3 - 5\phi_3(t))e^{\int_s^t (b(r)-a(r))\,dr}\,ds, & \underline{\beta}(b-a) > 0. \end{cases}$$

From this we arrive at the approximation

$$\phi(t,\lambda) = \sum_{i=0}^{2} \frac{\lambda^{2i+1}}{(2i+1)!}\phi_{2i+1}(t) + O(\lambda^7)$$

with the functions ϕ_1, ϕ_3, ϕ_5 computed above. The persistence of the zero solution under different perturbation functions is illustrated in Fig. 1.3.

A source for time-dependent models are treatment strategies for diseases.

Example 1.10 (HIV and T-cell Interaction, cf. [42]). The interaction between HIV and T-cells in the human immune system is described by the ODE

$$\begin{cases} \dot{T} = s_1(t) - \dfrac{s_2(t)V}{b_1 + V} - \mu T - kVT + r(t)T, \\ \dot{V} = \dfrac{gV}{b_2 + V} - cVT, \end{cases} \tag{1.10}$$

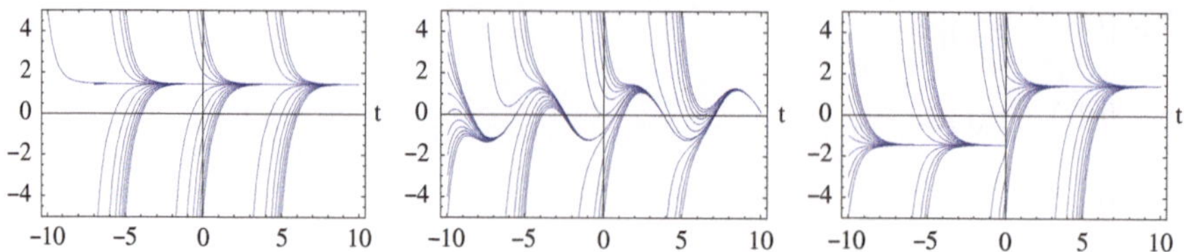

Fig. 1.3 Example 1.9: Solution portrait indicating the persistence of the trivial solution as constant solution (*left*, $c(t) \equiv 1$), periodic solution (*middle*, $c(t) = \sin t$) and as bounded entire solution (*right*, $c(t) = \operatorname{sgn} t$) for $\lambda = 1, \alpha = 1, \beta = 2$

where T is the uninfected CD4$^+$ T-cell, and V the HIV population. The terms $s_1(t)$ and $s_2(t)$ describe the source and production of T-cells and $r(t)$ is an interleukin treatment function during possibly aperiodic treatment intervals; the L^∞-functions $s_1, s_2, r : \mathbb{R} \to \mathbb{R}$ are assumed to be known.

To understand the dynamics of (1.10) we first consider the autonomous case

$$\begin{cases} \dot{T} = s_1 - \dfrac{s_2 V}{b_1 + V} - \mu T - kVT, \\ \dot{V} = \dfrac{gV}{b_2 + V} - cVT, \end{cases} \tag{1.11}$$

where both s_1 and s_2 are now assumed to be constant in time. The autonomous system (1.11) possesses the disease free equilibrium $(T_0, V_0) = (\frac{s_1}{\mu}, 0)$. From the corresponding Jacobian

$$\begin{pmatrix} -\mu, & -\dfrac{ks_1}{\mu} - \dfrac{s_2}{b_1} \\ 0, & \dfrac{g}{b_2} - \dfrac{cs_1}{\mu} \end{pmatrix}$$

we see that (T_0, V_0) is asymptotically stable for $\frac{g}{b_2} < \frac{cs_1}{\mu}$ and hyperbolic (with Morse index 1) for $\frac{g}{b_2} > \frac{cs_1}{\mu}$. The critical case $\frac{g}{b_2} = \frac{cs_1}{\mu}$ corresponds to a transcritical bifurcation of (T_0, V_0). Hence, in order to enforce that the virus free equilibrium (T_0, V_0) becomes asymptotically stable, we have to choose a treatment strategy such that the decay rate μ for the T-cells becomes small.

Now we proceed to the full nonautonomous equation (1.10). Then the virus free dynamics is given by the scalar linearly inhomogeneous problem

$$\dot{T} = s_1(t) + (r(t) - \mu)T, \tag{1.12}$$

which is stable under the upper Bohl exponent condition $\overline{\beta}(r - \mu) < 0$. Moreover, this assumption guarantees the existence of a unique globally bounded solution

$$T_*(t) = \int_{-\infty}^{t} e^{\int_s^t [r(\sigma)-\mu]\, d\sigma} s_1(s)\, ds$$

to (1.12). Clearly, $(\frac{s_1}{\mu}, 0)$ is not an equilibrium of (1.10) anymore, but it persists as the entire bounded solution

$$T_*(t) = \int_{-\infty}^{t} e^{\int_s^t [r(\sigma)-\mu]\,d\sigma} s_1(s)\,ds, \qquad V_*(t) = 0.$$

Its stability behavior is determined by the variational equation

$$\begin{pmatrix} \dot{T} \\ \dot{V} \end{pmatrix} = \begin{pmatrix} r(t) - \mu, & -\frac{s_2}{b_1} - kT_*(t) \\ 0, & \frac{g}{b_2} - cT_*(t) \end{pmatrix} \begin{pmatrix} T \\ V \end{pmatrix},$$

i.e., its corresponding dichotomy spectrum

$$\Sigma = \left[\underline{\beta}(r-\mu), \overline{\beta}(r-\mu)\right] \cup \left[\underline{\beta}(\tfrac{g}{b_2} - cT_*), \overline{\beta}(\tfrac{g}{b_2} - cT_*)\right].$$

Thus, an effective dosing function r must satisfy the Bohl exponent condition

$$\max\left\{\overline{\beta}(r-\mu), \overline{\beta}(\tfrac{g}{b_2} - cT_*)\right\} < 0,$$

since it guarantees that (T_*, V_*) is uniformly asymptotically stable.

For further time-varying (and higher dimensional) models see [40,41]. Moreover, related optimal control approaches to HIV-modelling using more complicated four- and higher dimensional equations, are studied in [27,43,60].

1.6 Attractors

In an autonomous system, the solutions depend only on the elapsed time $t - t_0$ since starting, so the limit relation $t - t_0 \to \infty$ either holds when $t \to \infty$ with t_0 fixed or as $t_0 \to -\infty$ with t fixed, so (yet to be defined) forward and pullback convergence are equivalent for an autonomous system.

Two types of attractors for nonautonomous systems are possible, depending on which of the above types of convergence is used. Moreover, unlike autonomous attractors, a nonautonomous attractor $\mathscr{A}$ for $(\mathfrak{D}_\lambda)$ consists of a family $(\mathscr{A}(t))_{t\in\mathbb{R}}$ of nonempty compact subsets which is invariant in the sense that

$$\varphi_\lambda(t, s, \mathscr{A}(s)) = \mathscr{A}(t) \quad \text{for all } t \in \mathbb{R}$$

and attracts bounded subsets $D \subseteq \mathbb{R}^d$ of initial values (t_0, x_0) (rather than just individual points), in the sense that

$$\operatorname{dist}(\varphi_\lambda(t, t_0, D), \mathscr{A}(t)) \to 0 \begin{cases} \text{as } t \to \infty \text{ with } t_0 \text{ fixed (forward case)}, \\ \text{as } t_0 \to -\infty \text{ with } t \text{ fixed (pullback case)}. \end{cases}$$

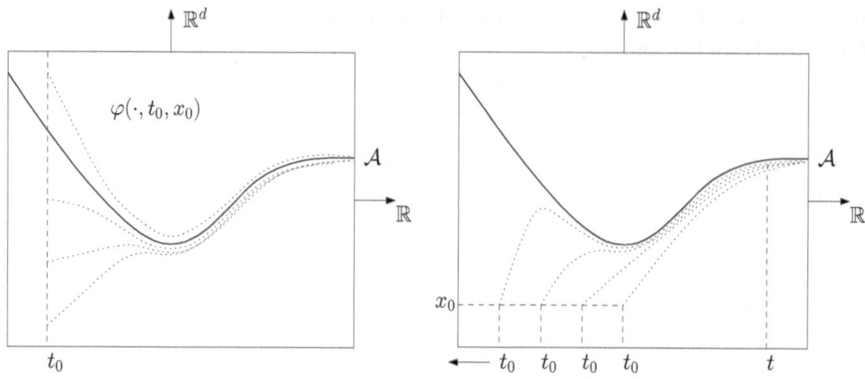

Fig. 1.4 Solution curves (*dotted*) and the attractor $\mathscr{A}$ with forward convergence $t \to \infty$ for t_0 fixed (*left*) and pullback convergence $t_0 \to -\infty$ for t fixed (*right*)

This gives, respectively, a *forward attractor* and a *pullback attractor*, which consist of entire solutions. In general, forward and pullback convergencies do not imply each other, but in special cases both may hold. Under appropriate dissipativity assumption on $(\mathfrak{D}_\lambda)$, pullback attractors exist as fiber-wise compact sets. See [12,44,46,52] for more information and Fig. 1.4 for an illustration.

We have already seen some examples of such attractors above without have explicitly said so (cf. Examples 1.4, 1.5 and 1.10). The following example is a very simple illustration of a situation that is common in the biological sciences, which are intrinsically nonautonomous.

Example 1.11 (Switching Systems). Consider two autonomous ODEs

$$\dot{x} = Ax + b_1, \qquad \dot{x} = Ax + b_2 \tag{1.13}$$

with a stable matrix $A \in \mathbb{R}^{d\times d}$, i.e., all its eigenvalues λ of A satisfy $\mathrm{Re}\lambda < 0$. Hence, the equations in (1.13) have its equilibria $-A^{-1}b_1$ and $-A^{-1}b_2$ as respective global attractor.

Let $s : \mathbb{R} \to \{b_1, b_2\}$ be a given piecewise continuous function and consider the nonautonomous ODE

$$\dot{x} = Ax + s(t) \tag{1.14}$$

formed by switching between the two autonomous systems in (1.13). Its explicit solution with initial value $x(t_0) = x_0$ is

$$\varphi(t, t_0, x_0) := x(t) = e^{A(t-t_0)}x_0 + \int_{t_0}^{t} e^{A(t-\tau)}s(\tau)\,d\tau. \tag{1.15}$$

The difference $\phi_1 - \phi_2$ of any two solutions with initial conditions $\phi_i(t_0) = x_i$ for $i = 1, 2$, satisfies the homogeneous ODE $\dot{x} = Ax$, so

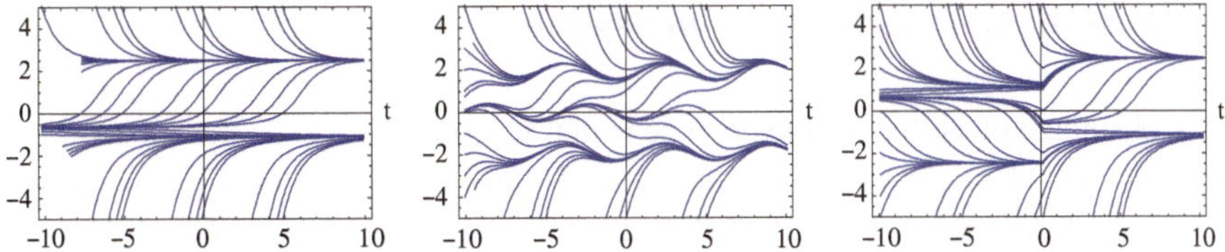

Fig. 1.5 Example 1.12: Solution portrait indicating the attractors $\mathscr{A}_\lambda$ of (1.16) with constants fibers $\mathscr{A}_\lambda(t)$ given by an interval (*left*, $c(t) \equiv 1$), periodic fibers $\mathscr{A}_\lambda(t)$ (*middle*, $c(t) = \sin t$) and temporally changing fibers $\mathscr{A}_\lambda(t)$ (*right*, $c(t) = \operatorname{sgn} t$) with $\lambda = \frac{1}{2}, \alpha = 1, \beta = 2$

$$\phi_1(t) - \phi_2(t) = e^{A(t-t_0)}(x_1 - y_2) \to 0 \quad \text{as } t \to \infty,$$

so all solutions converge together forward in time. What do they converge to? Taking the pullback limit $t_0 \to -\infty$ in (1.15) gives

$$\phi^*(t) := \int_{-\infty}^{t} e^{A(t-\tau)} s(\tau)\, d\tau,$$

which is an entire and bounded solution of the nonautonomous ODE (1.13).

This simple example has nonautonomous attractor in both pullback and forward senses, which consists of singleton subsets $\mathscr{A}(t) = \{\phi^*(t)\}$.

Example 1.12 (Neural Networks of Hopfield-type, cf. [59]). In Example 1.9 we saw that the zero solution to (1.9) for $\lambda = 0$ persists as an entire bounded solution for small stimuli in form of a continuous $c : \mathbb{R} \to \mathbb{R}$; the stability depends on the Bohl exponents of $b - a$. We now retreat to the simplified model

$$\dot{x} = f_\lambda(t, x) := -\alpha x + \beta \tanh x + \lambda c(t), \tag{1.16}$$

where the coefficient functions are constants fulfilling $0 < \alpha < \beta$. Then the unperturbed equation ($\lambda = 0$) has three equilibria $x_- < 0 < x_+$, with the trivial one being unstable and x_-, x_+ being asymptotically stable. Moreover, the interval $[x_-, x_+]$ is the global attractor. For $\lambda \neq 0$ (1.16) becomes nonautonomous and the equilibria x_-, x_+ persist as bounded entire solutions $\phi_\lambda^-, \phi_\lambda^+ : \mathbb{R} \to \mathbb{R}$, whose Taylor approximation in λ can be computed as in Example 1.9. Moreover, the global attractor of (1.16) is (cf. Fig. 1.5)

$$\mathscr{A}_\lambda = \left\{(t, x) \in \mathbb{R} \times \mathbb{R} : \phi_\lambda^-(t) \leq x \leq \phi_\lambda^+(t)\right\}.$$

Remark 1.3. We note that a pullback attractor $\mathscr{A} = \left\{(t, x) \in \mathbb{R} \times \mathbb{R}^d : x \in \mathscr{A}(t)\right\}$ contained in a uniformly bounded set (this means there exists a $R > 0$ such that $\mathscr{A}(t) \subset B_R(0)$ for all t) consist of the bounded entire solution of the system $(\mathfrak{D}_\lambda)$ (cf. [46]).

How Realistic Is Pullback Convergence?

An advantage of pullback convergence is that it provides a means for constructing the component sets of a pullback attractor and hence entire solutions. This obviously requires knowledge of the past history of the system. In some important systems involving the periodic or almost periodic forcing this past history is known.

There are, however, many modelling situations for which the past history is not known. Instead the system is known—in fact, prescribed—on future time intervals of the form $[t_0, \infty)$. Forward attractors can be easily modified to these situations. Pullback attractors can also be used if we invent an artificial past history, but the pullback attractor will then depend on which "history" we choose and it is not clear which history we should use.

A generalization of the theory of nonautonomous dynamical systems to nonautonomous semi-dynamical systems proposed in [46, 52] offers some insight here. Essentially, it contains all possible past histories of the driving system and the modified pullback attractor component sets are the accumulative effect of all of these past histories. In this sense, pullback attraction is still meaningful and useful for nonautonomous systems defined only for future time.

1.7 Bifurcation Theory

A satisfactory bifurcation theory for nonautonomous systems is still under development. Due to the lack of equilibria (or periodic solutions) for aperiodic time-variant problems $(\mathfrak{D}_\lambda)$, at least two approaches were investigated so far:

- *Attractor bifurcation*: Pullback attractors or repellers change their structure under varying parameters, i.e., become trivial or change their dimension (cf. [26,37,53] or [73–75]). In particular, this led to bifurcation patterns generalizing the classical counterparts of saddle-node, transcritical and pitchfork types.
- *Solution bifurcation*: The number of bounded entire solutions for $(\mathfrak{D}_\lambda)$ with a specific property changes, if parameters are varied. For instance, almost periodic solutions have been considered in [45] and [63] treat even more general classes. Inspired by the persistence of equilibria as globally defined bounded solutions under parametric perturbation, bifurcation results for such solutions have been obtained in [68, 69].

The following examples illuminate these different approaches:

Example 1.13 (A Time-dependent Logistic Model, cf. [56]). A nonlinear logistic model of population growth is given by

$$\dot{x} = \lambda x\,(\beta(t) - x)$$

with a constant $\lambda \in \mathbb{R}$ and a continuous function $\beta : \mathbb{R} \to [b_m, b_M]$, where $0 < b_m < b_M < \infty$. It is a Bernoulli ODE which is explicitly solvable. It has a positive entire solution

$$\phi_\lambda^*(t) = \frac{e^{\lambda \int_{-\infty}^t b(s)\,ds}}{\lambda \int_{-\infty}^t e^{\lambda \int_{-\infty}^\tau b(s)\,ds}\,d\tau} \quad \text{for all } \lambda > 0,$$

which can be determined by taking pullback convergence. For $\lambda < 0$ the system has a trivial nonautonomous attractor $\mathscr{A}_\lambda(t) \equiv \{0\}$ in both the pullback and forwards senses, which undergoes a transcritical attractor bifurcation at $\lambda = 0$ to

$$\mathscr{A}_\lambda(t) = [0, \phi_\lambda^*(t)]$$

for $\lambda > 0$, which is also pullback and forward attracting. It is also possible to verify this topological change of the attractor theoretically on basis of [75].

Example 1.14 (Neural Networks of Hopfield-type, cf. [59]). We return to the scalar model from our above Examples 1.9 and 1.12, but with different parameter constellations. Indeed, we focus on

$$\dot{x} = f_\lambda(t, x) := -a_\lambda(t)x + b_\lambda(t)\tanh x, \tag{1.17}$$

with continuous function $a_\lambda, b_\lambda : \mathbb{R} \to (0, \infty)$ and the trivial solution to (1.17). Furthermore, for continuous functions $\alpha, \beta : \mathbb{R} \to \mathbb{R}$ used below the upper and lower Bohl exponents of $\beta - \alpha$ are supposed to exist as finite numbers.

(a) *Pitchfork bifurcation*: At first we assume that

$$a_\lambda(t) := \gamma(t), \qquad b_\lambda(t) := \gamma(t) + \lambda\beta(t),$$

where $\alpha, \beta, \gamma : \mathbb{R} \to \mathbb{R}$ is continuous. The associate variational equation reads as $\dot{x} = \lambda\beta(t)x$ yielding

$$\Sigma_\lambda = [\underline{\beta}(\lambda\beta), \overline{\beta}(\lambda\beta)]$$

as dichotomy spectrum, which degenerates to the singleton $\{0\}$ for the critical parameter $\lambda = 0$. This setting implies a nonautonomous pitchfork bifurcation as understood in [75]: In case $\underline{\beta}(\lambda\beta) > 0$ it is supercritical, i.e., for $\lambda < 0$ the trivial solution is asymptotically stable, while it becomes unstable for $\lambda > 0$ and is embedded into a nontrivial attractor for (1.17). In case $\overline{\beta}(\lambda\beta) < 0$ the bifurcation is subcritical, i.e., a dual stability change occurs (cf. Fig. 1.6).

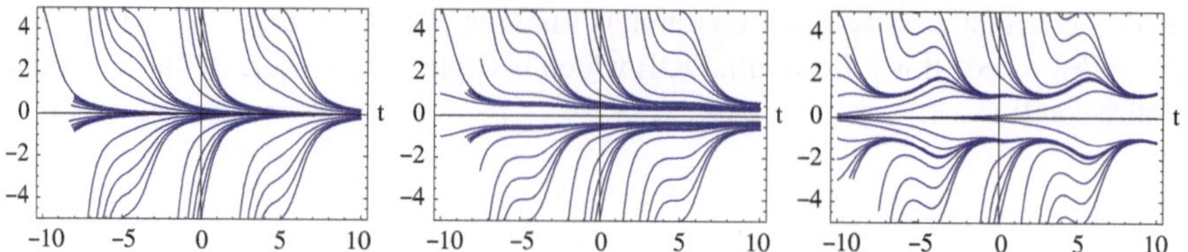

Fig. 1.6 Supercritical pitchfork bifurcation from Example 1.14(a): Solution portrait indicating the bifurcation of a trivial attractor $\mathscr{A}_\lambda = \mathbb{R} \times \{0\}$ (*left*, $\lambda < 0$) over the neutral situation (*middle*, $\lambda = 0$) into a nontrivial attractor (*right*, $\lambda > 0$) with $\gamma(t) = 1 + \sin|t|$, $\beta(t) \equiv 1$

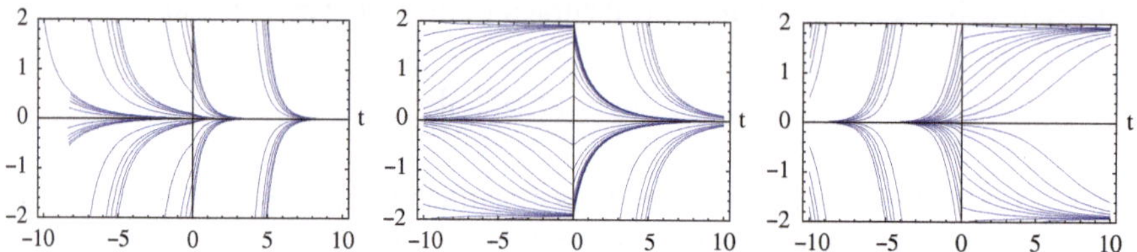

Fig. 1.7 Shovel bifurcation from Example 1.14(b): Solution portrait indicating the bifurcation of a 1-parameter family of bounded entire solutions. The trivial solution is the unique bounded entire solution and uniformly asymptotically stable ($\lambda < \frac{3}{2}$, *left*), 0 is asymptotically stable on $\mathbb{R}_+$ and embedded into a family of bounded entire solutions ($\lambda \in (-\frac{3}{2}, -\frac{1}{2})$, *middle*) and 0 is the unique bounded entire solution and unstable ($\lambda > -\frac{1}{2}$, *right*) with functions $\alpha(t) = 0.5\,\mathrm{sgn}\, t$, $\beta(t) \equiv 1$

(b) *Shovel bifurcation*: A different situation occurs when, for instance,

$$a_\lambda(t) := \alpha(t) - \lambda, \qquad b_\lambda(t) := \beta(t).$$

As variational equation we have $\dot{x} = [\lambda + \beta(t) - \alpha(t)]x$ and the dichotomy spectrum

$$\Sigma_\lambda = [\underline{\beta}(\beta - \alpha), \overline{\beta}(\beta - \alpha)] - \lambda.$$

This yields a shovel bifurcation as introduced in [69] and in particular two critical parameters:

- $\lambda_1^* = \overline{\beta}(\beta - \alpha)$: The trivial solution is asymptotically stable for $\lambda < \lambda_1^*$, while there exists a 1-parameter family of bounded entire solutions to (1.17) for $\lambda > \lambda_1^*$, i.e. a supercritical bifurcation.
- $\lambda_2^* = \underline{\beta}(\beta - \alpha)$: The trivial solution is asymptotically stable for $\lambda > \lambda_2^*$, and for $\lambda < \lambda_2^*$ there is a 1-parameter family of bounded entire solutions to (1.17), i.e. a subcritical bifurcation.

See Fig. 1.7 for an illustration.

The following section provides a geometrical tool to investigate bifurcation phenomena in higher dimensional problems.

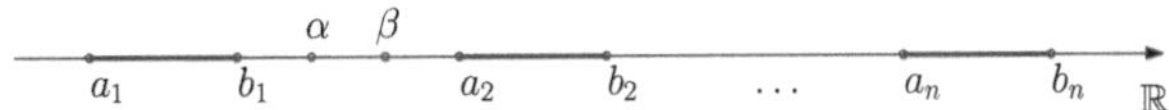

Fig. 1.8 Splitting of the dichotomy spectrum according to (1.18)

1.8 Integral Manifolds

Invariant manifolds are an important geometrical tool in dynamical systems to separate different domains of attraction or for dimension reductions (cf., for instance, Chap. 3) and bifurcation problems to capture the essential asymptotic behavior. The reason is based on the fact that after a decay of transients, the behavior of a differential equation modeling a biochemical system often stays on a low-dimensional surface which is an invariant manifold.

A corresponding theory for nonautonomous problems (1.3) is due to [5, 80] and one speaks of *integral manifolds* in this generalized context. They are associated to fixed reference solutions ϕ^* of $(\mathfrak{D}_\lambda)$, which are transformed to the trivial solution by means of the equation of perturbed motion (1.3).

Keeping the parameter λ fixed throughout, we introduce a nonautonomous version of an invariant manifold for (1.3) as follows: The essential assumption is a gap in the dichotomy spectrum $\Sigma_I(A_\lambda)$ for $(\mathfrak{L}_\lambda)$, i.e., there exists an $1 \leq i < n$ and an interval $(\alpha, \beta) \subseteq \mathbb{R}$ such that (cf. Fig. 1.8)

$$a_i < \alpha < \beta < b_{i+1}. \tag{1.18}$$

Thus, for $\gamma \in (\alpha, \beta)$ the scaled variational equation $\dot{x} = [A_\lambda(t) - \gamma \,\mathrm{id}]x$ has an exponential dichotomy with associated projector $P_i \in \mathbb{R}^{d \times d}$. We define projection-valued maps

$$P_i^+(t) := \Phi_\lambda(t,0) P_i \Phi_\lambda(0,t), \qquad P_i^-(t) := \Phi_\lambda(t,0)[\mathrm{id} - P_i]\Phi_\lambda(0,t)$$

and assume that $w_i^\pm : U \times I \to \mathbb{R}^d$ are continuously differentiable and satisfy

$$w_i^\pm(t,0) \equiv 0 \quad \text{on } I, \qquad \lim_{x \to 0} \left\| D_2 w_i^\pm(t,x) \right\| = 0 \quad \text{uniformly in } t \in I, \tag{1.19}$$

$$w_i^\pm(t,x) = w_i^\pm(t, P_i^\pm(t)x) \in R(P_i^\mp(t)) \tag{1.20}$$

for all $(t,x) \in I \times U$. Then the nonautonomous set given by the graph

$$\mathscr{W}_i^\pm := \left\{(\tau, \xi + w_i^\pm(\tau,\xi)) \in I \times \mathbb{R}^d : \xi \in R(P_i^\pm(\tau)) \cap U\right\} \tag{1.21}$$

is called a *local integral manifold* of the nonlinear ODE $(\mathfrak{D}_\lambda)$, if

$$(t_0, x_0) \in \phi^* + \mathscr{W}_i^\pm \quad \Rightarrow \quad (t, \varphi(t, t_0, x_0)) \in \phi^* + \mathscr{W}_i^\pm \quad \text{for all } t \in J_U(t_0, x_0)$$

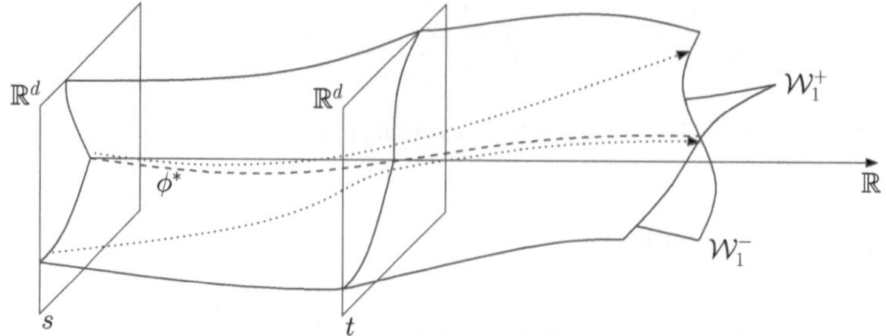

Fig. 1.9 Integral manifolds $\mathcal{W}_1^+, \mathcal{W}_i^- \subseteq \mathbb{R} \times \Omega$ for $(\mathfrak{D}_\lambda)$ associated to the solution ϕ^* (*dashed*) to a hyperbolic situation $\Sigma(A_\lambda) = [a_1, b_1] \cup [a_2, b_2]$ with $b_1 < 0 < a_2$. The *stable integral manifold* $\mathcal{W}_1^+$ consists of all solutions decaying exponentially to 0 in forward time, while the *unstable spectral manifold* $\mathcal{W}_1^-$ is formed of solutions with the corresponding asymptotics in backward time. Solution curves are indicated by *dotted lines*

holds, where $J_U(t_0, x_0) \subseteq I$ is the maximal existence interval for $\varphi_\lambda(\cdot, t_0, x_0)$ w.r.t. the tubular neighborhood $\phi^* + U$. One speaks of a C^m-integral manifold of $(\mathfrak{D}_\lambda)$, if the derivatives $D_2^n w_i^\pm$ exist and are continuous for $n \in \{1, \dots, m\}$.

Geometrically, conditions (1.19)–(1.20) imply that $\phi^* + \mathcal{W}_i^\pm$ contains the solution ϕ^* of $(\mathfrak{D}_\lambda)$, and $\mathcal{W}_i^\pm$ is fiber-wise tangent to the spectral manifolds

$$\mathcal{V}_i^\pm = \left\{(\tau, \xi) \in I \times \mathbb{R}^d : \xi \in R(P_i^\pm(\tau))\right\},$$

while (1.20) implies that each t-fiber $\mathcal{W}_i^\pm(t)$ is a graph over the intersection $R(P_i^\pm(t)) \cap U$, $t \in I$ (cf. Fig. 1.9).

Local integral manifolds satisfy the following nonlinear first order partial differential equation, called the *invariance equation*,

$$\begin{aligned} & A_\lambda(t) w_i^\pm(t, \xi) + P_i^\mp(t) F_\lambda(t, \xi + w_i^\pm(\xi, t)) \\ &= D_1 w_i^\pm(t, \xi) + D_2 w_i^\pm(t, \xi) \left(A_\lambda(t)\xi + P_i^\pm(t) F_\lambda(t, \xi + w_i^\pm(t, \xi))\right) \end{aligned} \tag{1.22}$$

for all $t \in I$, $\xi \in R(P_i^\pm(t)) \cap U$ such that $\xi + w_i^\pm(t, \xi) \in U_0$.

Remark 1.4 (Classical Hierarchy of Integral Manifolds). The sets $\phi^* + \mathcal{W}_i^+$ and $\phi^* + \mathcal{W}_i^-$ are known as *pseudo-stable* and *pseudo-unstable* integral manifolds of the solution ϕ^* to $(\mathfrak{D}_\lambda)$, respectively. Then

- for I unbounded above, $\phi^* + \mathcal{W}_i^+$ describes a *center-stable* integral manifold in case $\beta > 0$, a *stable* integral manifold in the hyperbolic situation $\alpha < 0 < \beta$ and a *strongly stable* integral manifold for $\beta < 0$.
- For I unbounded below, $\phi^* + \mathcal{W}_i^-$ is a *center-unstable* integral manifold in case $\alpha < 0$, an *unstable* integral manifold in the hyperbolic situation $\alpha < 0 < \beta$ and a *strongly unstable* integral manifold in case $\alpha > 0$.

This terminology corresponds to the autonomous situation in, e.g., [13].

Theorem 1.3 (Local Integral Manifolds, cf. [72]). *If $\phi^* : I \to \Omega$ is a solution to $(\mathfrak{D}_\lambda)$ such that* (1.18) *holds, then there exists a $\rho_0 > 0$ with:*

(a) for I unbounded above and under the gap condition

$$m\alpha < \beta, \tag{1.23}$$

the ODE $(\mathfrak{D}_\lambda)$ has a local pseudo-stable C^m-integral manifold $\phi^* + \mathscr{W}_i^+$,

(b) for I unbounded below and under the gap condition

$$\alpha < m\beta, \tag{1.24}$$

the ODE $(\mathfrak{D}_\lambda)$ has a local pseudo-unstable C^m-integral manifold $\phi^* + \mathscr{W}_i^-$,

(c) for the corresponding mapping $w_i^\pm : B_{\rho_0}(0) \times I \to \mathbb{R}^d$ from (1.21)*, there exist real numbers $\gamma_0, \dots, \gamma_m \geq 0$ such that*

$$\left\| D_2^n w_i^\pm(t,x) \right\| \leq \gamma_n \quad \text{for all } x \in B_{\rho_0}(0),\ t \in I,\ n \in \{0,\dots,m\}, \tag{1.25}$$

(d) if the right-hand side f_λ of $(\mathfrak{D}_\lambda)$, as well as the solution ϕ^* are periodic in t with period $\theta > 0$, then

$$w_i^\pm(t+\theta,x) = w_i^\pm(t,x) \quad \text{for all } x \in \Omega,\ t \in I,$$

and if $(\mathfrak{D}_\lambda)$ is autonomous and ϕ^* constant, then the mappings $w_i^\pm$ are independent of $t \in I$, i.e., the sets

$$\left\{\phi^* + \xi + w_i^\pm(\xi) \in \mathbb{R}^d : \xi \in R(P_i^\pm) \cap B_{\rho_0}(0)\right\}$$

are a locally invariant manifolds of $(\mathfrak{D}_\lambda)$.

Remark 1.5 (Reduction Principle). Center-unstable manifolds $\phi_i^* + \mathscr{W}_i^-$ are of particular importance, since they allow a reduction in the dimension in critical stability situations: First, the stability of ϕ^* is completely determined by the corresponding properties of the zero solution for the ODE reduced to $\mathscr{W}_i^-$. Second, bifurcating entire solutions near ϕ^* are contained in $\phi^* + \mathscr{W}_i^-$.

In general, the integral manifolds $\phi^* + \mathscr{W}_i^\pm$ are unknown. However, for a nonautonomous center manifold reduction it suffices to determine a Taylor approximation of the mappings $w_i^\pm$ in x. Thereto we make the ansatz

$$w_i^\pm(t,x) = \sum_{n=2}^{m} \frac{1}{n!} w_{i,n}^\pm(t) x^n + R_{i,m}^\pm(t,x) \tag{1.26}$$

with coefficient functions $w_{i,n}^\pm : I \to L_n(\mathbb{R}^d)$ given by $w_{i,n}^\pm(t) := D_1^n w_i^\pm(t,0)$ and a remainder $R_{i,m}^\pm$ satisfying $\lim_{x\to 0} \frac{R_{i,m}^\pm(t,x)}{\|x\|^m} = 0$. In addition, let us introduce the mappings $H_{i,n}^\pm : I \to L_n(\mathbb{R}^d)$,

$$H_{i,n}^{\pm}(t)x_1\cdots x_n := P_i^{\pm}(t)D_2^n F_\lambda(t,0)x_1\cdots x_n$$

$$+P_i^{\mp}(t)\sum_{j=2}^{n-1}\sum_{(N_1,\dots,N_j)\in P_j^{<}(n)} D_2^j F_\lambda(t,0)W_{\#N_1}^{\pm}(t)x_{N_1}\cdots W_{\#N_j}^{\pm}(t)x_{N_j}$$

$$-\sum_{\substack{(N_1,N_2)\in P_2(n)\\ 0<\#N_1<n-1\\ N_2\neq\emptyset}} g_{\#N_1+1}^{\pm}(t)x_{N_1}\cdot g_{\#N_2}^{\pm}(t)x_{N_2},$$

where we use the terminology from Sect. 1.5 and (neglecting the index i)

$$W_n^{\pm}(t) := D_2^n W^{\pm}(t,0), \quad W^{\pm}(t,x) := P_i^{\pm}(t)x + w_i^{\pm}(t,x),$$

$$g_n^{\pm}(t) := D_2^n g^{\pm}(t,0), \qquad g^{\pm}(t,x) := A_\lambda(t)P_i^{\pm}(t)x + P_i^{\pm}(t)F(t,W^{\pm}(t,x)).$$

One has $H_{i,2}^{\pm}(t) = P_i^{\mp}(t)D_2^2F_\lambda(t,0)$, and for $n\in\{3,\dots,m\}$, the values $H_{i,n}^{\pm}(t)$ only depend on $w_{i,2}^{\pm},\dots,w_{i,n-1}^{\pm}$. We obtain that the Taylor coefficients $w_{i,n}^{\pm}$ fulfill linear differential equations with inhomogeneities $H_{i,n}^{\pm}$ (for the precise form of this *homological equation*, see [72, Theorem 4.2]) and their unique bounded solutions can be determined recursively from

Theorem 1.4 (Taylor Approximation of Integral Manifolds, cf. [72]). *Given the mappings $w_i^{\pm}: U\times I\to\mathbb{R}^d$ introduced in Theorem 1.3, their Taylor coefficients $w_{i,n}^{+}: I\to L_n(\mathbb{R}^d)$ in the expansion* (1.26) *can be determined recursively from the respective* Lyapunov-Perron integrals

$$w_{i,n}^{+}(t) = -\int_t^{\infty}\Phi_\lambda(t,s)H_{i,n}^{+}(s)_{\Phi_\lambda(s,t)P_i^{+}(t)}\,ds,$$

$$w_{i,n}^{-}(t) = \int_{-\infty}^{t}\Phi_\lambda(t,s)H_{i,n}^{-}(s)_{\Phi_\lambda(s,t)P_i^{-}(t)}\,ds \quad \textit{for all } n\in\{2,\dots,m\},$$

with the abbreviation $X_T x_1\dots x_n := X(Tx_1,\dots,Tx_n)$ for a symmetric n-linear form $X\in L_n(\mathbb{R}^d,\mathbb{R}^d)$, a matrix $T\in\mathbb{R}^{d\times d}$ and vectors $x_1,\dots,x_n\in\mathbb{R}^n$.

The above integrals for $w_{i,n}^{\pm}$ provide an explicit formula. However, in concrete examples we recommend a direct and less formal approach:

Example 1.15 (Tumor Growth, cf. [82]). Let us return to the tumor growth model from Example 1.6 with parametric perturbations

$$\begin{cases} \dot{x}_1 = \dfrac{\lambda_0 x_1}{1 + \dfrac{\lambda_0}{\lambda_1} \sum_{j=1}^{4} x_j} - k_{pot} c_\lambda(t) x_1, \\ \dot{x}_2 = k_{pot} c_\lambda(t) x_1 - k x_2, \\ \dot{x}_3 = k(x_2 - x_3), \\ \dot{x}_4 = k(x_3 - x_4), \end{cases} \tag{$\mathfrak{T}_\lambda$}$$

i.e., we have $c_\lambda(t) = c_0 + \lambda c(t)$ with a bounded continuous function $c : \mathbb{R} \to \mathbb{R}$. The parameters are assumed to satisfy $c_0, k, k_{pot} > 0$. Restricting to the trivial solution to ($\mathfrak{T}_\lambda$), the variational equation (1.6) has the dichotomy spectrum

$$\Sigma_\lambda = \{-k\} \cup \left[\underline{\beta}(\lambda_0 - k_{pot} c_\lambda), \overline{\beta}(\lambda_0 - k_{pot} c_\lambda) \right].$$

For $\lambda = 0$ this reduces to $\Sigma_0 = \{-k, \lambda_0 - k_{pot} c_0\}$. Hence, the zero solution to ($\mathfrak{T}_0$) is asymptotically stable for $\lambda_0 < k_{pot} c_0$ and unstable for $\lambda_0 > k_{pot} c_0$. More interesting is the nonhyperbolic situation $\lambda_0 = k_{pot} c_0$ on which we focus now:

Denote the right-hand side of ($\mathfrak{T}_\lambda$) by $f_\lambda(t, x)$ and apply the linear transformation

$$\begin{pmatrix} y_0 \\ y_1 \\ y_2 \\ y_3 \\ y_4 \end{pmatrix} := \begin{pmatrix} 0 & 1\,0\,0 & 0 \\ \dfrac{k}{c_0 k_{pot}} & 0\,0\,0 & 0 \\ 1 & 0\,0\,0 & \dfrac{1}{k^2} \\ 1 & 0\,0\,\dfrac{1}{k} & 0 \\ 1 & 0\,1\,0 & 0 \end{pmatrix}^{-1} \begin{pmatrix} \lambda \\ x_1 \\ x_2 \\ x_3 \\ x_4 \end{pmatrix} = \begin{pmatrix} 0 & \dfrac{c_0 k_{pot}}{k} & 0\;0\;0 \\ 1 & 0 & 0\;0\;0 \\ 0 & -\dfrac{c_0 k_{pot}}{k} & 0\;0\;1 \\ 0 & -c_0 k_{pot} & 0\;k\;0 \\ 0 & -c_0 k k_{pot} & k^2\;0\;0 \end{pmatrix} \begin{pmatrix} \lambda \\ x_1 \\ x_2 \\ x_3 \\ x_4 \end{pmatrix}$$

to the five-dimensional ODE $\dot{\lambda} = 0$, $\dot{x} = f_\lambda(t, x)$. This readily implies

$$\dot{y} = Ay + F(t, y) \tag{1.27}$$

with the time-invariant linear part (in Jordan canonical form)

$$A := \begin{pmatrix} 0 & 0 & 0 & 0 & 0 \\ 0 & 0 & 0 & 0 & 0 \\ 0 & 0 & -k & 1 & 0 \\ 0 & 0 & 0 & -k & 1 \\ 0 & 0 & 0 & 0 & -k \end{pmatrix}$$

and the nonlinearity

$$F(t,y) := \begin{pmatrix} k_{pot}y_0\left(c_0\phi(y) - y_1c(t)\right) \\ 0 \\ -k_{pot}y_0y_1c(t) + c_0k_{pot}y_0\phi(y) + ky_2 - y_3 \\ -k_{pot}y_0y_1c(t) + c_0k_{pot}y_0\phi(y) + ky_3 - y_4 \\ -k_{pot}y_0y_1c(t) + c_0k_{pot}y_0\phi(y) + ky_4 \end{pmatrix},$$

where

$$\phi(y) := \frac{k^2\lambda_1}{k^2(c_0k_{pot}(3y_0 + y_2) + \lambda_1) + c_0kk_{pot}y_3 + c_0k_{pot}y_4 + k^3y_0} - 1.$$

Thanks to Theorem 1.3(b), the transformed equation (1.27) has a two-dimensional center-unstable manifold $\mathscr{W}_1^+ \subseteq \mathbb{R} \times \mathbb{R}^5$ given as graph of a mapping denoted as w. The ansatz

$$w(t, y_0, y_1) = \sum_{i=0}^{2} y_0^{2-i}y_1^i \begin{pmatrix} w_i^1(t) \\ w_i^2(t) \\ w_i^3(t) \end{pmatrix} + O\left(\sqrt{y_0^2 + y_1^2}^3\right)$$

in the invariance equation (1.22) yields the following homological equations

$$\dot{w}_0^1 = -kw_0^1 + \frac{3c_0^2k_{pot}^2 + c_0kk_{pot}}{\lambda_1} + w_0^2,$$

$$\dot{w}_0^2 = -kw_0^2 + \frac{3c_0^2kk_{pot}^2 + c_0k^2k_{pot}}{\lambda_1} + w_0^3,$$

$$\dot{w}_0^3 = -kw_0^3 + \frac{c_0k^2k_{pot}(3c_0k_{pot} + k)}{\lambda_1},$$

$$\dot{w}_1^1 = -kw_1^1 + k_{pot}c(t) + w_1^2, \qquad \dot{w}_2^1 = -kw_2^1 + w_2^2.$$

$$\dot{w}_1^2 = -kw_1^2 + kk_{pot}c(t) + w_1^3, \qquad \dot{w}_2^2 = -kw_2^2 + w_2^3,$$

$$\dot{w}_1^3 = -kw_1^3 + \frac{k^2(c_0k_{pot} + k)}{c_0}c(t), \qquad \dot{w}_2^3 = -kw_2^3$$

and their unique bounded entire solution is given by

$$\begin{pmatrix} w_0^1(t) \\ w_0^2(t) \\ w_0^3(t) \end{pmatrix} \equiv \begin{pmatrix} 3\frac{c_0k_{pot}(3c_0k_{pot}+k)}{k\lambda_1} \\ 2\frac{c_0k_{pot}(3c_0k_{pot}+k)}{\lambda_1} \\ \frac{c_0kk_{pot}(3c_0k_{pot}+k)}{\lambda_1} \end{pmatrix}, \qquad \begin{pmatrix} w_2^1(t) \\ w_2^2(t) \\ w_2^3(t) \end{pmatrix} \equiv 0 \quad \text{on } \mathbb{R},$$

as well as the successively given coefficients

$$w_1^3(t) = \frac{k^2(c_0k_{pot}+k)}{c_0}\int_{-\infty}^t e^{k(s-t)}c(s)\,ds,$$

$$w_1^2(t) = \int_{-\infty}^t e^{k(s-t)}kk_{pot}c(s) + w_1^3(s)\,ds,$$

$$w_1^1(t) = \int_{-\infty}^t e^{k(s-t)}k_{pot}c(s) + w_1^2(s)\,ds \quad \text{for all } t \in \mathbb{R}.$$

This finally shows that (1.27) reduced to $\mathscr{W}_1^-$ is given by the scalar equation

$$\dot{y}_0 = -k_{pot}\left(c(t)\lambda + \frac{c_0(k+3c_0k_{pot})}{\lambda_1}y_0\right)y_0 + O\left(y_0^3, y_0^2\lambda\right). \tag{1.28}$$

We thus observe a nonautonomous transcritical bifurcation of the trivial solution to both (1.27) and (1.28) in the sense of [75, Theorem 5.1]. In particular, depending on the Bohl exponents, the trivial solution is unstable for $\underline{\beta}(\lambda c) > 0$ and becomes asymptotically stable for $\overline{\beta}(\lambda c) < 0$.

1.9 Skew-Product, Control and Random Systems

Nonautonomous dynamics as presented so far was based on *processes* (or 2-*parameter semiflows*), i.e. continuous mappings $\varphi : \{(t,s,x) \in \mathbb{R}\times\mathbb{R}\times\Omega : s \le t\} \to \Omega$ satisfying

$$\varphi(t,t,x) = x, \quad \varphi(t,s,\varphi(s,\tau,x)) = \varphi(t,\tau,x) \quad \text{for all } \tau \le s \le t,\ x \in \Omega. \tag{1.29}$$

This has partly didactical reasons and is due to the fact that the general solution φ_λ to $(\mathfrak{D}_\lambda)$ fulfills (1.29) at least on maximal existence intervals.

Nevertheless, there is a further approach being theoretically important and flexible in applications at the same time (see [79, 83]). Indeed, a deterministic nonautonomous dynamical system can alternatively be formulated as a topological *skew-product flow* (θ,ϕ) consisting of a cocycle mapping ϕ on a state space X (a metric space) driven by an autonomous dynamical system θ acting on a base or parameter space P (also a metric space), which is particularly useful when the nonautonomity is due to periodic or almost periodic coefficients, e.g., [79, 83]. Specifically, $\theta = \{\theta_t : t \in \mathbb{R}\}$ is an *autonomous dynamical system* on P, i.e., a group of homeomorphisms under composition on P with the properties that

(α) $\theta_0(p) = p$ for all $p \in P$
(β) $\theta_{s+t}(p) = \theta_s(\theta_t(p))$ for all $s, t \in \mathbb{R}$
(γ) the mapping $(t, p) \mapsto \theta_t(p)$ is continuous,

and the *cocycle* mapping $\phi : \mathbb{R}^+ \times P \times X \to X$ satisfies

(a) $\phi(0, p, x) = x$ for all $(p, x) \in P \times X$
(b) $\phi(s + t, p, x) = \phi(s, \theta_t(p), \phi(t, p, x))$ for all $s, t \in \mathbb{R}^+$, $(p, x) \in P \times X$
(c) the mapping $(t, p, x) \mapsto \phi(t, p, x)$ is continuous.

Note that a skew-product flow (θ, ϕ) defines an autonomous semi-dynamical system on the product space $P \times X$. They include nonautonomous dynamical systems in the process formulation $\varphi(t, t_0, x_0)$ used above as a special case with a noncompact base $P = \mathbb{R}$, the shift operator $\theta_t(t_0) := t + t_0$ and the cocycle mapping $\phi(t, t_0, x_0) := \varphi(t + t_0, t_0, x_0)$, i.e., the parameter p is the initial time t_0.

A major advantage of the skew-product formulation appears when the parameter space P is compact, which arises in differential equations with periodic, almost periodic or almost automorphic forcing such as in the simplest situation

$$\dot{x} = -x + p(t).$$

Here, P is the *hull* of the inhomogeneity $p \in C(\mathbb{R}, \mathbb{R})$ defined by

$$P := \mathrm{cl}\,\{p(s + \cdot) : s \in \mathbb{R}\}$$

with the shift operator $\theta_t p(\cdot) := p(t + \cdot)$ and the closure taken in an appropriate topology.

Analogously, a set $\mathscr{A} = \{(p, A(p)) : p \in P\}$ with nonempty compact fibers $\mathscr{A}(p) \subseteq X$ is said to be *ϕ-invariant*, if $\phi(t, p, A(p)) = A(\theta_t(p))$ for all $t \in \mathbb{R}$ and $p \in P$ and *pullback attracting*, if

$$\mathrm{dist}_X(\phi(t, \theta_{-t}(r), D), A(p)) \to 0 \quad \text{as } t \to \infty$$

for appropriate bounded subsets $D \subseteq X$. See the monograph [52] for more details and examples.

Control Systems

A further class of intrinsically nonautonomous problems are *control systems*. While this chapter basically dealt with deterministic ODEs $(\mathfrak{D}_\lambda)$ where the parameter λ is a real number (or a tuple of them), control theory is concerned with the situation when

λ is a function in time—typically denoted as control function u from an appropriate set or function space U. In this field, the central question is to investigate how changing these functions within U affects the solutions or their long-term behavior. This gives rise to a number of different questions like controllability or "optimal control" not tackled here and we refer to e.g. [84] for further information.

We finally point out that also control systems allow a formulation as skew-product flows and refer to [17] for details.

Random Dynamical Systems

A *random dynamical system* is defined similarly, except that the base space P is now the sample space Ω of a probability space $(\Omega, \mathscr{F}, \mathbb{P})$ and continuity properties w.r.t. p are now replaced by measurability in $\omega \in \Omega$. In particular, a random attractor $\mathscr{A} = \{(\omega, A(\omega)) : \omega \in \Omega\}$ consists of nonempty compact fibers $\mathscr{A}(\omega) \subseteq X$ such that the set valued mapping $\omega \mapsto A(\omega)$ becomes $\mathscr{F}$-measurable. For details, we refer to [6, 14, 52] and the Chap. 2 by de Freitas and Sontag in this volume. In fact, de Freitas and Sontag propose a novel generalization of random dynamical systems that includes control and uncertainty as well as randomness.

Acknowledgements Peter E. Kloeden was partially supported by the DFG grant KL 1203/7-1, the Spanish Ministerio de Ciencia e Innovación project MTM2011-22411, the Consejería de Innovación, Ciencia y Empresa (Junta de Andalucía) under the Ayuda 2009/FQM314 and the Proyecto de Excelencia P07-FQM-02468.

References

1. B.M. Adams, H.T. Banks, J.E. Banks, J.D. Stark, Population dynamics models in plant-insect herbivore-pesticide interactions. Math. Biosci. **196**(1), 39–64 (2005)
2. L.Ya. Adrianova, *Introduction to Linear Systems of Differential Equations*. Translations of Mathematical Monographs, vol. 146 (AMS, Providence, 1995)
3. H. Amann, *Ordinary Differential Equations: An Introduction to Nonlinear Analysis*. Studies in Mathematics, vol. 13 (Walter De Gruyter, Berlin, 1990)
4. M. Anguiano, P.E. Kloeden, Nonautonomous SIR equations with diffusion. Commun. Pure Appl. Anal. **13**, 157–173 (2014)
5. B. Aulbach, T. Wanner, in *Integral Manifolds for Carathéodory Type Differential Equations in Banach Spaces*, ed. by B. Aulbach, F. Colonius. Six Lectures on Dynamical Systems (World Scientific, Singapore, 1996), pp. 45–119
6. M. Bachar, J. Batzel, S. Ditlevsen, *Stochastic Biomathematical Models*. Lecture Notes in Mathematics (Mathematical Biosciences Subseries), vol. 2058 (Springer, Berlin, 2013)
7. H.T. Banks, J.E. Banks, S.L. Joyner, J.D. Stark, Dynamic models for insect mortality due to exposure to insecticides. Math. Comput. Model. **48**(1–2), 316–332 (2008)
8. J. Baranyi, T.A. Roberts, P. McClure, A non-autonomous differential equation to model bacterial growth. Food Microbiol. **10**, 43–59 (1993)
9. M. Barenco, D. Tomescu, D. Brewer, R. Callard, J. Stark, M. Hubank, Ranked prediction of p53 targets using hidden variable dynamic modeling. Genome Biol. **7**(3) (2006)

10. B. Baeumer et al., Predicting the drug release kinetics of matrix tablets. Discrete Contin. Dyn. Syst. Ser. B **17**(2), 261–277 (2009)
11. C. Castillo-Chavez, B. Song, Dynamical models of tubercolosis and their applications. Math. Biosci. Eng. **1**(2), 361–404 (2004)
12. D.N. Cheban, P.E. Kloeden, B. Schmalfuß, The relationship between pullback, forward and global attractors of nonautonomous dynamical systems. Nonlin. Dynam. Syst. Theory **2**, 9–28 (2002)
13. S.-N. Chow, C. Li, D. Wang, *Normal Forms and Bifurcation of Planar Vector Fields* (Cambridge University Press, Cambridge, 1994)
14. I. Chueshov, *Monotone Random Systems Theory and Applications*. Lecture Notes in Mathematics, vol. 1779 (Springer, Berlin, 2002)
15. C.F. Clancy, M.J.A. O'Callaghan, T.C. Kelly, A multi-scale problem arising in a model of avian flu virus in a seabird colony. J. Phys. Conf. Ser. **55**, 45–54 (2006)
16. N.G. Cogan, Effects of persister formation on bacterial response to dosing. J. Theor. Biol. **238**(3), 694–703 (2006)
17. F. Colonius, W. Kliemann, *The Dynamics of Control* (Birkhäuser, Basel, 1999)
18. W.A. Coppel, *Dichotomies in Stability Theory*. Lecture Notes in Mathematics, vol. 629 (Springer, Berlin, 1978)
19. J.L. Daleckiĭ, M.G. Kreĭn, *Stability of Solutions of Differential Equations in Banach Space*. Translations of Mathematical Monographs, vol. 43 (AMS, Providence, 1974)
20. G. De Nicolao et al., *A Minimal Model Describing the Effect of Drug Administration on Tumor Growth Dynamics*. 14th Mediterranean Conference on Control and Automation (2006). doi:10.1109/MED.2006.328783
21. L.G. de Pillis, A. Radunskaya, The dynamics of an optimally controlled tumor model: a case study, Math. Comput. Model. **37**, 1221–1244 (2003)
22. L. Dieci, E.S. van Vleck, *Lyapunov and Other Spectra: A Survey*. Collected Lectures on the Preservation of Stability under Discretization (SIAM, Philadelphia, 2002), pp. 197–218
23. L. Dieci, E.S. van Vleck, Lyapunov and Sacker-Sell spectral intervals. J. Dyn. Differ. Equ. **19**(2), 265–293 (2007)
24. J. Dushoff, J.B. Plotkin, S.A. Levin, D.J.D. Earn, Dynamic resonance can account for seasonality of influenza epidemics. Proc. Natl. Acad. Sci. USA **101**(48), 16915–16916 (2004)
25. R. Eftimie, J.L. Bramson, D.J.D. Earn, Interaction between the immune system and cancer: a brief review of non-spatial mathematical models. Bull. Math. Biol. **73**, 2–32 (2011)
26. R. Fabbri, R.A. Johnson, F. Mantellini, A nonautonomous saddle-node bifurcation pattern. Stoch. Dyn. **4**(3), 335–350 (2004)
27. W. Garira, S.D. Musekwa, T. Shiri, Optimal control of combined therapy in a single strain HIV-1 model. Electron. J. Differ. Equ. **2005**(52), 1–22 (2005)
28. I. Győri, S. Michelson, J. Leith, Time-dependent subpopulation induction in heterogeneous tumors. Bull. Math. Biol. **50**(6,) 681–696 (1988)
29. P. Hahnfeldt, D. Panigrahy, J. Folkman, L.R. Hlatky, Tumor development under angiogenic signaling: a dynamical theory of tumor growth, treatment response, and postvascular dormancy. Cancer Res. **59**, 4770–4775 (1999)
30. G. Herzog, R. Redheffer, Nonautonomous SEIRS and Thron models for epidemiology and cell biology. Nonlinear Anal. Real World Appl. **4**, 33–44 (2004)
31. M.W. Hirsch, S. Smale, *Differential Equations, Dynamical Systems, and Linear Algebra* (Academic, Boston, 1974)
32. Z. Hu, P. Bi, W. Ma, S. Ruan, Bifurcations of an SIRS epidemic model with nonlinear incidence rate. Discrete Contin. Dyn. Syst. Ser. B **15**(3), 93–112 (2011)
33. T. Hüls, Homoclinic trajectories of non-autonomous maps. J. Differ. Equ. Appl. **17**(1), 9–31 (2011)
34. M. Imran, H.L. Smith, The pharmacodynamics of antibiotic treatment. Comput. Math. Methods Med. **7**(4), 229–263 (2006)
35. M. Imran, H.L. Smith, The dynamics of bacterial infection, innate immune response, and antibiotic treatment. Discrete Contin. Dyn. Syst. Ser. B **8**(1), 127–145 (2007)

36. B. Janssen, L. Révéesz, Analysis of the growth of tumor cell populations. Math. Biosci. **19**(1–2), 131–154 (1974)
37. R.A. Johnson, F. Mantellini, A nonautonomous transcritical bifurcation problem with an application to quasi-periodic bubbles. Discrete Contin. Dyn. Syst. **9**(1), 209–224 (2003)
38. G.P. Karev, A.S. Novozhilov, E.V. Koonin, Mathematical modeling of tumor therapy with oncolytic viruses: effects of parametric heterogeneity on cell dynamics. Biol. Direct **1** (2006)
39. J. Keener, J. Sneyd, *Mathematical Physiology*, vol. I & II, 2nd edn. (Springer, Heidelberg, 2009)
40. D. Kirschner, G.F. Webb, A model for treatment strategy in the chemotherapy of AIDS. Bull. Math. Biol. **58**(1), 367–390 (1996)
41. D. Kirschner, G.F. Webb, Understanding drug resistance for monotherapy treatment of HIV infection. Bull. Math. Biol. **59**(4), 763–785 (1997)
42. D. Kirschner, G.F. Webb, Immunotherapy of HIV-1 infection, J. Biol. Syst. **6**(1), 71–83 (1998)
43. D. Kirschner, S. Lenhart, S. Serbin, Optimal control of the chemotherapy of HIV. J. Math. Biol. **35**, 775–792 (1997)
44. P.E. Kloeden, Pullback attractors in nonautonomous difference equations. J. Differ. Equ. Appl. **6**(1), 33–52 (2000)
45. P.E. Kloeden, Pitchfork and transcritical bifurcations in systems with homogenous nonlinearities and an almost periodic time coefficient. Commun. Pure Appl. Anal. **1**(4), 1–14 (2002)
46. P.E. Kloeden, Pullback attractors for nonautonomous semidynamical systems. Stoch. Dyn. **3**(1), 101–112 (2003)
47. P.E. Kloeden, Nonautonomous attractors of switching systems. Dyn. Syst. **21**, 209–230 (2006)
48. P.E. Kloeden, V. Kozyakin, The dynamics of epidemiological systems with nonautonomous and random coefficients. Math. Eng. Sci. Aerosp. **2**(2), 105–118 (2011)
49. P.E. Kloeden, V. Kozyakin, Asymptotic behaviour of random Markov chains with tridiagonal generators. Bull. Aust. Math. Soc. **87**, 27–36 (2013)
50. P.E. Kloeden, V. Kozyakin, Asymptotic behaviour of random tridiagonal Markov chains in biological applications. Discrete Contin. Dyn. Syst. Ser. B **18**(2), 453–466 (2013)
51. P.E. Kloeden, C. Pötzsche, *Nonautonomous Bifurcation Scenarios in SIR Models*, Manuscript (2013)
52. P.E. Kloeden, M. Rasmussen, *Nonautonomous Dynamical Systems* (AMS, Providence, 2011)
53. P.E. Kloeden, S. Siegmund, Bifurcations and continuous transitions of attractors in autonomous and nonautonomous systems. Int. J. Bifurcat. Chaos **5**(2), 1–21 (2005)
54. P.E. Kloeden, C. Pötzsche, M. Rasmussen, Discrete-time nonautonomous dynamical systems, in *Stability and Bifurcation in Non-Autonomous Differential Equations*, ed. by R. Johnson, M.P. Pera. Lecture Notes in Mathematics, vol. 2065 (Springer, Berlin, 2012)
55. M.Y. Li, J.R. Graef, L. Wang, J. Karsai, Global dynamics of a SEIR model with varying total population size. Math. Biosci. **160**, 191–213 (1999)
56. R.M. Lopez, B.R. Morin, S.K. Suslov, *Logistic models with time-dependent coefficients and some of their applications* (2011, preprint)
57. A. Makroglou, J. Li, Y.K. Kuang, Mathematical models and software tools for the glucose-insulin regulatory system and diabeties: an overview. Appl. Numer. Math. **56**, 559–573 (2006)
58. S. Michelson, B.E. Miller, A.S. Glicksmann, J. Leith, Tumor micro-ecology and competitive interactions. J. Theor. Biol. **128**, 233–246 (1987)
59. S. Mohamad, K. Gopalsamy, Neuronal dynamics in the time varying environments: continuous and discrete time models. Discrete Contin. Dyn. Syst. **6**(4), 841–860 (2000)
60. H. Moore, W. Gu, A mathematical model for treatment-resistant mutations of HIV. Math. Biosci. Eng. **2**(2), 363–380 (2005)
61. L. Moreau, E.D. Sontag, M. Arcak, Feedback tuning of bifurcations. Syst. Control Lett. **50**(3), 229–239 (2003)
62. J.D. Murray, *Mathematical Biology: I. An Introduction*. Interdisciplinary Applied Mathematics, vol. 17, 3rd edn. (Springer, Berlin, 2001)
63. C. Núñez, R. Obaya, A non-autonomous bifurcation theory for deterministic scalar differential equations. Discrete Contin. Dyn. Syst. Ser. B **9**(3–4), 701–730 (2008)

64. K.J. Palmer, Exponential dichotomies for almost periodic equations. Proc. Am. Math. Soc. **101**, 293–298 (1987)
65. P. Palumbo, W. Clausen, S. Panunzi, A. De Gaetano, Linear periodic models of subcutaneous insulin absorption. HERMIS **6**, 60–79 (2005)
66. C. Pötzsche, Exponential dichotomies of linear dynamic equations on measure chains under slowly varying coefficients, J. Math. Anal. Appl. **289**, 317–335 (2004)
67. C. Pötzsche, Robustness of hyperbolic solutions under parametric perturbations. J. Differ. Equ. Appl. **15**(8–9), 803–819 (2009)
68. C. Pötzsche, Nonautonomous bifurcation of bounded solutions I: a Lyapunov-Schmidt approach. Discrete Contin. Dyn. Syst. Ser. B **14**(2), 739–776 (2010)
69. C. Pötzsche, Nonautonomous bifurcation of bounded solutions II: a shovel bifurcation pattern. Discrete Contin. Dyn. Syst. Ser. A **31**(1), 941–973 (2011)
70. C. Pötzsche, Nonautonomous continuation of bounded solutions. Commun. Pure Appl. Anal. **10**(3), 937–961 (2011)
71. C. Pötzsche, Bifurcations in nonautonomous dynamical systems: Results and tools in discrete time, in *Proceedings of the Workshop on "Future Directions in Difference Equations", Vigo, Spain, 2011*, ed. by E. Liz. Colección Congresos, no. 69, Servizo de Publicacións de Universidade de Vigo, 13–17 June 2011, pp. 163–212
72. C. Pötzsche, M. Rasmussen, Taylor approximation of integral manifolds. J. Dyn. Differ. Equ. **18**(2), 427–460 (2006)
73. M. Rasmussen, Towards a bifurcation theory for nonautonomous difference equation. J. Differ. Equ. Appl. **12**(3–4), 297–312 (2006)
74. M. Rasmussen, *Attractivity and Bifurcation for Nonautonomous Dynamical Systems*. Lecture Notes in Mathematics, vol. 1907 (Springer, Berlin, 2007)
75. M. Rasmussen, Nonautonomous bifurcation patterns for one-dimensional differential equations. J. Differ. Equ. **234**, 267–288 (2007)
76. R.K. Sachs, L.R. Hlatky, P. Hahnfeldt, Simple ODE models of tumor growth and anti-angiogenic or radiation treatment. Math. Comput. Model. **33**, 1297–1305 (2001)
77. R.J. Sacker, G.R. Sell, A spectral theory for linear differential systems. J. Differ. Equ. **27**, 320–358 (1978)
78. W.M. Schaffer, T.V. Bronnikova, Parametric dependence in model epidemics. I: contact-related parameters. J. Biol. Dyn. **1**(2), 183–195 (2007)
79. G.R. Sell, *Topological Dynamics and Differential Equations* (Van Nostrand Reinhold, London, 1971)
80. G.R. Sell, The structure of a flow in the vicinity of an almost periodic motion. J. Differ. Equ. **27**(3), 359–393 (1978)
81. S. Siegmund, Dichotomy spectrum for nonautonomous differential equations. J. Dyn. Differ. Equ. **14**(1), 243–258 (2002)
82. M. Simeoni, et al, Predictive pharmacokinetic-pharmacodynamic modeling of tumor growth kinetics in xenograft models after administration of anticancer agents. Cancer Res. **64**, 1094–1101 (2004)
83. W. Shen, Y. Yi, *Almost Automorphic and Almost Periodic Dynamics in Skew-product Semi-flows*. Memoirs of the AMS, vol. 647 (AMS, Providence, 1998)
84. E.D. Sontag, *Mathematical Control Theory*. Texts in Applied Mathematics, vol. 6, 2nd edn. (Springer, New York, 1998)
85. E.D. Sontag, Some new directions in control theory inspired by systems biology. Syst. Biol. **1**(1), 9–18 (2004)
86. E.D. Sontag, Molecular systems biology and control. Europ. J. Control **11**, 1–40 (2005)
87. H.R. Thieme, Uniform weak implies uniform strong persistence for non-autonomous semi-flows. Proc. Am. Math. Soc. **127**(8), 2395–2403 (1999)
88. H.R. Thieme, Uniform persistence and permanence for non-autonomous semiflows in population biology. Math. Biosci. **166**, 173–201 (2000)
89. M. Turelli, Random environments and stochastic calculus. Theor. Popul. Biol. **12**, 140–178 (1977)

90. H. Wang, J. Li, Y.K. Kuang, Mathematical modeling and qualitative analysis of insulin therapies. Math. Biosci. **210**, 17–33 (2007)
91. T. Zhang, Z. Teng, On a nonautonomous SEIRS model in epidemiology. Bull. Math. Biol. **69**, 2537–2559 (2007)
92. X.-Q. Zhao, Persistence in almost periodic predator-prey reaction-diffusion equations, in *Dynamical Systems and their Application in Biology*, ed. by S. Ruan, G.S.K. Wolkowicz, J. Wu. Fields Institute Communications (AMS, Providence, 2003), pp. 259–268
93. M. Zhien, B. Song, T.G. Hallam, The threshold of survival for systems in a fluctuating environment. Bull. Math. Biol. **53**(3), 311–323 (1989)

Chapter 2
Random Dynamical Systems with Inputs

Michael Marcondes de Freitas and Eduardo D. Sontag

Abstract This work introduces a notion of random dynamical systems with inputs, providing several basic definitions and results on equilibria and convergence. It also presents a "converging input to converging state" ("CICS") result, a concept that plays a key role in the analysis of stability of feedback interconnections, for monotone systems.

Keywords Pullback convergence • Random dynamical systems • Stochastic dynamics

2.1 Introduction

In the late 1980s, Ludwig Arnold conceived an elegant and deep approach to the foundations of random dynamics [3]. His paradigm of a *random dynamical system* (RDS for short) is based on an ultimately simple idea: view an RDS as consisting of two ingredients, a stochastic but autonomous "noise process," and a classical dynamical system that is driven by this process. The noise process is described by a measure-preserving dynamical system. It is typically probabilistic, representing for example environmental perturbations, internal variability, randomly fluctuating parameters, model uncertainty, or measurement errors. But the formalism allows for deterministic periodic or almost-periodic driving processes as well. The resulting theory, developed since by many authors, provides a seamless integration of classical ergodic theory with modern dynamical systems, giving a theoretical framework parallel to classical smooth and topological dynamics (stability, attractors, bifurcation theory, and so forth), while allowing one to treat ina unified way the most

M. Marcondes de Freitas · E.D. Sontag (✉)
Department of Mathematics, Rutgers University, Piscataway, NJ, USA
e-mail: marcfrei@math.rutgers.edu; sontag@math.rutgers.edu

P.E. Kloeden and C. Pötzsche (eds.), *Nonautonomous Dynamical Systems in the Life Sciences*, Lecture Notes in Mathematics 2102, DOI 10.1007/978-3-319-03080-7_2,

important classes of dynamical systems with randomness—random differential or difference equations (basically, deterministic systems with randomly changing parameters), or stochastic ordinary and partial differential equations (white noise or, more generally, martingale-driven systems as studied in the Itô calculus). The main goal of this chapter is to propose a new RDS-based formalism for random control systems, that is, systems with inputs (and outputs), which we abbreviate RDSI (or RDSIO).

Why Systems with Inputs and Outputs?

Our motivation for studying RDS with inputs and outputs arises from the need to provide foundations for a constructive theory of interconnections and feedback for stochastic systems, one that will eventually generalize successful and widely applied deterministic approaches to the analysis and design of dynamic networks [17, 19, 20]. To motivate this need and in order to set the stage for our definitions, let us start by recalling the basic paradigm of (deterministic) control theory. We use for concreteness ordinary differential equations. (For a more abstract general dynamical systems approach, see [30], as well as the definition of RDSIO's in this chapter.) The objects of study are systems with inputs and outputs:

$$\begin{aligned} \dot{x}_1(t) &= f_1(x_1(t), \ldots, x_n(t), u_1(t), \ldots, u_m(t)) \\ &\vdots \\ \dot{x}_n(t) &= f_n(\underbrace{x_1(t), \ldots, x_n(t)}_{\text{states}}, \underbrace{u_1(t), \ldots, u_m(t)}_{\text{inputs}}) \end{aligned}$$

supplemented by a set of output variables $y_1, \ldots, y_p$ that are functions of the state vector x:

$$y_j(t) = h_j(x(t)), \quad j = 1, \ldots, p.$$

The inputs $u_i(t)$ may be viewed as controls, forcing functions, external signals, or stimuli, depending on the context. The outputs y_j represent responses, typically a partial read-out of the system state vector $(x_1, \ldots, x_n)$. Such a formalism, which originated in the analysis of engineering systems, is also natural in biology. Cells are not autonomous systems; they process external information, provided by physical (UV or other radiation, mechanical, temperature) or chemical (drugs, growth factors, hormones, nutrients) inputs. They also produce signals which we may view as outputs, such as chemical signals sent to other cells, commands to motors that move flagella or pseudopods, or the internal activation of transcription factors which may be monitored by measurement technologies. Thus, the control-theory formalism—in contrast to dynamical-systems theory, which deals with isolated systems—is not only reasonable, but natural in biology.

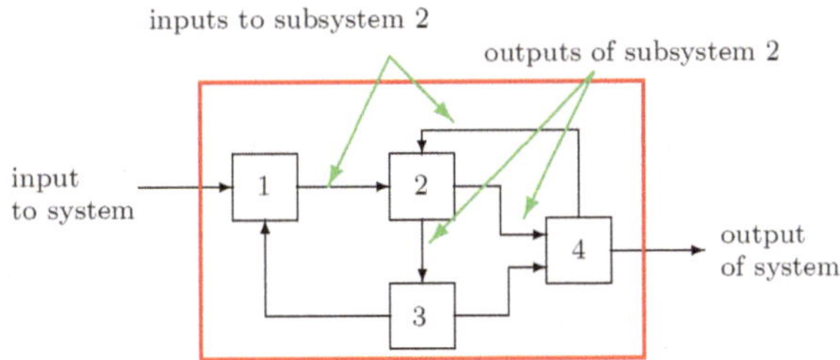

Fig. 2.1 A system viewed as an interconnection of subsystems with inputs and outputs

There is also a somewhat different reason for considering systems with inputs and outputs. Cells can be seen as composed of a large number of subsystems, networks of proteins, RNA, DNA, and metabolites involved in various processes such as cell growth and maintenance, division, and death. Indeed, one of the important themes in current molecular biology [9,15,22] is that of understanding cell behavior in terms of cascades and feedback interconnections of elementary "modules." The hope is that one should be able to decompose large systems into, hopefully simpler, subsystems, and then study the emergent properties of interconnections. Diagrammatically, one might represent this situation by a graph as in Fig. 2.1, which shows an overall system as composed of four subsystems. In Fig. 2.1, there are inputs and outputs for the overall system. However, even if the entire system were autonomous (no arrows into or out of the large box), in order to be able to define such interconnections, one must necessarily consider subsystems that admit time-dependent input signals and which produce output signals. Thus, the control theoretic formalism is a necessity even in the analysis of autonomous systems, when using a decomposition-based approach. Observe that, if the behavior of subsystems is subject to random effects, then it is imperative to allow inputs to be random when studying subsystems: for example, the subsystem "2" in Fig. 2.1 has inputs that depend on subsystems "1" and "4" and thus, if these are described by random processes, the inputs to "2" are also random processes.

As an illustration of how these ideas play out in the deterministic case, consider an inhibitory or activating cyclic structure

$$\begin{aligned} \dot{x}_1 &= f_1(x_n, x_1) \\ \dot{x}_2 &= f_2(x_1, x_2) \\ &\vdots \\ \dot{x}_n &= f_n(x_{n-1}, x_n), \end{aligned}$$

as diagrammed in the left panel of Fig. 2.2. This is the "Goodwin model" of gene expression, and appears as well in many other models in mathematical biology (e.g. [14, 25]). It has been much studied mathematically, notably by Mallet-Paret and others [13,24,27,28], which among other major results, established a Poincaré-Bendixson theorem which tightly characterizes Ω-limit sets for such systems in

Fig. 2.2 A cyclic system (*left*), built by feedback from a cascade of n systems (*right*)

terms of periodic orbits and heteroclinic connections among equilibria. In the present context, we wish to view the system as built out of n components x_i, $i = 1, \ldots, n$. These components form a "cascade" or "series interconnection" when the feedback connection is ignored (right panel of Fig. 2.2). This view has been very successful when combined with tools from passivity theory [2], input-to-state (ISS) stability [29], and monotone systems with inputs and outputs [1]. To be more concrete, suppose, for example, that the system has the following special form, with each x_i scalar:

$$\begin{aligned}
\dot{x}_1 &= b_1\kappa_1(x_n) - a_1 x_1 \\
\dot{x}_2 &= b_2\kappa_2(x_1) - a_2 x_2 \\
&\vdots \\
\dot{x}_n &= b_n\kappa_n(x_{n-1}) - a_n x_n,
\end{aligned}$$

where the a_i's and b_i's are (for the moment) positive constants. The functions $\kappa_i(x_{i-1})$ represent the way in which the previous state in the cycle affects the given state. "Opening up" the feedback loop amounts to studying the system:

$$\begin{aligned}
\dot{x}_1 &= b_1\kappa_1(u) - a_1 x_1 \\
\dot{x}_2 &= b_2\kappa_2(x_1) - a_2 x_2 \\
&\vdots \\
\dot{x}_n &= b_n\kappa_n(x_{n-1}) - a_n x_n,
\end{aligned}$$

in which now u represents an external input. We may, in turn, view this open system as an interconnection of n subsystems

$$\dot{x} = b_i\kappa_i(u) - a_i x.$$

The hope is to be able to conclude something interesting about the overall system by the following two steps: (1) study the "open" system by recursively interconnecting the systems $\dot{x} = b_i\kappa_i(u) - a_i x$ until the whole system is obtained, and then (2) study the effect of "closing the loop" with feedback to recover the original system. The key property needed in the first step, at least in order to recursively study stability, is the CICS property: the state $x_i(t)$ should converge to an equilibrium provided that the input $u(t)$ converges to a limit. Obviously, in this simple example CICS is trivially true (assuming that κ is continuous), since we just have a forced linear system, easily solved in closed form using variation of parameters. However, for general nonlinear systems, CICS fails even for systems which are globally asymptotically stable with respect to constant inputs. This motivated, for deterministic systems, the introduction of the notions of ISS [29] and of monotone systems with inputs [1], either of which allows one to obtain CICS types of theorems, and these approaches coupled with what are generically called "small-gain theorems" (essentially, asking that the feedback loop results in a contraction in an appropriate sense) allow one to complete the program (step 2). In this work, we focus exclusively on the CICS problem (step 1) for stochastic systems, and leave the study of small-gain theorems for follow-up work.

Stochastic extensions of deterministic theory should take full advantage of the power of ergodic theory. Suppose, continuing with the above simple example, that we have the scalar linear system $\dot{x} = bu - ax$, where a and b are now not constants but are randomly varying, $a = a(\omega)$, $b = b(\omega)$. Randomness might model the effect of cell-to-cell variability in essential enzymes, or physical factors such as temperature or pH. If $a(\omega) \leq -\lambda < 0$ for all ω (and b is, for example, bounded), then stability will not be an issue. However, it may be that the only possible assumption is that the expected value of $a(\omega)$ is negative, but $a(\omega)$ might take zero, or even positive, values (for example, a might be a difference between an autocatalytic term of production and a degradation/dilution term). Then, ergodic theory is needed in order to establish results on almost-sure stability (or convergence to steady-state probability distributions). We feel, therefore, that an RDS-based theory is most natural in this context.

Much work has been done on random control systems, but not employing an RDS axiomatic approach. This includes the papers [11, 26] on stochastic stabilization, as well as the papers [7, 8, 31] on feedback stabilization using noise to state stability analogs of input to state stability. We believe, however, that an RDS approach is a useful addition to the literature, for the reasons mentioned above. Also very relevant is an extension [6] of RDS to allow (*deterministic*) inputs that are themselves generated by a dynamical system (in the terminology of regulation and disturbance rejection, one would say that inputs are generated by an "exosystem").

Outline of Chapter

We first review the classical RDS theory. This material is not new; however, with an eye to generalizations, we reformulate it in a slightly different language. We next define our new concept of RDSI (and RDSIO), which extends the notion of RDS to systems in which there is an external input or forcing function, which is itself a stochastic process. A major contribution of this work lies upon the precise formulation of this concept, particularly the way in which the stochastic argument of the input is shifted in the semigroup (cocycle) property. Note that stochasticity of inputs is essential if one is to develop a theory of interconnected subsystems, as an input to one system in such an interconnection is typically obtained by using a combination of outputs (necessarily random) of other subsystems. After establishing several basic results that provide a foundation for further study, we turn to the question of "converging input to converging state" (CICS) properties. Specifically, recent work by Chueshov [5] introduced the class of monotone RDS (without inputs), a theory that provides us with the concepts needed to pursue the generalization of the latter to RDSI. Thus, we introduce also a class of monotone RDSI, and are able to formulate and prove a CICS theorem for monotone systems. A follow-up of this work will provide a small-gain theorem for monotone RDSI, generalizing [1], which follows from the CICS tools developed here. Separate work in progress deals with generalizations of ISS. Space prevents giving many examples, so we limit ourselves to a simple linear RODE (a pathwise random ODE). In principle, however, our setup also allows one to study more complicated objects including stochastic differential equations as in the Itô calculus. (A good reference for RODE's and SDE's in the context of RDS is the original book by Arnold [3]; see also [18]).

Other chapters in this volume deal with concepts closely related to those discussed in this chapter. Linear systems with inputs are considered, for example, in Chap. 1, Example 1.4, when viewing the transcription factor activity $f(t)$ as an input. Pullback limits are discussed in Example 1.7 of that same chapter, and especially at the end of Sect. 1.6, where the significance of this concept is discussed. Cascade flows (semi-direct products, skew-product flows) are described in Sect. 1.9. The mass-action kinetics model of the JAK-STAT signal transduction pathway described in Chap. 9, (9.7), can also be interpreted as a cascade closed under the feedback of x_4 into the first coordinate. It is in fact a monotone system. Finally, the base model given in Chap. 8, Sect. 8.2.6 for hepatitis C virus viral kinetics in chronically infected patients, can be interpreted as a closed-loop system. More specifically, it can be viewed as the closed-loop obtained from a monotone stochastic RDS (with cone $\mathbb{R}_{\geq 0} \times \mathbb{R}_{\geq 0} \times \mathbb{R}_{\leq 0}$, and when the term $T(t)$ in the equation for $I(t)$ is viewed as an input), closed under "negative" feedback, when setting this input again to $T(t)$.

2.2 Random Dynamical Systems

We first review the random dynamical systems framework of Arnold [3]. Along the way we introduce a couple of pieces of terminology not found in [3], to facilitate the discussion. Suppose given a *measure preserving dynamical system*[1] (*MPDS*)

$$\theta = (\Omega, \mathscr{F}, \mathbb{P}, \{\theta_t\}_{t \in \mathscr{T}});$$

[1] Arnold [3, p. 635] and Chueshov [5, p. 10, Definition 1.1.1] refer to such an object primarily as a *metric dynamical system*. We find *measure preserving*, which Arnold also uses as a synonym, less confusing and more informative.

that is, a probability space $(\Omega, \mathscr{F}, \mathbb{P})$, a topological group $(\mathscr{T}, +)$, and a measurable flow $\{\theta_t\}_{t\in\mathscr{T}}$ of measure preserving maps $\Omega \to \Omega$ satisfying (T1)–(T3):

(T1) $(t,\omega) \mapsto \theta_t\omega$, $(t,\omega) \in \mathscr{T} \times \Omega$, is $(\mathscr{B}(\mathscr{T}) \otimes \mathscr{F})$-measurable,
(T2) $\theta_{t+s} = \theta_t \circ \theta_s$ for every $t, s \in \mathscr{T}$ (semigroup property),
(T3) $\mathbb{P} \circ \theta_t = \mathbb{P}$ for each $t \in \mathscr{T}$ (measure preserving[2]).

In this work $\mathscr{T}$ will always refer to either $\mathbb{R}$ or $\mathbb{Z}$, depending on whether one is talking about continuous or discrete time, respectively. In either case $\mathscr{T}_{\geq 0}$ refers to the nonnegative elements of $\mathscr{T}$. We will occasionally need to make measure-theoretic considerations about $\mathscr{T}$ or Borel subsets of it. If $\mathscr{T} = \mathbb{R}$, that is, in continuous time, then we tacitly equip any Borel subset of $\mathscr{T}$ with the measure induced by the Lebesgue measure on $\mathbb{R}$. If $\mathscr{T} = \mathbb{Z}$, or in discrete time, then we think of the counting measure in $\mathbb{Z}$. When $\mathscr{T} = \mathbb{Z}$, it follows from (T2) that θ is completely determined by $\theta_1 = \theta(1,\cdot)$. In that case we will abuse the notation and use the same θ to denote both the underlying MPDS and θ_1.

In the context of a given MPDS θ, a set $B \in \mathscr{F}$ is said to be θ-invariant if $\theta_t(B) = B$ for all $t \in \mathscr{T}$. We say that an MPDS θ is *ergodic* (*under* $\mathbb{P}$) if, whenever $B \in \mathscr{F}$ is θ-invariant, then we have either $\mathbb{P}(B) = 0$ or $\mathbb{P}(B) = 1$.

Let X be a metric space constituting the measurable space $(X, \mathscr{B})$ when equipped with the σ-algebra $\mathscr{B}$ of Borel subsets of X. A (*continuous*) *random dynamical system* (*RDS*) *on* X is a pair (θ, φ) in which θ is an MPDS and

$$\varphi : \mathscr{T}_{\geq 0} \times \Omega \times X \longrightarrow X$$

is a (*continuous*) *cocycle over* θ; that is, a $(\mathscr{B}(\mathscr{T}_{\geq 0}) \otimes \mathscr{F} \otimes \mathscr{B})$-measurable map such that

(S1) $\varphi(t,\omega) := \varphi(t,\omega,\cdot) : X \to X$ is continuous for each $t \in \mathscr{T}_{\geq 0}$, $\omega \in \Omega$,
(S2) $\varphi(0,\omega) = \mathrm{id}_X$ for each $\omega \in \Omega$, and (cocycle property)

$$\varphi(t+s,\omega) = \varphi(t,\theta_s\omega) \circ \varphi(s,\omega), \quad \forall s,t \in \mathscr{T}_{\geq 0},\ \forall \omega \in \Omega.$$

The cocycle property generalizes the semigroup property of deterministic dynamical systems. More specifically, RDS's include deterministic dynamical systems as the special case in which Ω is a singleton.

Example 2.1 (RDS's Generated by Random Linear Differential Equations). Given an MPDS θ, suppose $A: \Omega \to \mathbb{R}^{n\times n}$ is a random $n \times n$ real matrix such that, for each $\omega \in \Omega$,

---

[2] Property (T3) is normally [32, Definition 1.1] stated as

$$\mathbb{P}(\theta_t^{-1}(B)) = \mathbb{P}(B), \quad \forall B \in \mathscr{F},\ \forall t \in \mathscr{T}.$$

But since it follows from (T2) that θ_t is invertible with $\theta_t^{-1} = \theta_{-t}$ for each $t \in \mathscr{T}$, the two formulations are equivalent in this context.

$$t \longmapsto \|A(\theta_t\omega)\|, \quad t \geq 0,$$

is locally essentially bounded. For each $\omega \in \Omega$, let

$$\Xi(\cdot,\cdot,\omega)\colon \mathbb{R} \times \mathbb{R} \to \mathbb{R}^{n\times n}$$

be the *fundamental matrix solution*[3] of the linear differential equation

$$\dot{\xi} = A(\theta_t\omega)\xi, \quad t \in \mathbb{R}; \tag{2.1}$$

that is, for each fixed $s \in \mathbb{R}$, $\Xi(s,\cdot,\omega)$ is the unique absolutely continuous $\mathbb{R} \to \mathbb{R}^{n\times n}$ map such that

$$\Xi(s,s,\omega) = I_n := \begin{bmatrix} 1 & 0 & \cdots & 0 \\ 0 & 1 & \cdots & 0 \\ \vdots & \vdots & \ddots & \vdots \\ 0 & 0 & \cdots & 1 \end{bmatrix}$$

and

$$\frac{d}{dt}\Xi(s,t,\omega) = A(\theta_t\omega)\Xi(s,t,\omega)$$

for almost all $t \in \mathbb{R}$.

Let

$$\begin{aligned} \Phi\colon \mathbb{R}_{\geq 0} \times \Omega \times \mathbb{R}^n &\longrightarrow \mathbb{R}^n \\ (t,\omega,x) &\longmapsto \Xi(0,t,\omega)\cdot x \end{aligned}.$$

Then $\Phi(0,\omega,x) = x$ for every $(\omega,x) \in \Omega \times \mathbb{R}^n$ and

$$\frac{d}{dt}\Phi(t,\omega,x) = A(\theta_t\omega)\Phi(t,\omega,x)$$

for almost all $t \geq 0$. Moreover, $\Phi(t,\omega,\cdot)\colon \mathbb{R}^n \to \mathbb{R}^n$ is continuous for each fixed $(t,\omega) \in \mathbb{R}_{\geq 0} \times \Omega$, and it can be shown using existence and uniqueness of solutions for (2.1) that Φ has the cocycle property:

$$\Phi(t+s,\omega,x) = \Phi(t,\theta_s\omega,\Phi(t,\omega,x)), \quad \forall (t,\omega,x) \in \mathbb{R}_{\geq 0} \times \Omega \times \mathbb{R}^n.$$

Thus (θ,Φ) constitutes an RDS, referred to as the RDS *generated* by the (*homogeneous, linear*) *random differential equation* (*RDE*) (2.1).

---

[3]The reason we are introducing the fundamental matrix solution as a function of $(s,t) \in \mathbb{R} \times \mathbb{R}$ rather than a function of just $t \in \mathbb{R}$ (for each fixed $\omega \in \Omega$) will become clear in Example 2.3. This notation will make it easier to discuss the rate of growth of the fundamental matrix solution.

In this work, we use linear (or affine) systems as a case study to illustrate the theory developed. Such systems (and their discrete counterparts) may be interpreted as "switched linear systems," and include classes of systems of great interest in applications such as iterated function systems. Throughout the remainder of the chapter, we will be building upon the example above. Thus for any random matrix A as in Example 2.1, the symbols "Ξ" and "Φ" will be reserved to carry the meanings established in the example. We shall need the following two properties of the fundamental matrix solution:

(F1) $\Xi(0,t,\omega)\cdot\left(\Xi(0,s,\omega)\right)^{-1} = \Xi(s,t,\omega)$, for all $(s,t,\omega) \in \mathbb{R}\times\mathbb{R}\times\Omega$, and
(F2) $\Xi(s,t,\theta_\sigma\omega) = \Xi(\sigma+s,\sigma+t,\omega)$, for all $(s,t,\omega) \in \mathbb{R}\times\mathbb{R}\times\Omega$, for all $\sigma \in \mathbb{R}$.

These properties also follow from uniqueness of solutions.

2.2.1 Trajectories, Equilibria and θ-Stationary Processes

In the context of RDS's, the analogue to points in the state space X for a deterministic system are random variables $\Omega \to X$, that is, $\mathscr{B}$-measurable maps $\Omega \to X$. We denote the set of all random variables on a metric space X by $X^{\Omega}_{\mathscr{B}}$. We refer to a $(\mathscr{B}(\mathscr{T}_{\geq 0}) \otimes \mathscr{F})$-measurable map $q : \mathscr{T}_{\geq 0} \times \Omega \to X$ as a *θ-stochastic process*[4] *on* X, and denote $q_t := q(t,\cdot)$ for each $t \in \mathscr{T}_{\geq 0}$. The set of all θ-stochastic processes on a metric space X is denoted by $\mathscr{S}^X_\theta$.

Let (θ,φ) be an RDS. Given $x \in X^{\Omega}_{\mathscr{B}}$, we define the *(forward) trajectory starting at* x to be the θ-stochastic process $\xi^x \in \mathscr{S}^X_\theta$ defined by

$$\xi^x_t(\omega) := \varphi(t,\omega,x(\omega)), \quad (t,\omega) \in \mathscr{T}_{\geq 0} \times \Omega. \tag{2.2}$$

The *pullback trajectory starting at* x is in turn defined to be the θ-stochastic process $\check{\xi}^x : \mathscr{T}_{\geq 0} \times \Omega \to X$ defined by

$$\check{\xi}^x_t(\omega) := \varphi(t,\theta_{-t}\omega,x(\theta_{-t}\omega)), \quad (t,\omega) \in \mathscr{T}_{\geq 0} \times \Omega. \tag{2.3}$$

More generally, the *pullback* of a θ-stochastic process $q \in \mathscr{S}^X_\theta$ is the θ-stochastic process $\check{q} \in \mathscr{S}^X_\theta$ defined by

$$\check{q}_t(\omega) := q_t(\theta_{-t}\omega), \quad (t,\omega) \in \mathscr{T}_{\geq 0} \times \Omega.$$

So the pullback trajectory starting at x is simply the pullback of the forward trajectory starting at x. We will always use the accent ˇ to indicate the pullback of the θ-stochastic process being accented.

---

[4]A "θ-stochastic process" is indeed a stochastic process in the traditional sense. We use the prefix "θ-" to emphasize the underlying probability space, as well as the time semigroup.

We slightly modify the standard notion of equilibrium for RDS's (see, for instance, [5, p. 38, Definition 1.7.1]) to allow for the defining property to hold only almost everywhere, as opposed to everywhere. So an *equilibrium* of an RDS (θ, φ) is a random variable $x \in X^{\Omega}_{\mathscr{B}}$ such that

$$\xi^x_t(\omega) = \varphi(t, \omega, x(\omega)) = x(\theta_t \omega), \quad \forall t \in \mathscr{T}_{\geq 0}, \ \forall \omega \in \tilde{\Omega},$$

for some θ-invariant $\tilde{\Omega} \subseteq \Omega$ of full measure.[5] It is often not necessary to specify the said $\tilde{\Omega}$. So we say "for θ-almost all $\omega \in \Omega$" and write

$$\text{`}\ \tilde{\forall} \omega \in \Omega\ \text{'}$$

to mean "for all $\omega \in \tilde{\Omega}$, for some θ-invariant set $\tilde{\Omega} \subseteq \Omega$ of full measure."

In view of the notion of pullback convergence with which we will be working (see Sect. 2.2.3), it is more natural to think of the concept of equilibrium in terms of pullback trajectories. Observe that a random variable $x \in X^{\Omega}_{\mathscr{B}}$ is an equilibrium of the RDS (θ, φ) if, and only if

$$\check{\xi}^x_t(\omega) = \varphi(t, \theta_{-t}\omega, x(\theta_{-t}\omega)) = x(\omega), \quad \forall t \in \mathscr{T}_{\geq 0}, \ \tilde{\forall} \omega \in \Omega.$$

The remaining of this section is devoted to interpreting the concept of equilibrium for an RDS in terms of a shift operator in the set $\mathscr{S}^X_\theta$ of all θ-stochastic processes on X. For each $s \in \mathscr{T}_{\geq 0}$, let

$$\begin{aligned} \rho_s : \mathscr{S}^X_\theta &\longrightarrow \mathscr{S}^X_\theta \\ q &\longmapsto \rho_s(q) \end{aligned} \tag{2.4}$$

be defined by

$$(\rho_s(q))_t(\omega) := q_{t+s}(\theta_{-s}\omega), \quad (t, \omega) \in \mathscr{T}_{\geq 0} \times \Omega. \tag{2.5}$$

Definition 2.1 (θ-Stationary Process). A θ-stochastic process $\bar{q} \in \mathscr{S}^X_\theta$ is said to be θ-*stationary* if

$$(\rho_s(\bar{q}))_t(\omega) = \bar{q}_t(\omega),$$

for all $s, t \in \mathscr{T}_{\geq 0}$, for θ-almost all $\omega \in \Omega$.

We use the prefix "θ-" in "θ-stationary" to emphasize the dependence on the underlying MPDS θ. Using the characterization of θ-stationary processes given in

[5] That is, $\theta_t \tilde{\Omega} = \tilde{\Omega}$ for all $t \in \mathscr{T}$, and $\mathbb{P}(\tilde{\Omega}) = 1$.

Lemma 2.1 below, it is not difficult to show that a θ-stationary θ-stochastic process $\bar{q}$ is indeed *stationary* in the traditional stochastic processes sense:

$$\mathbb{P}(\bar{q}_{t_1} \in A_1, \dots, \bar{q}_{t_k} \in A_k) = \mathbb{P}(\bar{q}_{t_1+h} \in A_1, \dots, \bar{q}_{t_k+h} \in A_k)$$

for all $A_1, \dots, A_k \in \mathscr{F}$, for any $t_1, \dots t_k, h \geq 0$ (see, for instance, [23, Sect. 1.3]).

Lemma 2.1. *The θ-stochastic process $\bar{q} \in \mathscr{S}_\theta^X$ is θ-stationary if and only if there exists a random variable $q \in X_{\mathscr{B}}^\Omega$ such that*

$$\bar{q}_t(\omega) = q(\theta_t \omega), \quad \forall t \in \mathscr{T}_{\geq 0}, \ \tilde{\forall} \omega \in \Omega. \tag{2.6}$$

Proof. (Sufficiency) Suppose that (2.6) holds for some $q \in X_{\mathscr{B}}^\Omega$. Pick any $s \in \mathscr{T}_{\geq 0}$. For any $t \in \mathscr{T}_{\geq 0}$ and θ-almost all $\omega \in \Omega$,

$$(\rho_s(\bar{q}))_t(\omega) = \bar{q}_{t+s}(\theta_{-s}\omega) = q(\theta_{t+s}\theta_{-s}\omega) = q(\theta_t\omega) = \bar{q}_t(\omega).$$

So $\bar{q}$ is θ-stationary.

(Necessity) Suppose that $\bar{q} \in \mathscr{S}_\theta^X$ is θ-stationary and define $q \in X_{\mathscr{B}}^\Omega$ by

$$q(\omega) := \bar{q}_0(\omega), \quad \omega \in \Omega. \tag{2.7}$$

We have

$$\bar{q}_{t+s}(\theta_{-s}\omega) = (\rho_s(q))_t(\omega) = \bar{q}_t(\omega), \quad \forall s, t \in \mathscr{T}_{\geq 0}, \ \tilde{\forall} \omega \in \Omega.$$

Setting $t = 0$ and renaming s as t we then have

$$\bar{q}_t(\theta_{-t}\hat{\omega}) = \bar{q}_0(\hat{\omega}) = q(\hat{\omega}), \quad \forall t \in \mathscr{T}_{\geq 0}, \ \tilde{\forall} \hat{\omega} \in \Omega.$$

Given any $\omega \in \tilde{\Omega}$ and any $t \in \mathscr{T}_{\geq 0}$, we may apply this property with $\hat{\omega} = \theta_t \omega$ due to the θ-invariance of $\tilde{\Omega}$, thus obtaining

$$\bar{q}_t(\omega) = q(\theta_t \omega).$$

Therefore (2.6) holds. □

Note that the random variable q associated to $\bar{q}$ is unique up to a θ-invariant set of measure zero. Indeed, it is determined θ-almost everywhere by (2.7). Thus, we have:

Corollary 2.1. *Given an RDS (θ, φ) over a metric space X and a random state $x \in X_{\mathscr{B}}^\Omega$, the following three properties are equivalent:*

(1) x is an equilibrium;
(2) the trajectory ξ^x, as defined in (2.2), is θ-stationary;
(3) the map $t \mapsto \xi_t^x \in X_{\mathscr{B}}^\Omega$, $t \in \mathscr{T}_{\geq 0}$, is constant.

We will always use an overbar to denote the θ-stationary θ-stochastic process $\bar{q}$ associated with a given random variable q.

2.2.2 *Perfection of Crude Cocycles*

We briefly review the theory of perfection of crude cocycles discussed in Arnold's [3, Sect. 1.2]. It is customary for the definition of an RDS to require that the cocycle property of φ in (S2) holds for every $s, t \in \mathscr{T}_{\geq 0}$ and *every* $\omega \in \Omega$. If we want to emphasize this fact we shall say that φ is a *perfect cocycle* (over the underlying MPDS θ).

Definition 2.2 (Crude Cocycle). We say that $\varphi: \mathscr{T}_{\geq 0} \times \Omega \times X \to X$ is a *crude cocycle (over* θ*)* if it is a $(\mathscr{B}(\mathscr{T}) \otimes \mathscr{F} \otimes \mathscr{B})$-measurable map satisfying (S1) and

(S2′) $\varphi(0, w) = \mathrm{id}_X$ for each $\omega \in \Omega$, and for every $s \in \mathscr{T}_{\geq 0}$, there exists a subset $\Omega_s \subseteq \Omega$ of full measure such that

$$\varphi(t+s, \omega) = \varphi(t, \theta_s \omega) \circ \varphi(s, \omega), \quad \forall t \in \mathscr{T}_{\geq 0}, \ \forall \omega \in \Omega_s.$$

The Ω_s's need not be θ-invariant.

As Arnold points out, there are circumstances where this flexibility in the requirements for a cocycle is desirable. For instance, the flow of a stochastic differential equation is only guaranteed to be a crude cocycle [3, Sect. 2.3]. Another example will come up below after we introduce random dynamical systems with inputs. Consider (deterministic) controlled dynamical systems. Such systems yield a (deterministic) dynamical system when restricted to a constant input. One would expect a sensible extension of the concept to random dynamical systems to have an analogous property. However we shall see in the proof of Lemma 2.3 in the next section that the restriction of the flow of an RDS with inputs to a θ-stationary input is not necessarily a perfect cocycle.

In this work we deal only with random dynamical systems (with inputs) evolving in locally compact, connected subsets of $\mathbb{R}^n$. We will informally refer to such systems as *finite dimensional*. It turns out that crude cocycles evolving in these spaces can be perfected in a very reasonable sense.

Definition 2.3 (Indistinguishable Cocycles). Let θ be an MPDS and $\varphi, \psi: \mathscr{T}_{\geq 0} \times \Omega \times X \to X$ crude cocycles over θ. If there exists a subset $N \in \mathscr{F}$ such that $\mathbb{P}(N) = 0$ and

$$\{\omega \in \Omega;\ \varphi(t, \omega) \neq \psi(t, \omega), \text{ for some } t \in \mathscr{T}_{\geq 0}\} \subseteq N,$$

then φ and ψ are said to be *indistinguishable*.

Proposition 2.1. *Let $\theta = (\mathscr{F}, \Omega, \mathbb{P}, (\theta_t)_{t\in\mathscr{T}})$ be an MPDS with $\mathscr{T} = \mathbb{Z}$ or $\mathscr{T} = \mathbb{R}$. Suppose $\varphi\colon \mathscr{T}_{\geq 0} \times \Omega \times X \to X$ is a crude cocycle over θ evolving in a locally compact, locally connected, Hausdorff topological space X. Then there exists a perfect cocycle $\psi\colon \mathscr{T}_{\geq 0} \times \Omega \times X \to X$ such that φ and ψ are indistinguishable.*

Proof. See Arnold [3, Theorem 1.2.1] for the discrete case, which actually holds with weaker hypotheses and yields stronger conclusions. For the continuous case, see Arnold [3, Theorem 1.2.2 and Corollary 1.2.4]. □

2.2.3 *Pullback Convergence*

We work with the notion of pullback convergence developed in the literature and canonized in the works of Arnold and Chueshov [3,5]. As with equilibria, we relax the notion to require only that pointwise convergence happens θ-almost everywhere.

Definition 2.4 (Pullback Convergence). A θ-stochastic process $\xi \in \mathscr{S}_\theta^X$ is said to *converge to a random variable* $\xi_\infty \in X_{\mathscr{B}}^\Omega$ *in the pullback sense* if

$$\check{\xi}_t(\omega) := \xi_t(\theta_{-t}\omega) \longrightarrow \xi_\infty(\omega) \quad \text{as} \quad t \to \infty,$$

for θ-almost all $\omega \in \Omega$.

Proposition 2.2. *Let (θ, φ) be an RDS evolving on a metric space X. Suppose there exists a random initial state $x \in X_{\mathscr{B}}^\Omega$ and a map $x_\infty : \Omega \to X$ such that*

$$\check{\xi}_t^x(\omega) = \varphi(t, \theta_{-t}\omega, x(\theta_{-t}\omega)) \longrightarrow x_\infty(\omega) \quad as \quad t \to \infty, \quad \tilde{\forall}\omega \in \Omega. \tag{2.8}$$

Then x_∞ is an equilibrium.

Proof. For each $t \in \mathscr{T}_{\geq 0}$, the map $\omega \mapsto \varphi(t, \theta_{-t}\omega, x(\theta_{-t}\omega))$, $\omega \in \Omega$, is measurable, since it is the composition of measurable maps:

$$\omega \longmapsto \theta_{-t}\omega \longmapsto x(\theta_{-t}\omega),$$

$$(\theta_{-t}\omega, x(\theta_{-t}\omega)) \longmapsto \varphi(t, \theta_{-t}\omega, x(\theta_{-t}\omega)).$$

So it follows from [21, Chap. 11, Sect. 1, Property M7 on page 248] that x_∞ is measurable. (If $\mathscr{T}$ is continuous time, just pick a subsequence $(t_n)_{n\in\mathbb{N}}$ going to infinity.)

In addition, for each $\omega \in \Omega$ such that the limit in (2.8) exists, and each $\tau \in \mathscr{T}_{\geq 0}$, we have

$$\lim_{t\to\infty} \varphi(t - \tau, \theta_{\tau-t}\omega, x(\theta_{\tau-t}\omega)) = x_\infty(\omega)$$

also. By θ-invariance, the limit in (2.8) exists for $\theta_\tau\omega$ as well. Hence

$$\begin{aligned} x_\infty(\theta_\tau\omega) &= \lim_{t\to\infty} \varphi(t, \theta_{-t}\theta_\tau\omega, x(\theta_{-t}\theta_\tau\omega)) \\ &= \lim_{t\to\infty} \varphi(\tau, \theta_{t-\tau}\theta\tau - t\omega, \varphi(t-\tau, \theta_{\tau-t}\omega, x(\theta_{\tau-t}\omega))) \\ &= \varphi(\tau, \omega, x_\infty(\omega)) \end{aligned}$$

by continuity (property (S1) in the definition of an RDS). □

2.3 RDS's with Inputs and Outputs

We now define a new concept. It extends the notion of RDS's to systems in which there is an external input or forcing function. A contribution of this work is the precise formulation of this concept, particularly the way in which the argument of the input is shifted in the semigroup (cocycle) property.

As in the previous section, given a metric space U, we equip it with its Borel σ-algebra $\mathscr{B}(U)$ and denote by $U^{\Omega}_{\mathscr{B}}$ the set of Borel measurable maps $\Omega \to U$. Let $\mathscr{S}^U_\theta$ be the set of all θ-stochastic processes $\mathscr{T}_{\geq 0} \times \Omega \to U$. Given $u, v \in \mathscr{S}^U_\theta$ and $s \in \mathscr{T}_{\geq 0}$, we define $u \Diamond_s v$: $\mathscr{T}_{\geq 0} \times \Omega \to U$ by

$$(u\Diamond_s v)_\tau(\omega) = \begin{cases} u_\tau(\omega),\ 0 \leq \tau < s \\ v_{\tau-s}(\theta_s\omega),\ s \leq \tau \end{cases}, \quad \tau \in \mathscr{T}_{\geq 0},\ \omega \in \Omega.$$

We say that a subset $\mathscr{U} \subseteq \mathscr{S}^U_\theta$ is a *set of θ-inputs* if $u \Diamond_s v \in \mathscr{U}$ for any $u, v \in \mathscr{U}$ and any $s \in \mathscr{T}_{\geq 0}$. In other words, a set of θ-inputs is a subset of $\mathscr{S}^U_\theta$ which is closed under concatenation.

Given $\tilde{u} \in U$, we denote by $c(\tilde{u})$ the trivial θ-stochastic process defined by $(c(\tilde{u}))_t(\omega) := \tilde{u}$ for every $t \in \mathscr{T}_{\geq 0}$ and every $\omega \in \Omega$.

Definition 2.5 (Random Dynamical Systems with Inputs). A *random dynamical system with inputs* (*RDSI*) is a triple $(\theta, \varphi, \mathscr{U})$ consisting of an MPDS

$$\theta = (\Omega, \mathscr{F}, \mathbb{P}, \{\theta_t\}_{t\in\mathscr{T}}),$$

a set of θ-inputs $\mathscr{U} \subseteq \mathscr{S}^U_\theta$, and a map

$$\varphi : \mathscr{T}_{\geq 0} \times \Omega \times X \times \mathscr{U} \to X$$

satisfying

(I1) $\varphi(\cdot, \cdot, \cdot, u)$: $\mathscr{T}_{\geq 0} \times \Omega \times X \to X$ is $(\mathscr{B}(\mathscr{T}_{\geq 0}) \otimes \mathscr{F} \otimes \mathscr{B})$-measurable for each fixed $u \in \mathscr{U}$;

(I1′) the map $\tilde{\varphi}\colon \mathcal{T}_{\geq 0} \times \Omega \times X \times U \to X$ defined by

$$\tilde{\varphi}(t,\omega,x,\tilde{u}) := \varphi(t,\omega,x,c(\tilde{u})), \quad (t,\omega,x,\tilde{u}) \in \mathcal{T}_{\geq 0} \times \Omega \times X \times U,$$

is $(\mathcal{B}(\mathcal{T}_{\geq 0}) \otimes \mathcal{F} \otimes \mathcal{B} \otimes \mathcal{B}(U))$-measurable;

(I2) $\varphi(t,\omega,\cdot,u) : X \to X$ is continuous, for each fixed $(t,\omega,u) \in \mathcal{T}_{\geq 0} \times \Omega \times \mathcal{U}$;

(I3) $\varphi(0,\omega,x,u) = x$ for each $(\omega,x,u) \in \Omega \times X \times \mathcal{U}$;

(I4) given $s,t \in \mathcal{T}_{\geq 0}$, $\omega \in \Omega$, $x \in X$, $u,v \in \mathcal{U}$, if

$$\varphi(s,\omega,x,u) = y$$

and

$$\varphi(t,\theta_s\omega,y,v) = z,$$

then

$$\varphi(s+t,\omega,x,u\diamondsuit_s v) = z;$$

(I5) and given $t \in \mathcal{T}_{\geq 0}$, $\omega \in \Omega$, $x \in X$, and $u,v \in \mathcal{U}$, if $u_\tau(\omega) = v_\tau(\omega)$ for almost all $\tau \in [0,t)$, then $\varphi(t,\omega,x,u) = \varphi(t,\omega,x,v)$.

We refer to the elements $u \in \mathcal{U}$ as *θ-inputs*, or simply *inputs*. Whenever we talk about an RDSI $(\theta,\varphi,\mathcal{U})$, we tacitly assume the notation laid above, unless otherwise specified.

(I1), (I1′) and (I2) are regularity conditions. (I3) means that nothing has "happened" if one is still at time $t = 0$. (I4) generalizes the cocycle property and (I5) states that the evolution of an RDS subject to an input u is, so to speak, independent of "irrelevant" random input values.

Remark 2.1. Notice that for each $s,t \in \mathcal{T}_{\geq 0}$, $x \in X$, $\omega \in \Omega$,

$$\varphi(t+s,\omega,x,u) \;=\; \varphi(t,\theta_s\omega,\varphi(s,\omega,x,u),\rho_s(u)), \quad \forall u \in \mathcal{U},$$

where $\rho_s\colon \mathcal{S}^U_\theta \to \mathcal{S}^U_\theta$ is defined by (2.5)[6]:

$$(\rho_s(u))_t(\theta_s\omega) \equiv u_{t+s}(\omega). \tag{2.9}$$

This follows from (I4) with $v = \rho_s(u)$, which then yields $u\diamondsuit_s v = u$. □

[6]We will use the same notation ρ_s for the shift operator $\mathcal{S}^V_\theta \to \mathcal{S}^V_\theta$ defined by (2.5), irrespective of the underlying metric space V. Since the domain of any θ-stochastic process is always $\mathcal{T}_{\geq 0}\times\Omega$, this will not be a source of confusion.

The shift operator ρ_s has a physical interpretation. The right-hand side is the input as interpreted by an observer of the RDSI φ who started at time $t_1 = 0$, while the left-hand side is how someone who started observing the system at time $t_2 = s$ would describe it at time t $(+\, t_2)$. Following this interpretation, a θ-stationary input would then be an input which is observed to be just the same, regardless of when one started observing it.

Example 2.2 (RDSI's Generated by Random Differential Linear Equations with Inputs). This generalizes Example 2.1. Given an MPDS θ, suppose that $A\colon \Omega \to \mathbb{R}^{n\times n}$ and $B\colon \Omega \to \mathbb{R}^{n\times k}$ are random real matrices such that, for each $\omega \in \Omega$,

$$t \longmapsto \|A(\theta_t\omega)\|, \quad t \geq 0, \quad \text{and} \quad t \longmapsto \|B(\theta_t\omega)\|, \quad t \geq 0,$$

are locally essentially bounded. Let $U := \mathbb{R}^k$ and let $\mathscr{S}^U_\infty \subseteq \mathscr{S}^U_\theta$ be the set of θ-inputs consisting of all θ-stochastic processes $u \in \mathscr{S}^U_\theta$ such that

$$t \longmapsto |u_t(\omega)|, \quad t \geq 0,$$

is locally essentially bounded for each $\omega \in \Omega$. We consider the *random differential equation with inputs (RDEI)*

$$\dot{\xi} = A(\theta_t\omega)\xi + B(\theta_t\omega)u_t(\omega), \quad t \geq 0,\ \omega \in \Omega,\ u \in \mathscr{S}^U_\infty. \tag{2.10}$$

Let $\Xi\colon \mathbb{R} \times \mathbb{R} \times \Omega \to \mathbb{R}^{n\times n}$ be the fundamental matrix solution of the homogeneous, linear RDE

$$\dot{\xi} = A(\theta_t\omega)\xi, \quad t \geq 0,$$

and let (θ, Φ) be the RDS generated by the same equation (see Example 2.1). For each fixed $(\omega, u) \in \Omega \times \mathscr{S}^U_\infty$, define

$$\Psi(\cdot, \omega, u)\colon \mathbb{R}_{\geq 0} \to \mathbb{R}^n$$

by

$$\Psi(t, \omega, u) := \int_0^t \Xi(\sigma, t, \omega)B(\theta_\sigma\omega)u_\sigma(\omega)\, d\sigma, \quad t \geq 0.$$

Finally, define

$$\begin{aligned} \varphi\colon \mathbb{R}_{\geq 0} \times \Omega \times \mathbb{R}^n \times \mathscr{S}^U_\infty &\longrightarrow \mathbb{R}^n \\ (t, \omega, x, u) &\longmapsto \Phi(t, \omega, x) + \Psi(t, \omega, u) \end{aligned}.$$

Fixing $(\omega, x, u) \in \Omega \times \mathbb{R}^n \times \mathscr{S}^U_\infty$ arbitrarily, and differentiating $\varphi(t, \omega, x, u)$ with respect to t, we get

$$\begin{aligned}\frac{d}{dt}\varphi(t,\omega,x,u) &= A(\theta_t\omega)\Phi(t,\omega,x) + \Xi(t,t,\omega)B(\theta_t\omega)u_t(\omega)\\ &\quad + A(\theta_t\omega)\int_0^t \Xi(\sigma,t\omega)B(\theta_\sigma\omega)u_\sigma(\omega)\,d\sigma\\ &= A(\theta_t\omega)\Psi(t,\omega,u) + B(\theta_t\omega)u_t(\omega), \quad \forall t \geq 0.\end{aligned}$$

Thus $t \to \varphi(t,\omega,x,u)$, $t \geq 0$, is a solution of (2.10) with initial state

$$\varphi(0,\omega,x,u) = \Phi(0,\omega,x) + \Psi(0,\omega,u) = x.$$

In fact, $(\theta,\varphi,\mathscr{S}^U_\infty)$ is an RDSI. Indeed, (I1) and (I1′) follow from the fact that the limit of a sequence of measurable functions is measurable. Properties (I2) and (I3) follow directly from the analogous properties of Φ. And (I4) and (I5) follow from uniqueness of solutions applied for each fixed $\omega \in \Omega$—one basically verifies that both sides of each equation we want to prove to be true, when looked at as functions of t, define solutions of the same differential equation with the same initial condition. We refer to $(\theta,\varphi,\mathscr{S}^U_\infty)$ as the RDSI *generated* by the RDEI (2.10).

We also introduce a notion of outputs.

Definition 2.6 (Random Dynamical System with Inputs and Outputs). A *random dynamical system with inputs and outputs* (*RDSIO*) is a quadruple $(\theta,\varphi,\mathscr{U},h)$, such that $(\theta,\varphi,\mathscr{U})$ is an RDSI, and

$$h : \Omega \times X \to Y$$

is an $(\mathscr{F}\otimes\mathscr{B})$-measurable map into a metric space Y such that $h(\omega,\cdot)$ is continuous for each $\omega \in \Omega$. In this context we call h an *output function* and Y an *output space.*

It may sometimes be useful to refer to a *random dynamical system with outputs* (*RDSO*) only, by which we mean a triple (θ,φ,h) where (θ,φ) is an RDS and h is an output function.

The Ω-component in the domain of output functions is important. It allows for the concept to model uncertainties in the readout as well. We will return to systems with outputs further down, in the context of RDSIO's which can be realized as cascades of RDSO's and RDSIO's.

2.3.1 *Pullback Trajectories*

Let $(\theta,\varphi,\mathscr{U},h)$ be an RDSIO with output space Y. Given $x \in X^\Omega_\mathscr{B}$ and $u \in \mathscr{U}$, we define the *(forward) trajectory starting at x and subject to u* to be the θ-stochastic process $\xi^{x,u} \in \mathscr{S}^X_\theta$ defined by

$$\xi^{x,u}_t(\omega) := \varphi(t,\omega,x(\omega),u), \quad (t,\omega) \in \mathscr{T}_{\geq 0} \times \Omega.$$

We then define the *pullback trajectory starting at x and subject to u* to be the θ-stochastic process $\breve{\xi}^{x,u} \in \mathscr{S}_\theta^X$ defined by

$$\breve{\xi}_t^{x,u}(\omega) := \xi_t^{x,u}(\theta_{-t}\omega) = \varphi(t, \theta_{-t}\omega, x(\theta_{-t}\omega), u), \quad (t,\omega) \in \mathscr{T}_{\geq 0} \times \Omega.$$

The *(forward) output trajectory corresponding to initial state x and input u* is defined to be the θ-stochastic process $\eta^{x,u} \in \mathscr{S}_\theta^Y$, where

$$\eta_t^{x,u}(\omega) := h(\theta_t\omega, \varphi(t,\omega,x(\omega),u)) = h(\theta_t\omega, \xi_t^{x,u}(\omega)), \quad (t,\omega) \in \mathscr{T}_{\geq 0} \times \Omega,$$

while the *pullback output trajectory corresponding to initial state x and input u* is analogously defined to be the θ-stochastic process $\breve{\eta}^{x,u} \in \mathscr{S}_\theta^Y$, where

$$\begin{aligned} \breve{\eta}_t^{x,u}(\omega) &:= \eta_t^{x,u}(\theta_{-t}\omega) \\ &= h(\omega, \varphi(t, \theta_{-t}\omega, x(\theta_{-t}\omega), u)) \\ &= h(\omega, \breve{\xi}_t^{x,u}(\omega)), \qquad\qquad (t,\omega) \in \mathscr{T}_{\geq 0} \times \Omega. \end{aligned}$$

For RDSI's the definitions of forward and pullback trajectories are the same and we also use the notations $\xi^{x,u}$ and $\breve{\xi}^{x,u}$. For RDSO's the definitions are analogous, except that they of course do not depend on any inputs. So forward and pullback trajectories are defined as for RDS's and we also use the notations ξ^x and $\breve{\xi}^x$, respectively. We denote the forward and pullback output trajectories corresponding to initial state x by η^x and $\breve{\eta}^x$, respectively:

$$\eta_t^x(\omega) := h(\theta_t\omega, \varphi(t,\omega,x(\omega))) = h(\theta_t\omega, \xi_t^x(\omega))$$

and

$$\breve{\eta}_t^x(\omega) := h(\omega, \varphi(t, \theta_{-t}\omega, x(\theta_{-t}\omega))) = h(\omega, \breve{\xi}_t^x(\omega))$$

for every $(t,\omega) \in \mathscr{T}_{\geq 0} \times \Omega$.

Note that the input u is not shifted in the argument of φ in the pullback, while at first one might intuitively think it should have been. There are several reasons this is so. First notice that

$$\breve{\xi}_t^{x,u}(\omega) = \xi_t^{x,u}(\theta_{-t}\omega), \quad \forall (t,\omega) \in \mathscr{T}_{\geq 0} \times \Omega.$$

So $\breve{\xi}^{x,u}$ is just the pullback of the θ-stochastic process $\xi^{x,u}$, as it should be the case. However we are more concerned with what happens in the context of cascades and feedback interconnections of RDSIO's. But before we get to that we first discuss discrete RDSIO's. This will further motivate axioms (I1)–(I5) in the definition of an RDSI, provide—and completely characterize—a whole class of examples, and provide the framework for said discussion of pullback trajectories and cascades.

We say that an RDSI (or RDSIO) is *discrete* when $\mathcal{T} = \mathbb{Z}$. We first note that, just like RDS's [3, Sect. 2.1], RDSI's also have their flows completely determined by their state at time $t = 1$.

Theorem 2.1 (Characterization of Discrete RDSI's). *For every discrete RDSI*

$$(\theta, \varphi, \mathcal{U}),$$

there exists a unique map $f: \Omega \times X \times U \to X$ such that

(G1) $f: \Omega \times X \times U \to X$ *is* $(\mathcal{F} \otimes \mathcal{B} \otimes \mathcal{B}(U))$*-measurable,*
(G2) $f(\omega, \cdot, \tilde{u}): X \to X$ *is continuous for each* $(\omega, \tilde{u}) \in \Omega \times U$,

and

$$\varphi(n+1, \omega, x, u) = f(\theta_n \omega, \varphi(n, \omega, x, u), u_n(\omega)), \tag{2.11}$$

for every $(n, \omega, x, u) \in \mathcal{T}_{\geq 0} \times \Omega \times X \times \mathcal{U}$.

Conversely, given an MPDS θ, a set of θ-inputs $\mathcal{U}$ and a map

$$f: \Omega \times X \times U \to X$$

satisfying (G1) *and* (G2)*, define* $\varphi: \mathcal{T}_{\geq 0} \times \Omega \times X \times \mathcal{U} \to X$ *recursively by*

$$\varphi(0, \omega, x, u) := x, \quad (\omega, x, u) \in \Omega \times X \times \mathcal{U}, \tag{2.12}$$

and (2.11)*. Then $(\theta, \varphi, \mathcal{U})$ is an RDSI.*

We refer to the map f as the generator *of the RDSI $(\theta, \varphi, \mathcal{U})$.*

Proof. Define f by setting

$$f(\omega, x, \tilde{u}) := \varphi(1, \omega, x, c(\tilde{u})), \quad (\omega, x, \tilde{u}) \in \Omega \times X \times U.$$

Then (G1) and (G2) follow directly from (I1′) and (I2), respectively. Equation (2.11) follows from (I4) (see Remark 2.1) and (I5):

$$\begin{aligned}
\varphi(n+1, \omega, x, u) &= \varphi(1, \theta_n \omega, \varphi(n, \omega, x, u), \rho_n(u)) \\
&= \varphi(1, \theta_n \omega, \varphi(n, \omega, x, u), c((\rho_n(u))_0(\theta_n \omega))) \\
&= f(\theta_n \omega, \varphi(n, \omega, x, u), (\rho_n(u))_0(\theta_n \omega)) \\
&= f(\theta_n \omega, \varphi(n, \omega, x, u), u_n(\omega))
\end{aligned}$$

for any $(n, \omega, x, u) \in \mathcal{T}_{\geq 0} \times \Omega \times X \times \mathcal{U}$. Uniqueness follows from (I3) and (I5), together with the computations above performed backwards for $t = 0$.

Now suppose f satisfies (G1) and (G2), and that φ is defined recursively by (2.12) and (2.11). For (I1), pick any $u \in \mathcal{U}$. One first shows using induction on n

that

$$\varphi(n,\cdot,\cdot,u) = f(\theta_{n-1}\cdot,\varphi(n-1,\cdot,\cdot,u),u_{n-1}(\cdot)) \tag{2.13}$$

is $(\mathscr{F} \otimes \mathscr{B})$-measurable for each $n \in \mathbb{Z}_{>0}$. Indeed, at $n = 1$ we have

$$\varphi(1,\cdot,\cdot,u) = f(\theta_{1-1}\cdot,\varphi(1-1,\cdot,\cdot,u),u_{1-1}(\cdot)) = f(\cdot,\cdot,u_0(\cdot)),$$

which is $(\mathscr{F} \otimes \mathscr{B})$-measurable, since f satisfies (G1) and u_0 is $\mathscr{F}$-measurable. Now (2.13) gives us the inductive step, since the right hand side is a composition of measurable functions and, hence, itself measurable. Now pick any $A \in \mathscr{B}$. We then have

$$\varphi(\cdot,\cdot,\cdot,u)^{-1}(A) = \bigcup_{n=0}^{\infty}\{n\} \times \varphi(n,\cdot,\cdot,u)^{-1}(A) \in 2^{\mathbb{Z}_{\geq 0}} \otimes \mathscr{F} \otimes \mathscr{B},$$

since it is a countable union of $(2^{\mathbb{Z}_{\geq 0}} \otimes \mathscr{F} \otimes \mathscr{B})$-measurable sets. Thus (I1) holds. One can prove (I1′) in the same way by noting that

$$\tilde{\varphi}^{-1}(A) = \bigcup_{n=0}^{\infty}\{n\} \times \tilde{\varphi}(n,\cdot,\cdot,\cdot)^{-1}(A)$$

for each $A \in \mathscr{B}$, and that

$$\tilde{\varphi}(n,\cdot,\cdot,\cdot) = f(\theta_{n-1}\cdot,\tilde{\varphi}(n-1,\cdot,\cdot,\cdot),\cdot)$$

is $(\mathscr{F} \otimes \mathscr{B} \otimes \mathscr{B}(U))$-measurable for each $n \in \mathbb{Z}_{>0}$.

Property (I2) follows from (G2), (2.12) and (2.11), again by induction on $n \in \mathbb{Z}_{\geq 0}$. Indeed, at $n = 0$, $\varphi(0,\omega,\cdot,u)$ is continuous for every $\omega \in \Omega$ and every $u \in \mathscr{U}$. So once (I2) has been proved for a certain value of $n \in \mathbb{Z}_{\geq 0}$, we conclude that

$$\varphi(n+1,\omega,\cdot,u) = f(\theta_n\omega,\varphi(n,\omega,\cdot,u),u_n(\omega))$$

is continuous for any $\omega \in \Omega$ and any $u \in \mathscr{U}$ as well.

Property (I3) follows from (2.12).

Before proving (I4) we first prove (I5) by induction on $n \in \mathbb{Z}_{\geq 0}$. Fix $\omega \in \Omega$, $x \in X$. Equation (2.12) gives us the base of the induction. Now assume (I5) holds for a certain value of $n \in \mathbb{Z}_{\geq 0}$. If $u, v \in \mathscr{U}$ are such that $u_j(\omega) = v_j(\omega)$ for $j = 0,1,\ldots,n$, then $\varphi(n,\omega,x,u) = \varphi(n,\omega,x,v)$ by the induction hypothesis. So it follows from (2.11) that

$$\begin{aligned}\varphi(n+1,\omega,x,u) &= f(\theta_n\omega,\varphi(n,\omega,x,u),u_n(\omega))\\ &= f(\theta_n\omega,\varphi(n,\omega,x,v),v_n(\omega))\\ &= \varphi(n+1,\omega,x,v).\end{aligned}$$

This proves (I5).

It remains to prove (I4). For each arbitrarily fixed $p \in \mathbb{Z}_{\geq 0}$, we use induction on $n \in \mathbb{Z}_{\geq 0}$. For $n = 0$, (I4) holds in virtue of (I3) and (I5). For any $\omega \in \Omega$, we have $u_j(\omega) = (u \Diamond_p v)_j(\omega)$ for $j = 0, \dots, p-1$. Therefore

$$\varphi(0, \theta_p\omega, \varphi(p, \omega, x, u), v) = \varphi(p, \omega, x, u) = \varphi(0 + p, \omega, x, u \Diamond_p v),$$

for any $x \in X$. Now suppose (I4) holds for some $n \in \mathbb{Z}_{\geq 0}$. Given $\omega \in \Omega$ and $x \in X$, set $y := \varphi(n, \theta_p\omega, x, u)$. Then

$$\begin{aligned}
\varphi(n+1, \theta_p\omega, y, v) &= f(\theta_n\theta_p\omega, \varphi(n, \theta_p\omega, y, v), v_n(\theta_p\omega)) \\
&= f(\theta_{n+p}\omega, \varphi(n+p, \omega, x, u \Diamond_p v), (u \Diamond_p v)_{n+p}(\omega)) \\
&= \varphi(n+p+1, \omega, x, u \Diamond_p v).
\end{aligned}$$

This completes the proof that $(\theta, \varphi, \mathcal{U})$ is an RDSI. □

Observe that we did not need (I1) in order to prove the first half of the theorem. So we could have in principle dropped this axiom from the definition of an RDSI and an analogous result would still hold. We remind the reader that (I1) was nevertheless used in showing that RDSI's restricted to θ-stationary inputs are RDS's (see Lemma 2.3 below).

From the construction of the generator f of an RDSI $(\theta, \varphi, \mathcal{U})$, it is clear how the dependence of the flow φ at time $n \in \mathbb{Z}_{\geq 0}$ and subject to $\omega \in \Omega$ on the input u is really through the value $u_n(\omega)$ of the input u. So when one shifts the Ω-argument ω of φ in the pullback trajectory to $\theta_{-n}\omega$, there is no need to change the input, since $\varphi(n, \theta_{-n}\omega, x(\theta_{-n}\omega), u)$ depends on $u_n(\theta_{-n}\omega)$ already. This is our second reason for defining the pullback trajectories of systems with inputs like so.

We now discuss the third and most important reason this is the mathematically sensible way of defining pullback trajectories for RDSI's. Let (θ, ψ) be a discrete RDS evolving on the state space $Z = X_1 \times X_2$:

$$\psi: \mathbb{Z}_{\geq 0} \times \Omega \times (X_1 \times X_2) \longrightarrow (X_1 \times X_2).$$

Let $g: \Omega \times Z \to Z$ be the generator of (θ, ψ). Suppose g can be written as

$$g(\omega, (x_1, x_2)) \equiv \begin{pmatrix} f_1(\omega, x_1) \\ f_2(\omega, x_2, h_1(\omega, x_1)) \end{pmatrix}, \tag{2.14}$$

where $f_1: \Omega \times X_1 \to X_1$ is the generator of some RDSO (θ, φ_1, h_1) with output space Y_1, and $f_2: \Omega \times X_2 \times U_2 \to X_2$ is the generator of some RDSI $(\theta, \varphi_2, \mathcal{U}_2)$ with input space $U_2 = Y_1$. Let $\pi_2: X_1 \times X_2 \to X_2$ be the projection onto the second coordinate. We use η_1 to denote the output trajectories of (θ, φ_1, h_1), ξ for the state trajectories of ψ, and ξ_2 for the state trajectories of $(\theta, \varphi_2, \mathcal{U}_2)$.

Theorem 2.2 (Projection of Pullback Equals Pullback of Projection). *For any random initial state*

$$z = (x_1, x_2) \in Z^{\Omega}_{\mathscr{B}(Z)} = (X_1)^{\Omega}_{\mathscr{B}(X_1)} \times (X_2)^{\Omega}_{\mathscr{B}(X_2)},$$

the following two identities hold:

(1) $\psi(n, \omega, z(\omega)) \equiv \begin{pmatrix} \varphi_1(n, \omega, x_1(\omega)) \\ \varphi_2(n, \omega, x_2(\omega), (\eta_1)^{x_1}) \end{pmatrix}$, *and*

(2) $\pi_2(\check{\xi}^z_n(\omega)) \equiv (\check{\xi}_2)_n^{x_2,(\eta_1)^{x_1}}(\omega)$.

Proof. (1) For each fixed $\omega \in \Omega$ and $z \in Z^{\Omega}_{\mathscr{B}(Z)}$, we use induction on $n \in \mathbb{Z}_{\geq 0}$. At $n = 0$ we have

$$\psi(0, \omega, z(\omega)) = z(\omega) = \begin{pmatrix} x_1(\omega) \\ x_2(\omega) \end{pmatrix} = \begin{pmatrix} \varphi_1(0, \omega, x_1(\omega)) \\ \varphi_2(0, \omega, x_2(\omega), (\eta_1)^{x_1}) \end{pmatrix}.$$

Now suppose that (1) holds for some $n \in \mathbb{Z}_{\geq 0}$. Since

$$h_1(\theta_n\omega, \varphi_1(n, \omega, x_1(\omega))) = (\eta_1)_n^{x_1}(\omega)$$

by definition, it follows that

$$\begin{aligned} \psi(n+1, \omega, z(\omega)) &= g(\theta_n\omega, \psi(n, \omega, z(\omega))) \\ &= \begin{pmatrix} f_1(\theta_n\omega, \varphi_1(n, \omega, x_1(\omega))) \\ f_2(\theta_n\omega, \varphi_2(n, \omega, x_2(\omega), (\eta_1)^{x_1}), (\eta_1)_n^{x_1}(\omega)) \end{pmatrix} \\ &= \begin{pmatrix} \varphi_1(n+1, \omega, x_1(\omega)) \\ \varphi_2(n+1, \omega, x_2(\omega), (\eta_1)^{x_1}) \end{pmatrix}. \end{aligned}$$

This completes the induction.

(2) We prove by induction that (2) holds, for each $n \in \mathbb{Z}_{\geq 0}$, for all random initial states $z = (x_1, x_2) \in Z^{\Omega}_{\mathscr{B}(Z)}$, and all $\omega \in \Omega$. At $n = 0$ we have

$$\begin{aligned} \pi_2(\check{\xi}^z_0(\omega)) &= \pi_2(\psi(0, \omega, (x_1(\omega), x_2(\omega)))) \\ &= x_2(\omega) \\ &= \varphi_2(0, \omega, x_2(\omega), (\eta_1)^{x_1}) \\ &= (\check{\xi}_2)_0^{x_2,(\eta_1)^{x_1}}. \end{aligned}$$

Now assume (2) has been proved to hold for all integer values of n up to some $n_0 \geq 0$, for all random initial states $z = (x_1, x_2) \in Z^{\Omega}_{\mathscr{B}(Z)}$ and all $\omega \in \Omega$. Given $z = (x_1, x_2) \in Z^{\Omega}_{\mathscr{B}(Z)}$, define $\hat{z} = (\hat{x}_1, \hat{x}_2) \in Z^{\Omega}_{\mathscr{B}(Z)}$ by

$$\begin{aligned}\hat{z}(\omega) &= g(\theta_{-1}\omega, z(\theta_{-1}\omega)) \\ &:= \begin{pmatrix} f_1(\theta_{-1}\omega, x_1(\theta_{-1}\omega)) \\ f_2(\theta_{-1}\omega, x_2(\theta_{-1}\omega), h_1(\theta_{-1}\omega, x_1(\theta_{-1}\omega))) \end{pmatrix}, \quad \omega \in \Omega.\end{aligned} \tag{2.15}$$

We have $(\eta_1)^{\hat{x}_1} = \rho_1((\eta_1)^{x_1})$ by Lemma 2.2 below, and also

$$h_1(\theta_{-(n_0+1)}\omega, x_1(\theta_{-(n_0+1)}\omega)) = (\eta_1)_0^{x_1}(\theta_{-(n_0+1)}\omega), \quad \omega \in \Omega.$$

Fix $\omega \in \Omega$ arbitrarily and denote $\hat{\omega} := \theta_{-(n_0+1)}\omega$. Then

$$\begin{aligned}\pi_2(\check{\xi}^z_{n_0+1}(\omega)) &= \pi_2(\psi(n_0+1, \hat{\omega}, z(\hat{\omega}))) \\ &= \pi_2(\psi(n_0, \theta_{-n_0}\omega, \psi(1, \hat{\omega}, z(\hat{\omega})))) \\ &= \pi_2(\psi(n_0, \theta_{-n_0}\omega, g(\hat{\omega}, z(\hat{\omega})))) \\ &= \pi_2(\psi(n_0, \theta_{-n_0}\omega, \hat{z}(\theta_{-n_0}\omega))) \\ &= \pi_2(\check{\xi}^{\hat{z}}_{n_0}(\omega)) \\ &= (\check{\xi}_2)_{n_0}^{\hat{x}_2,(\eta_1)^{\hat{x}_1}}(\omega)\end{aligned}$$

by the induction hypothesis. Now

$$\begin{aligned}(\check{\xi}_2)_{n_0}^{\hat{x}_2,(\eta_1)^{\hat{x}_1}}(\omega) &= \varphi_2(n_0, \theta_{-n_0}\omega, \hat{x}_2(\theta_{-n_0}\omega), (\eta_1)^{\hat{x}_1}) \\ &= \varphi_2(n_0, \theta_{-n_0}\omega, f_2(\hat{\omega}, x_2(\hat{\omega}), (\eta_1)_0^{x_1}(\hat{\omega})), (\eta_1)^{\hat{x}_1}) \\ &= \varphi_2(n_0, \theta_{-n_0}\omega, \varphi_2(1, \hat{\omega}, x_2(\hat{\omega}), (\eta_1)^{x_1}), \rho_1((\eta_1)^{\hat{x}_1})) \\ &= \varphi_2(n_0+1, \theta_{-(n_0+1)}\omega, x_2(\theta_{-(n_0+1)}\omega), (\eta_1)^{x_1}) \\ &= (\check{\xi}_2)_{n_0+1}^{x_2,(\eta_1)^{x_1}}(\omega).\end{aligned}$$

So

$$\pi_2(\check{\xi}^z_{n_0+1}(\omega)) = (\check{\xi}_2)_{n_0+1}^{x_2,(\eta_1)^{x_1}}(\omega).$$

Since $z = (x_1, x_2) \in Z^{\Omega}_{\mathscr{B}(Z)}$ and $\omega \in \Omega$ were arbitrary, this completes the inductive step. □

The left hand side of (2) in the proposition above is the projection over the second coordinate of the pullback trajectory starting at $z = (x_1, x_2)$ of the RDS (θ, ψ). The right hand side is the pullback trajectory of the RDSI $(\theta, \varphi_2, \mathscr{U}_2)$ starting at x_2 and subject to the input $(\eta_1)^{x_1}$, the output trajectory of (θ, φ_1, h_1) starting at x_1. Theorem 2.2 then says that they coincide. An analogous result holds in continuous time for systems generated by random differential equations. These provide the motivation for the definition of cascades of systems with inputs and outputs, an introductory discussion of which is carried out in Sect. 2.4.2.

We now state and prove the technical lemma referred to in the proof of item (2) in Theorem 2.2:

Lemma 2.2. *Let* $f : \Omega \times X \to X$ *be the generator of a discrete RDSO* (θ, φ, h). *Given* $x \in X^{\Omega}_{\mathscr{B}}$, *let* $\hat{x} \in X^{\Omega}_{\mathscr{B}}$ *be defined by*

$$\hat{x}(\omega) := f(\theta_{-1}\omega, x(\theta_{-1}\omega)), \quad \omega \in \Omega.$$

Then $\eta^{\hat{x}} = \rho_1(\eta^x)$.

Proof. Indeed, we have

$$\begin{aligned}
\eta^{\hat{x}}_n(\omega) &= h(\theta_n\omega, \varphi(n, \omega, \hat{x}(\omega))) \\
&= h(\theta_n\omega, \varphi(n, \omega, f(\theta_{-1}\omega, x(\theta_{-1}\omega)))) \\
&= h(\theta_n\omega, \varphi(n, \omega, \varphi(1, \theta_{-1}\omega, x(\theta_{-1}\omega)))) \\
&= h(\theta_{n+1}\theta_{-1}\omega, \varphi(n+1, \theta_{-1}\omega, x(\theta_{-1}\omega))) \\
&= \eta^x_{n+1}(\theta_{-1}\omega) \\
&= (\rho_1(\eta^x))_n(\omega),
\end{aligned}$$

for every $n \in \mathbb{Z}_{\geq 0}$ and every $\omega \in \Omega$. □

2.3.2 θ-Stationary Inputs

The concept of RDSI subsumes that of an RDS, as we shall see below. Denote the subset of $\mathscr{S}^U_\theta$ consisting of θ-stationary inputs by $\bar{\mathscr{S}}^U_\theta$. We identify $\bar{\mathscr{S}}^U_\theta$ and $U^{\Omega}_{\mathscr{B}}$ via Lemma 2.1.

Let $(\theta, \varphi, \mathscr{U})$ be a RDSI, and suppose that $\bar{u} \in \mathscr{U} \cap \bar{\mathscr{S}}^U_\theta$ is some θ-stationary input. Consistent with the convention that an overbar is used to indicate the θ-stationary process associated with a given random variable, we remove the bar to denote the random variable associated with a given θ-stationary process. So we denote by u the random variable in $U^{\Omega}_{\mathscr{B}}$ associated via Lemma 2.1 with $\bar{u}$. We then define

$$\varphi_u := \varphi(\cdot, \cdot, \cdot, \bar{u}) : \mathscr{T}_{\geq 0} \times \Omega \times X \longrightarrow X.$$

Lemma 2.3. φ_u *is a crude cocycle.*

Proof. It follows from condition (I1) and [12, p. 65, Proposition 2.34] that φ_u is measurable. From (I2), $\varphi_u(t, \omega, \cdot)$ is continuous for each $(t, \omega) \in \mathscr{T}_{\geq 0} \times \Omega$, yielding (S1). From (I3), we know that $\varphi_u(0, \omega, \cdot) = \mathrm{id}_X$ for every $\omega \in \Omega$. So to verify (S2′) it remains to prove that φ_u satisfies the "crude cocycle property." Let $\tilde{\Omega} \subseteq \Omega$ be a θ-invariant subset of full measure such that

$$(\rho_s(\bar{u}))_t(\omega) = \bar{u}_t(\omega), \quad \forall s, t \in \mathscr{T}_{\geq 0}, \ \forall \omega \in \tilde{\Omega}. \tag{2.16}$$

Fix arbitrarily $\omega \in \tilde{\Omega}$. For any $s, t \in \mathscr{T}_{\geq 0}$, we have $\theta_s\omega \in \tilde{\Omega}$ by θ-invariance, and so it follows from (2.16) and (I5) that

$$\varphi(t, \theta_s\omega, \varphi_u(s, \omega, x), \rho_s(\bar{u})) = \varphi(t, \theta_s\omega, \varphi_u(s, \omega, x), \bar{u}).$$

It then follows from (I4)—see Remark 2.1—that

$$\begin{aligned}
\varphi_u(t + s, \omega, x) &= \varphi(t + s, \omega, x, \bar{u}) \\
&= \varphi(t, \theta_s\omega, \varphi(s, \omega, x, \bar{u}), \rho_s(\bar{u})) \\
&= \varphi(t, \theta_s\omega, \varphi_u(s, \omega, x), \bar{u}) \\
&= \varphi_u(t, \theta_s\omega, \varphi_u(s, \omega, x)).
\end{aligned}$$

So (S2′) is satisfied with $\Omega_s := \tilde{\Omega}$ for every $s \in \mathscr{T}_{\geq 0}$. □

Proposition 2.3. *If X is a locally compact and locally connected, Hausdorff topological space, then φ_u can be perfected.*

Proof. This follows straight from Proposition 2.1. □

Note that, since $\tilde{\Omega}$ in the proof of Lemma 2.3 is θ-invariant, so is its complement in Ω, namely $\Omega \backslash \tilde{\Omega}$. So Proposition 2.3 could have also been proved directly by redefining φ_u to take an arbitrarily fixed value of $x_0 \in X$ on the set

$$\mathscr{T}_{\geq 0} \times (\Omega \backslash \tilde{\Omega}) \times X.$$

Whenever the state space X is such that φ_u can be perfected, we shall assume that φ_u has already been replaced by an indistinguishable perfection and then refer to the resulting RDS (θ, φ_u).

2.3.3 Tempered Random Sets

Recall that, given a topological space X, a multifunction $D: \Omega \to 2^X$ is said to be a *random set* if

$$D^{-1}(U) := \{\omega \in \Omega;\ D(\omega) \cap U \neq \varnothing\} \in \mathscr{F}$$

for every open set $U \subseteq X$ (see [16, Chap. 2]). In this work, we shall be concerned exclusively with so-called *Polish spaces*; that is, separable topological spaces generated by a metric with respect to which they are complete. In such spaces, the definition above is known [16, p. 142, Proposition 1.4] to be equivalent to the requirement that

$$\omega \longmapsto \operatorname{dist}(x, D(\omega)) := \inf_{y \in D(\omega)} d(x, y), \quad \omega \in \Omega,$$

defines a Borel-measurable[7] map $\Omega \to \bar{\mathbb{R}}_{\geq 0}$ for each $x \in X$.

Definition 2.7 (Tempered Random Variables). A nonnegative, Borel-measurable function $r\colon \Omega \to \mathbb{R}_{\geq 0}$ is said to be a *tempered random variable* (*with respect to the underlying MPDS* θ) if, for every $\gamma > 0$,

$$\sup_{s \in \mathcal{T}} r(\theta_s \omega)\, \mathrm{e}^{-\gamma |s|} < \infty, \quad \tilde{\forall} \omega \in \Omega.$$

We denote the family of nonnegative, tempered (with respect to θ) random variables $\Omega \to \mathbb{R}_{\geq 0}$ by $(\mathbb{R}_{\geq 0})^{\Omega}_{\theta}$.

Observe that we do not require the bound to be independent of $\omega \in \Omega$. In fact, if it were, then r would have been essentially bounded. More precisely, suppose that, for some $\gamma > 0$, there exists a $K_\gamma \geq 0$ such that

$$\sup_{s \in \mathcal{T}} r(\theta_s \omega)\, \mathrm{e}^{-\gamma |s|} \leq K_\gamma, \quad \tilde{\forall} \omega \in \Omega.$$

Then

$$0 \leq r(\omega) \leq \sup_{s \in \mathcal{T}} r(\theta_s \omega)\, \mathrm{e}^{-\gamma |s|} \leq K_\gamma, \quad \tilde{\forall} \omega \in \Omega.$$

So r is actually essentially bounded.

Definition 2.8 (Tempered Random Set). Let (X, d) be a metric space. A random set $D\colon \Omega \to 2^X$ is said to be *tempered* (*with respect to* θ) if there exist $x_0 \in X$ and a nonnegative tempered random variable $r\colon \Omega \to \mathbb{R}_{\geq 0}$ such that

$$D(\omega) \subseteq \{x \in X;\ d(x, x_0) \leq r(\omega)\}, \quad \forall \omega \in \Omega. \tag{2.17}$$

A Borel-measurable map $v\colon \Omega \to X$ is said to be a *tempered random variable* (*with respect to* θ) if the random singleton defined by $\omega \mapsto \{v(\omega)\}$, $\omega \in \Omega$, is a tempered random set.

We denote the family of tempered (with respect to θ) random sets $\Omega \to 2^X$ by $(2^X)^{\Omega}_{\theta}$. Likewise, the family of tempered (with respect to θ) random variables $\Omega \to X$ is denoted by X^{Ω}_{θ}.

Lemma 2.4. *Suppose* θ *is an MPDS,* $(X, \|\cdot\|)$ *is a normed space over* $\mathbb{R}$*, and let* $R_1, R_2 \in X^{\Omega}_{\theta}$, $r \in \mathbb{R}^{\Omega}_{\theta}$*, and* $c \in \mathbb{R}$*. Then*

[7] Our convention is that $\inf \varnothing := +\infty$.

(1) $R_1 + R_2$ *is tempered.*
(2) cR_1 *is tempered.*
(3) rR_1 *is tempered; in particular, the product of two real-valued tempered random variables is tempered.*

Proof. (1) Indeed, for any $\gamma > 0$ and any $\omega \in \tilde{\Omega}$, we have

$$\begin{aligned}\sup_{s\in\mathcal{T}} \|(R_1 + R_2)(\theta_s\omega)\|e^{-\gamma|s|} &\leq \sup_{s\in\mathcal{T}} \|R_1(\theta_s\omega)\|e^{-\gamma|s|} + \sup_{s\in\mathcal{T}} \|R_2(\theta_s\omega)\|e^{-\gamma|s|} \\ &< \infty,\end{aligned}$$

where we write $(R_1 + R_2)(\theta_s\omega)$ for $R_1(\theta_s\omega) + R_2(\theta_s\omega)$. So both $R_1 + R_2$ is tempered.

(2) follows from (3), which we now prove. Given $\gamma > 0$ and $\omega \in \tilde{\Omega}$, apply the definition of tempered random variable for $\gamma/2$:

$$\begin{aligned}\sup_{s\in\mathcal{T}} \|r(\theta_s\omega)R_1(\theta_s\omega)\|e^{-\gamma|s|} &= \sup_{s\in\mathcal{T}} |r(\theta_s\omega)|e^{-\frac{\gamma}{2}|s|}\|R_1(\theta_s\omega)\|e^{-\frac{\gamma}{2}|s|} \\ &\leq \left(\sup_{s\in\mathcal{T}} |r(\theta_s\omega)|e^{-\frac{\gamma}{2}|s|}\right)\left(\sup_{s\in\mathcal{T}} \|R_1(\theta_s\omega)\|e^{-\frac{\gamma}{2}|s|}\right) \\ &< \infty.\end{aligned}$$

Thus rR_1 is tempered. □

In other words, X^{Ω}_{θ} is a real vector space, and also a module over the ring of real-valued tempered random variables.

We now introduce concepts of convergence and continuity taking into account the notion of temperedness just introduced.

Definition 2.9 (Tempered Convergence). Suppose θ is an MPDS and (X, d) is a metric space. We say that a net $(\xi_\alpha)_{\alpha\in A}$ in $X^{\Omega}_{\mathcal{B}}$ converges in the *tempered sense* to a random variable $\xi_\infty \in X^{\Omega}_{\mathcal{B}}$ if there exists a nonnegative, tempered random variable $r\colon \Omega \to \mathbb{R}_{\geq 0}$ and an $\alpha_0 \in A$ such that

(1) $\xi_\alpha(\omega) \to \xi_\infty(\omega)$ as $\alpha \to \infty$ for θ-almost all $\omega \in \Omega$, and
(2) $d(\xi_\alpha(\omega), \xi_\infty(\omega)) \leq r(\omega)$ for all $\alpha \succcurlyeq \alpha_0$, for θ-almost all $\omega \in \Omega$.

In this case we denote $\xi_\alpha \to_\theta \xi_\infty$ (as $\alpha \to \infty$).

Definition 2.10 (Tempered Continuity). Suppose θ is an MPDS and X, U are metric spaces. A map $\mathcal{K}\colon \mathcal{U} \subseteq U^{\Omega}_{\mathcal{B}} \to X^{\Omega}_{\mathcal{B}}$ is said do be *tempered continuous* if $\mathcal{K}(u_\alpha) \to_\theta \mathcal{K}(u_\infty)$ for every net $(u_\alpha)_{\alpha\in A}$ in $\mathcal{U}$ such that $u_\alpha \to_\theta u_\infty$ for some $u_\infty \in \mathcal{U}$.

We close this subsection with the definition of several asymptotic behavior concepts. Let X be a metric space. Given $\xi \in \mathcal{S}^X_\theta$ and $\tau \geq 0$, we call the multifunction $\beta^\tau_\xi\colon \Omega \to 2^X\backslash\{\varnothing\}$, defined by

$$\beta_\xi^\tau(\omega) := \{\xi_t(\theta_{-t}\omega);\ t \geq \tau\}, \quad \omega \in \Omega,$$

the *tail (from moment* τ) of the pullback trajectories of ξ. If a θ-stochastic process $\xi \in \mathscr{S}_\theta^X$ is such that there exists a $\tau_\xi \geq 0$ such that $\beta_\xi^\tau(\omega)$ is precompact for all $\tau \geq \tau_\xi$, for θ-almost all $\omega \in \Omega$, then we say that ξ is *eventually precompact.* We denote the subset of all eventually precompact θ-stochastic processes $\xi \in \mathscr{S}_\theta^X$ by $\mathscr{K}_\theta^X$. A θ-stochastic process $\xi \in \mathscr{S}_\theta^X$ is said to be *tempered* if there exists a tempered random set $D \in (2^X)_\theta^\Omega$ such that

$$\beta_\xi^\tau(\omega) \subseteq D(\omega), \quad \forall \tau \geq 0,\ \tilde{\forall}\omega \in \Omega; \tag{2.18}$$

in other words,

$$\xi_t(\theta_{-t}\omega) = \check{\xi}_t(\omega) \in D(\omega), \quad \forall t \geq 0,\ \tilde{\forall}\omega \in \Omega. \tag{2.19}$$

Any $D \in (2^X)_\theta^\Omega$ for which the relation above holds is called a *rest set*. The subset of $\mathscr{S}_\theta^X$ consisting of all tempered θ-stochastic processes $\xi \in \mathscr{S}_\theta^X$ is denoted by $\mathscr{V}_\theta^X$. Observe that, in virtue of θ-invariance, condition (2.19) is equivalent to

$$\xi_t(\omega) \in D(\theta_t\omega), \quad \forall t \geq 0,\ \tilde{\forall}\omega \in \Omega.$$

We further motivate the concept of temperedness just introduced. The idea is to have a term to talk about θ-stochastic processes which, as far as their oscillatory behavior is concerned, look somewhat like a θ-stationary process generated by a tempered random variable. Since this pertains to long-term behavior, this property should be preserved by shifting or concatenating tempered stochastic processes. Indeed, it is not difficult to show that (1) θ-stationary processes generated by tempered random variables are tempered, (2) $\rho_s(u)$ is tempered for any tempered u, and (3) $u \Diamond_s v$ is tempered for any tempered u, v.

Definition 2.11 (Tempered RDSI). An RDSI $(\theta, \varphi, \mathscr{U})$ is said to be *tempered* if the trajectories $\xi^{x,u}$ are tempered for every tempered initial state $x \in X_\theta^\Omega$ and every tempered input $u \in \mathscr{U}$.

2.3.4 *Input to State Characteristics*

Let $(\theta, \varphi, \mathscr{U})$ be an RDSI and suppose that $\bar{u} \in \mathscr{U}$ is a θ-stationary process, with generating random variable u (refer to Lemma 2.1). Any equilibrium ξ of the RDS (θ, φ_u) will be referred to as an *equilibrium associated to* $\bar{u}$ (or *to* u). The set of all equilibria associated to $\bar{u}$ (or to u) is denoted as $\mathscr{E}(\bar{u})$ (we may also write $\mathscr{E}(u)$). So an element $\xi \in \mathscr{E}(\bar{u})$ is a random variable $\Omega \to X$ such that

$$\varphi_u(t, \theta_{-t}\omega, \xi(\theta_{-t}\omega)) = \xi(\omega), \quad \forall t \geq 0, \quad \tilde{\forall}\omega \in \Omega. \tag{2.20}$$

When we have a "proper" RDS (θ, φ), we write simply $\mathscr{E}$ for the set of equilibria of (θ, φ).

For deterministic systems—when Ω is a singleton and we may identify the set of θ-inputs $\mathscr{U}$ with the input space U—, if the set $\mathscr{E}(\bar{u})$ consists of a single, globally attracting equilibrium, then the mapping $u \mapsto \mathscr{E}(\bar{u})$, $u \in U$, is the object called the "input to state characteristic" in the literature on monotone i/o systems. For systems with outputs, composition with the output map h provides the "input to output" characteristic [1]. One of the contributions of this work is the extension of these concepts to RDSI's and RDSIO's.

In this section we introduce the notion of input to state characteristics for RDSI's and discuss a class of examples. Systems with outputs will be considered in greater detail in the next section. For reasons which will be illustrated in Example 2.3 and become clearer in the proof of Theorem 2.3 (CICS), further conditions on the convergence of the states are needed.

Definition 2.12 (I/S Characteristic). An RDSI $(\theta, \varphi, \mathscr{U})$ is said to have an *input to state (i/s) characteristic* $\mathscr{K}\colon U_\theta^\Omega \to X_\theta^\Omega$ if

$$U_\theta^\Omega \subseteq \mathscr{U}$$

and

$$\breve{\xi}_t^{x,u} \longrightarrow_\theta \mathscr{K}(u) \quad \text{as} \quad t \to \infty,$$

for every $x \in X_\theta^\Omega$, for every $u \in U_\theta^\Omega$.

Example 2.3 below illustrates the concepts of tempered RDSI (Definition 2.11) and i/s characteristics (Definition 2.12 above). Temperedness features in said example will be a special case (with $p = 1$ or $p = \infty$) of the general result below.

Proposition 2.4. *Suppose $r\colon \Omega \to \mathbb{R}_{\geq 0}$ is a tempered random variable. For each $\gamma > 0$ and each $p \in [1, \infty]$, the map*

$$\omega \longmapsto \|r(\theta.\omega)\, \mathrm{e}^{-\gamma|\cdot|}\|_{L^p(\mathbb{R})}, \quad \omega \in \Omega,$$

is a tempered random variable. Moreover, temperedness bounds are uniform in $p \in [1, \infty]$; that is, for each $\gamma > 0$ and each $\delta > 0$,

$$\sup_{p \in [1,\infty]} \sup_{s \in \mathbb{R}} \|r(\theta.\theta_s\omega)\, \mathrm{e}^{-\gamma|\cdot|}\|_{L^p(\mathbb{R})}\, \mathrm{e}^{-\delta|s|} < \infty, \quad \tilde{\forall}\omega \in \Omega.$$

Proof. For each $\mu > 0$, set

$$K_{\mu,\omega} := \sup_{s \in \mathbb{R}} r(\theta_s\omega)\, \mathrm{e}^{-\mu|s|}$$

for every $\omega \in \Omega$ such that the supremum above is finite. Since r is tempered by assumption, this will be true for θ-almost all $\omega \in \Omega$.

Fix arbitrarily $\gamma > 0$ and choose any $\delta > 0$. We consider two different cases.

(Case $1 \le p < \infty$) Setting $m := \min\{\gamma, \delta\} > 0$ and using the triangle inequality we obtain

$$\begin{aligned}
\|r(\theta.\theta_s\omega)\,\mathrm{e}^{\gamma|\cdot|}\,\|_{L^p(\mathbb{R})}\,\mathrm{e}^{-\delta|s|} &= \left(\int_{-\infty}^{\infty} |r(\theta_{t+s}\omega)\,\mathrm{e}^{-\gamma|t|-\delta|s|}\,|^p\,dt\right)^{1/p} \\
&\le \left(\int_{-\infty}^{\infty} |r(\theta_{t+s}\omega)\,\mathrm{e}^{-m|t+s|}\,|^p\,dt\right)^{1/p} \\
&\le K_{\frac{m}{2},\omega}\left(\int_{-\infty}^{\infty} \mathrm{e}^{-\frac{pm}{2}|t+s|}\,dtsw\right)^{1/p} \\
&= K_{\frac{m}{2},\omega}\left(\frac{4}{pm}\right)^{1/p},
\end{aligned}$$

which is finite for all $s \in \mathbb{R}$, for θ-almost all $\omega \in \Omega$. In fact, since the map

$$p \longmapsto K_{\frac{m}{2},\omega}\left(\frac{4}{pm}\right)^{1/p}, \quad 1 \le p < \infty, \tag{2.21}$$

is continuous in p and

$$\lim_{p\to\infty} K_{\frac{m}{2},\omega}\left(\frac{4}{pm}\right)^{1/p} = 1,$$

we then know that the map in (2.21) is bounded. Thus

$$M_{\gamma,\delta,\omega} := \sup_{p\in[1,\infty)}\sup_{s\in\mathbb{R}} \|r(\theta.\theta_s\omega)\,\mathrm{e}^{-\gamma|\cdot|}\,\|_{L^p(\mathbb{R})}\,\mathrm{e}^{-\delta|s|} < \infty, \quad \tilde{\forall}\omega \in \Omega.$$

(Case $p = \infty$) The trick is basically the same as before. We have

$$\begin{aligned}
\|r(\theta.\theta_s\omega)\,\mathrm{e}^{\gamma|\cdot|}\,\|_{L^\infty(\mathbb{R})}\,\mathrm{e}^{-\delta|s|} &= \sup_{t\in\mathbb{R}} r(\theta_{t+s}\omega)\,\mathrm{e}^{-\gamma|t|-\delta|s|} \\
&\le \sup_{t\in\mathbb{R}} r(\theta_{t+s}\omega)\,\mathrm{e}^{-m|t+s|} \\
&= K_{m,\omega},
\end{aligned}$$

which is finite for all $s \in \mathbb{R}$, for θ-almost all $\omega \in \Omega$.

Combining both cases we conclude that

$$\sup_{p\in[1,\infty]}\sup_{s\in\mathbb{R}}\|r(\theta.\theta_s\omega)\,\mathrm{e}^{-\gamma|\cdot|}\|_{L^p(\mathbb{R})}\,\mathrm{e}^{-\delta|s|}=\max\{M_{\gamma,\delta,\omega},K_{m,\omega}\},$$

which is finite for θ-almost all $\omega\in\Omega$. Since $\gamma,\delta>0$ were chosen arbitrarily, this completes the proof. □

Example 2.3 (I/S Characteristics for RDSI's Generated by Linear RDEI's). Consider the RDSI $(\theta,\varphi,\mathcal{S}^U_\infty)$ from Example 2.2, generated by the RDEI

$$\dot{\xi}=A(\theta_t\omega)\xi+B(\theta_t\omega)u_t(\omega),\quad t\ge 0,\quad u\in\mathcal{S}^U_\infty,\tag{2.22}$$

where $X=\mathbb{R}^n$, $U=\mathbb{R}^k$, and $A\colon\Omega\to\mathbb{R}^{n\times n}$ and $B\colon\Omega\to\mathbb{R}^{n\times k}$ are random matrices such that

$$t\longmapsto A(\theta_t\omega),\quad t\ge 0,\quad\text{and}\quad t\longmapsto B(\theta_t\omega),\quad t\ge 0,$$

are locally essentially bounded for every $\omega\in\Omega$. Now suppose in addition that A, B are such that

(L1) B is tempered and
(L2) there exist a $\lambda>0$ and a nonnegative, tempered random variable $\gamma\in(\mathbb{R}_{\ge})^{\Omega}_{\theta}$ such that the fundamental matrix solution Ξ of the homogeneous part of (2.22) satisfies

$$\|\Xi(s,s+r,\omega)\|\le\gamma(\theta_s\omega)\,\mathrm{e}^{-\lambda r},\quad\forall s\in\mathbb{R},\quad\forall r\ge 0,\quad\tilde{\forall}\omega\in\Omega.$$

Then $(\theta,\varphi,\mathcal{S}^U_\infty)$ is tempered (in the sense of Definition 2.11) and has a continuous input to state characteristic $\mathcal{K}\colon U^{\Omega}_{\theta}\to X^{\Omega}_{\theta}$ (refer to Definition 2.12). We will prove this in several steps, indicated below.

Construction of $\mathcal{K}\colon U^{\Omega}_{\theta}\to X^{\Omega}_{\theta}$. We first claim that the limit

$$\lim_{t\to\infty}\check{\xi}^{x,\bar{u}}_t(\omega)=\int_{-\infty}^{0}\Xi(\sigma,0,\omega)B(\theta_\sigma\omega)u(\theta_\sigma\omega)\,d\sigma\tag{2.23}$$

exists for each $x\in X^{\Omega}_{\theta}$ and each $u\in U^{\Omega}_{\theta}$, for θ-almost $\omega\in\Omega$. Let Φ and Ψ be as in Example 2.2, so that we may write

$$\varphi(t,\omega,x,u)\equiv\Phi(t,\omega,x)+\Psi(t,\omega,u).$$

So it is enough to show that

$$\lim_{t\to\infty}\Phi(t,\theta_{-t}\omega,x(\theta_{-t}\omega))=0,\quad\forall x\in X^{\Omega}_{\theta},\quad\tilde{\forall}\omega\in\Omega,\tag{2.24}$$

and that

$$\lim_{t\to\infty} \Psi(t, \theta_{-t}\omega, \bar{u}) = \int_{-\infty}^{0} \Xi(\sigma, 0, \omega) B(\theta_\sigma\omega) u(\theta_\sigma\omega)\, d\sigma, \quad \forall u \in U_\theta^\Omega, \quad \tilde{\forall}\omega \in \Omega. \tag{2.25}$$

Fix arbitrarily $x \in X_\theta^\Omega$ and let $\omega \in \Omega$ be such that

$$K_{\omega,\frac{\lambda}{2},x} := \sup_{s\in\mathbb{R}} \gamma(\theta_s\omega)|x(\theta_s\omega)|\, \mathrm{e}^{-\frac{\lambda}{2}|s|} < \infty, \tag{2.26}$$

where $\lambda > 0$ and γ nonnegative and tempered are given by (L2). Combining (L2) and (2.26), we obtain

$$\begin{aligned} |\Phi(t, \theta_{-t}\omega, x(\theta_{-t}\omega))| &= |\Xi(0, t, \theta_{-t}\omega) \cdot x(\theta_{-t}\omega)| \\ &\le \gamma(\theta_{-t}\omega)\, \mathrm{e}^{-\lambda t}\, |x(\theta_{-t}\omega)| \\ &= \left(\gamma(\theta_{-t}\omega)|x(\theta_{-t}\omega)|\, \mathrm{e}^{-\frac{\lambda}{2}|-t|}\right) \mathrm{e}^{-\frac{\lambda}{2}t} \\ &\le K_{\omega,\frac{\lambda}{2},x}\, \mathrm{e}^{-\frac{\lambda}{2}t}, \quad \forall t \ge 0. \end{aligned}$$

Hence

$$|\Phi(t, \theta_{-t}\omega, x(\theta_{-t}\omega))| \longrightarrow 0 \quad \text{as} \quad t \to \infty.$$

Since $K_{\omega,\frac{\lambda}{2},x}$ is finite for θ-almost all $\omega \in \Omega$—recall that, by Lemma 2.4(3), the product of two tempered random variables is tempered—, this holds θ-almost everywhere. So since $x \in X_\theta^\Omega$ was chosen arbitrarily, this proves (2.24).

Now fix arbitrarily $u \in U_\theta^\Omega$. Then by (F2) and a change of variables,

$$\begin{aligned} \Psi(t, \theta_{-t}\omega, \bar{u}) &= \int_0^t \Xi(\sigma, t, \theta_{-t}\omega) B(\theta_{\sigma-t}\omega) u(\theta_{\sigma-t}\omega)\, d\sigma \\ &= \int_0^t \Xi(\sigma - t, 0, \omega) B(\theta_{\sigma-t}\omega) u(\theta_{\sigma-t}\omega)\, d\sigma \\ &= \int_{-t}^0 \Xi(\sigma, 0, \omega) B(\theta_\sigma\omega) u(\theta_\sigma\omega)\, d\sigma, \quad \forall (t, \omega) \in \mathbb{R}_{\ge 0} \times \Omega. \end{aligned}$$

In virtue of (L2), for each $\omega \in \Omega$ such that

$$L_{\omega,\frac{\lambda}{2},u} := \sup_{s\in\mathbb{R}} \gamma(\theta_s\omega)\|B(\theta_s\omega)\| \cdot |u(\theta_s\omega)|\, \mathrm{e}^{-\frac{\lambda}{2}|s|} < \infty, \tag{2.27}$$

we have

$$\begin{aligned} |\Xi(\sigma, 0, \omega) B(\theta_\sigma\omega) u(\theta_\sigma\omega)| &\le \gamma(\theta_\sigma\omega)\, \mathrm{e}^{-\lambda|\sigma|}\, \|B(\theta_\sigma\omega)\| \cdot |u(\theta_\sigma\omega)| \\ &\le L_{\omega,\frac{\lambda}{2},u}\, \mathrm{e}^{-\frac{\lambda}{2}|\sigma|}, \quad \forall \sigma \in \mathbb{R}. \end{aligned}$$

Since

$$\sigma \longmapsto L_{\omega,\frac{\lambda}{2},u}\,\mathrm{e}^{-\frac{\lambda}{2}|\sigma|}, \quad \sigma \in \mathbb{R},$$

is integrable on $(-\infty, \infty)$, so is

$$\sigma \longmapsto \Xi(\sigma, 0, \omega)B(\theta_\sigma\omega)u(\theta_\sigma\omega), \quad \sigma \in \mathbb{R}.$$

In particular, it follows from dominated convergence that the limit

$$\begin{aligned} \lim_{t\to\infty}\Psi(t, \theta_{-t}\omega, \bar{u}) &= \lim_{t\to\infty}\int_{-t}^{0}\Xi(\sigma, 0, \omega)B(\theta_\sigma\omega)u(\theta_\sigma\omega)\,d\sigma \\ &= \int_{-\infty}^{0}\Xi(\sigma, 0, \omega)B(\theta_\sigma\omega)u(\theta_\sigma\omega)\,d\sigma \end{aligned}$$

exists. Finally, observe that, for each $u \in U_\theta^\Omega$, $L_{\omega,\frac{\lambda}{2},u}$ as defined in (2.27) is finite for θ-almost all $\omega \in \Omega$. This establishes (2.25). We have then proved that (2.23) holds for each $x \in X_\theta^\Omega$ and each $u \in U_\theta^\Omega$, for θ-almost all $\omega \in \Omega$.

Define $\mathscr{K}: U_\theta^\Omega \to X_\mathscr{B}^\Omega$ by

$$(\mathscr{K}(u))(\omega) := \int_{-\infty}^{0}\Xi(\sigma, 0, \omega)B(\theta_\sigma\omega)u(\theta_\sigma\omega)\,d\sigma, \quad \tilde{\forall}\omega \in \Omega.$$

It remains to show that $\mathscr{K}(U_\theta^\Omega) \subseteq X_\theta^\Omega$. Indeed, fix $u \in U_\theta^\Omega$ arbitrarily. It follows from the computations above that

$$\begin{aligned} |(\mathscr{K}(u))(\omega)| &\le \int_{-\infty}^{0}\gamma(\theta_\sigma\omega)\|B(\theta_\sigma\omega)\|\cdot|u(\theta_\sigma\omega)|\,\mathrm{e}^{-\lambda|\sigma|}\,d\sigma \\ &\le \int_{-\infty}^{\infty}\gamma(\theta_\sigma\omega)\|B(\theta_\sigma\omega)\|\cdot|u(\theta_\sigma\omega)|\,\mathrm{e}^{-\lambda|\sigma|}\,d\sigma \\ &= \|(\gamma\|B\|\cdot|u|)(\theta.\omega)\,\mathrm{e}^{-\lambda|\cdot|}\|_{L^1(\mathbb{R})}, \quad \tilde{\forall}\omega \in \Omega. \end{aligned}$$

From Proposition 2.4,

$$\omega \longmapsto \|(\gamma\|B\|\cdot|u|)(\theta.\omega)\,\mathrm{e}^{-\lambda|\cdot|}\|_{L^1(\mathbb{R})}, \quad \omega \in \Omega,$$

is tempered. Thus $\mathscr{K}(u)$ is also tempered.

$\mathscr{K}$ is an i/s characteristic. To show that $\mathscr{K}$ is an i/s characteristic, it remains to show that the convergence in both (2.24) and (2.25) is tempered.

Fix $x \in X_\theta^\Omega$ arbitrarily. From the estimates above, we have

$$\begin{aligned} |\Phi(t, \theta_{-t}\omega, x(\theta_{-t}\omega))| &\le \gamma(\theta_{-t}\omega)|x(\theta_{-t}\omega)|\,\mathrm{e}^{-\lambda t} \\ &\le \sup_{s\in\mathbb{R}}\gamma(\theta_s\omega)|x(\theta_s\omega)|\,\mathrm{e}^{-\lambda|s|} \\ &= \|(\gamma|x|)(\theta.\omega)\,\mathrm{e}^{-\lambda|\cdot|}\|_{L^\infty(\mathbb{R})}, \quad \forall t \ge 0, \quad \tilde{\forall}\omega \in \Omega. \end{aligned}$$

It follows from Proposition 2.4 (applied with $p = \infty$) that

$$\omega \longmapsto \|(\gamma|x|)(\theta.\omega)\,\mathrm{e}^{-\lambda|\cdot|}\,\|_{L^\infty(\mathbb{R})}, \quad \omega \in \Omega,$$

is tempered. We conclude that the convergence in (2.24) is tempered.

Similarly, for any arbitrarily fixed $u \in U_\theta^\Omega$, we have

$$\begin{aligned}|\Psi(t, \theta_{-t}\omega, \bar{u}) - (\mathscr{K}(u))(\omega)| &= \left|\int_{-\infty}^{-t} \Xi(\sigma, 0, \omega) B(\theta_\sigma\omega) \cdot u(\theta_\sigma\omega)\, d\sigma\right| \\ &\le \int_{-\infty}^{\infty} |\Xi(\sigma, 0, \omega) B(\theta_\sigma\omega) \cdot u(\theta_\sigma\omega)|\, d\sigma \\ &= \|(\gamma\|B\| \cdot |u|)(\theta.\omega)\,\mathrm{e}^{-\lambda|\cdot|}\,\|_{L^1(\mathbb{R})}\end{aligned}$$

for all $t \ge 0$ and θ-almost all $\omega \in \Omega$. As we saw above, the rightmost term in these inequalities is a tempered random variable. So the convergence in (2.25) is also tempered.

$\mathscr{K}$ is continuous. Suppose that $u_\alpha \to_\theta u_\infty \in U_\theta^\Omega$ for some net $(u_\alpha)_{\alpha\in A}$ in U_θ^Ω. Let $\alpha_0 \in A$ and $r \in (\mathbb{R}_{\ge 0})_\theta^\Omega$ be such that

$$|u_\alpha(\omega) - u_\infty(\omega)| \le r(\omega), \quad \forall\alpha \ge \alpha_0, \quad \tilde{\forall}\omega \in \Omega.$$

Then

$$\begin{aligned}&|(\mathscr{K}(u_\alpha))(\omega) - (\mathscr{K}(u_\infty))(\omega)| \\ &= \left|\int_{-\infty}^{0} \Xi(\sigma, 0, \omega) B(\theta_\sigma\omega) \cdot (u_\alpha(\theta_\sigma\omega) - u_\infty(\theta_\sigma\omega))\, d\sigma\right| \\ &\le \int_{-\infty}^{\infty} \|\Xi(\sigma, 0, \omega) B(\theta_\sigma\omega)\| \cdot r(\theta_\sigma\omega)\, d\sigma\end{aligned}$$

for every $\alpha \ge \alpha_0$, for θ-almost all $\omega \in \Omega$. As above, we can combine (L2), the temperedness of γ, B and r, Lemma 2.4(3) and Proposition 2.4 to conclude that

$$\omega \longmapsto \|\Xi(\sigma, 0, \omega) B(\theta_\sigma\omega)\| \cdot r(\theta_\sigma\omega), \quad \omega \in \Omega,$$

is integrable for θ-almost all $\omega \in \Omega$, and that the map

$$\omega \longmapsto \int_{-\infty}^{\infty} \|\Xi(\sigma, 0, \omega) B(\theta_\sigma\omega)\| \cdot r(\theta_\sigma\omega)\, d\sigma, \quad \omega \in \Omega,$$

is tempered. In particular, since

$$|u_\alpha(\omega) - u_\infty(\omega)| \longrightarrow 0 \quad \text{as} \quad \alpha \to \infty, \quad \tilde{\forall}\omega \in \Omega,$$

it follows from dominated convergence that the map

$$|(\mathscr{K}(u_\alpha))(\omega) - (\mathscr{K}(u_\infty))(\omega)| \longrightarrow 0 \quad \text{as} \quad \alpha \to \infty, \quad \tilde{\forall}\omega \in \Omega,$$

as well. This shows that $\mathscr{K}(u_\alpha) \to_\theta \mathscr{K}(u_\infty)$. Since $u_\infty \in U_\theta^\Omega$ and the net $(u_\alpha)_{\alpha \in A}$ converging to it were arbitrary, this shows $\mathscr{K}$ is continuous.

φ is tempered. The argument here goes along the same lines. Fix arbitrarily any tempered input $u \in \mathscr{S}_\infty^U$ and any tempered initial state $x \in X_\theta^\Omega$. When we were showing that $\mathscr{K}$ is an i/s characteristic above, we saw that

$$|\Phi(t, \theta_{-t}\omega, x(\theta_{-t}\omega))| \le r_1(\omega), \quad \forall t \ge 0, \quad \tilde{\forall}\omega \in \Omega,$$

where $r_1 \colon \Omega \to \mathbb{R}_{\ge 0}$ is a tempered random variable defined by

$$r_1(\omega) := \|(\gamma|x|)(\theta.\omega)\, \mathrm{e}^{-\lambda|\cdot|}\|_{L^\infty(\mathbb{R})}, \quad \omega \in \Omega.$$

Now let $D \in (2^U)_\theta^\Omega$ be a (tempered) rest set for u. Let $r \in (\mathbb{R}_{\ge 0})_\theta^\Omega$ be such that

$$D(\omega) \subseteq \{u \in U;\ \|u\| \le r(\omega)\}, \quad \forall \omega \in \Omega.$$

Thus indeed

$$\|u_t(\theta_{-t}\omega)\| \le r(\omega), \quad \forall t \ge 0, \quad \tilde{\forall}\omega \in \Omega.$$

Then

$$\begin{aligned}
|\Psi(t, \theta_{-t}\omega, u)| &= \left| \int_0^t \Xi(\sigma, t, \theta_{-t}\omega) B(\theta_{\sigma - t}\omega) u_\sigma(\theta_{-t}\omega)\, d\sigma \right| \\
&\le \int_0^t \|\Xi(\sigma, t, \theta_{-t}\omega) B(\theta_{\sigma - t}\omega)\| \cdot |u_\sigma(\theta_{-\sigma}\theta_{\sigma - t}\omega)|\, d\sigma \\
&\le \int_0^t \|\Xi(\sigma - t, 0, \omega) B(\theta_{\sigma - t}\omega)\| \cdot r(\theta_{\sigma - t}\omega)\, d\sigma \\
&\le \int_{-\infty}^{\infty} \|\Xi(\sigma, 0, \omega) B(\theta_\sigma\omega)\| \cdot r(\theta_\sigma\omega)\, d\sigma, \quad \forall t \ge 0, \quad \tilde{\forall}\omega \in \Omega.
\end{aligned}$$

The argument repeatedly applied above shows that the map $r_2 \colon \Omega \to \mathbb{R}_{\ge 0}$ defined by

$$r_2(\omega) := \int_{-\infty}^{\infty} \|\Xi(\sigma, 0, \omega) B(\theta_\sigma\omega)\| \cdot r(\theta_\sigma\omega)\, d\sigma, \quad \omega \in \Omega,$$

is tempered. Now $r_1 + r_2$ is tempered and we have

$$|\check{\xi}_t^{x,u}(\omega)| = |\varphi(t, \theta_{-t}, x(\theta_{-t}\omega), u)| \le r_1(\omega) + r_2(\omega), \quad \forall t \ge 0, \quad \tilde{\forall}\omega \in \Omega.$$

This proves that $\xi^{x,u}$ is tempered. Since u tempered and x tempered were chosen arbitrarily, this completes the proof that φ is a tempered cocycle.

Remark 2.2. If $\|A(\cdot)\| \in L^1(\Omega, \mathscr{F}, \mathbb{P})$, the largest eigenvalue $\overline{\lambda}(\cdot)$ of the Hermitian part of $A(\cdot)$ is such that

$$\mathbb{E}\overline{\lambda} := \int_{\Omega} \overline{\lambda}(\omega)\, d\mathbb{P}(\omega) < 0,$$

and the underlying MPDS θ is ergodic, then it follows from [5, p. 60, Theorem 2.1.2] that (L2) holds with $\lambda := -(\mathbb{E}\overline{\lambda} + \varepsilon)$ for any choice of $\varepsilon \in (0, -\mathbb{E}\overline{\lambda})$. □

2.4 Monotone RDSI's

Suppose that $(X, \leq)$ is a partially ordered space. For any $a, b \in X^{\Omega}_{\mathscr{B}}$, we write $a \leq b$ to mean that $a(\omega) \leq b(\omega)$ for θ-almost all $\omega \in \Omega$. Similarly, for any $p, q \in \mathscr{S}^X_\theta$, we write $p \leq q$ to mean that $p(t,\omega) \leq q(t,\omega)$ for all $t \geq 0$, for θ-almost all $\omega \in \Omega$. Observe that this convention naturally induces partial orders in $X^{\Omega}_{\mathscr{B}}$ and $\mathscr{S}^X_\theta$.

Definition 2.13 (Monotone RDSI). An RDSI $(\theta, \varphi, \mathscr{U})$ is said to be *monotone* if the underlying state and input spaces are partially ordered spaces $(X, \leq_X)$, $(U, \leq_U)$, and

$$\varphi(\cdot,\cdot,x(\cdot),u) \leq_X \varphi(\cdot,\cdot,z(\cdot),v)$$

whenever $x, z \in X^{\Omega}_{\mathscr{B}}$ and $u, v \in \mathscr{U}$ are such that $x \leq_X z$ and $u \leq_U v$.

In particular, if

$$\varphi(t,\omega,x,u) \leq_X \varphi(t,\omega,z,v)$$

holds for every $t \geq 0$, every $\omega \in \Omega$, and every $x, z \in X$ and $u, v \in \mathscr{U}$ such that $x \leq_X z$ and $u \leq_U v$, then it follows that $(\theta, \varphi, \mathscr{U})$ is monotone as per definition above.

Most of the time the underlying partially ordered space will be clear from the context. So unless there is any risk of confusion, we shall often drop the indices in "$\leq_X$" and "$\leq_U$," and write simply "$\leq$."

Proposition 2.5. *If an RDSI $(\theta, \varphi, \mathscr{U})$ is monotone and has an i/s characteristic $\mathscr{K}: U^{\Omega}_\theta \to X^{\Omega}_\theta$, then $\mathscr{K}$ is order-preserving; in other words, if $u, v \in U^{\Omega}_\theta$ and $u \leq v$, then $\mathscr{K}(u) \leq \mathscr{K}(v)$.*

Proof. The proof is straightforward, and we emphasize its main purpose of pointing out a subtlety in Definition 2.13 which might have otherwise gone overlooked. Pick any $u, v \in U^{\Omega}_\theta$ such that $u \leq v$, and fix $x \in X^{\Omega}_\theta$ arbitrarily. Then $x \leq x$, and $\bar{u} \leq \bar{v}$.

By Definition 2.13, there exists a θ-invariant subset of full-measure $\tilde{\Omega} \subseteq \Omega$ such that

$$\varphi(t, \omega, x(\omega), \bar{u}) \leq \varphi(t, \omega, x(\omega), \bar{v}), \quad \forall t \geq 0, \ \forall \omega \in \tilde{\Omega}.$$

Thus

$$\varphi(t, \theta_{-t}\omega, x(\theta_{-t}\omega), \bar{u}) \leq \varphi(t, \theta_{-t}\omega, x(\theta_{-t}\omega), \bar{v}), \quad \forall t \geq 0, \ \forall \omega \in \tilde{\Omega},$$

in view of the θ-invariance of $\tilde{\Omega}$. The result then follows by taking the limit as $t \to \infty$ on both sides of the inequality above for each fixed $\omega \in \tilde{\Omega}$. (Recall that, from the definition of i/s characteristic, such limits exist for θ-almost all $\omega \in \Omega$.) □

2.4.1 Converging Input to Converging State

The "converging input to converging state" result below was first stated and proved for deterministic and finite-dimensional "monotone control systems" by Angeli and Sontag [1, Proposition V.5(2)]. In [10, Theorem 1], Enciso and Sontag explore normality to extend the result to infinite-dimensional systems. Replacing the geometric properties in [10] by minihedrality and adding a compactness assumption it is possible to extend this result to monotone RDSI's.

Recall that a (closed) *cone* in a vector space X is a subset $X_+ \subseteq X$ such that $X_+ + X_+ \subseteq X$, $cX_+ \subseteq X_+$ for every $c \geq 0$, and $X_+ \cap (-X_+) = \varnothing$. The cone X_+ induces a partial order $\leq_X$ in X, defined by

$$x \leq_X y \quad \Leftrightarrow \quad y - x \in X_+.$$

A cone is said to be *solid* if it has nonempty interior, and *minihedral* if every finite subset has a supremum. If X is a normed space, then X_+ is said to be *normal* if there exists a constant $k \geq 0$ such that $0 \leq x \leq y$ implies $\|x\| \leq k\|y\|$.

Theorem 2.3 (Random CICS). *Suppose that X and U are separable Banach spaces, partially ordered by solid, normal, minihedral cones $X_+ \subseteq X$ and $U_+ \subseteq U$, respectively. Let $(\theta, \varphi, \mathscr{U})$ be a tempered, monotone RDSI with state space X and input space U, and suppose that φ has a continuous i/s characteristic $\mathscr{K} : U_\theta^\Omega \to X_\theta^\Omega$. If $u \in \mathscr{U}$ and $u_\infty \in U_\theta^\Omega$ are such that*

(i) u is tempered and eventually precompact, and
(ii) $\breve{u}_t \longrightarrow_\theta u_\infty$ as $t \to \infty$,

then

$$\breve{\xi}_t^{x,u} \longrightarrow_\theta \mathscr{K}(u_\infty) \quad as \quad t \to \infty, \quad \forall x \in X_\theta^\Omega. \tag{2.28}$$

In other words, if the pullback trajectories of u are eventually precompact and converge to u_∞ in the tempered sense, then the pullback trajectories of φ subject to u and starting at any tempered random state x will converge to $\mathscr{K}(u_\infty)$ in the tempered sense as well.

Proof. Fix arbitrarily $x \in X_\theta^\Omega$. From (i), u is tempered. Since φ is assumed to be tempered, the θ-stochastic process $\xi^{x,u}$ is also tempered (see Definition 2.11). In particular,

$$\|\check{\xi}_t^{x,u}(\cdot) - (\mathscr{K}(u_\infty))(\cdot)\| \le \|\check{\xi}_t^{x,u}(\cdot)\| + \|(\mathscr{K}(u_\infty))(\cdot)\|,$$

which is in turn bounded by a nonnegative tempered random variable for large enough values of $t \ge 0$. Thus in order to prove the tempered convergence in (2.28), it remains to show the pointwise convergence; in other words, we need only show that

$$\check{\xi}_t^{x,u}(\omega) \longrightarrow \big(\mathscr{K}(u_\infty)\big)(\omega) \quad \text{as} \quad t \to \infty, \quad \tilde{\forall}\omega \in \Omega. \tag{2.29}$$

This will require some setting up.

Since U_+ is solid and normal, it follows from temperedness and Proposition [5, p. 89, Proposition 3.2.2] that there exist a tempered random variable $v\colon \Omega \to \operatorname{int} U_+$ and a $t_u \ge 0$ such that

$$\check{u}_t(\omega) \in [-v(\omega), v(\omega)], \quad \forall t \ge t_u, \quad \tilde{\forall}\omega \in \Omega.$$

Moreover, $[-v, v]$ is a random closed set by Proposition [5, p. 88, Proposition 3.2.1](1); in particular, it is a random set. So $[-v, v]$ is indeed a tempered random set—temperedness follows from normality. In view of the assumption (i) that u is eventually precompact, by picking a larger t_u, if necessary, we may assume without loss of generality that $\beta_u^{t_u}(\omega)$ is precompact for θ-almost all $\omega \in \Omega$.

Let $(a_\tau)_{\tau \ge t_u}$ and $(b_\tau)_{\tau \ge t_u}$ be, respectively, *lower* and *upper tails* of the pullback trajectories of u:

$$a_\tau(\omega) := \inf_{t \ge \tau} u_t(\theta_{-t}\omega) = \inf \beta_u^\tau(\omega), \quad \tau \ge t_u,$$

and

$$b_\tau(\omega) := \sup_{t \ge \tau} u_t(\theta_{-t}\omega) = \sup \beta_u^\tau(\omega), \quad \tau \ge t_u,$$

for each $\omega \in \Omega$ such that $\beta_u^{t_u}(\omega)$ is precompact. It follows from the hypotheses that U is separable and U_+ is minihedral that the lower and upper tails of the pullback trajectories of u are well-defined, and the maps $\omega \mapsto a_\tau(\omega)$, $\omega \in \Omega$, and $\omega \mapsto b_\tau(\omega)$, $\omega \in \Omega$, are measurable for each $\tau \ge t_u$ (see [5, Theorem 3.2.1, p. 90]). For each $\tau \ge t_u$, we have $a_\tau, b_\tau \in [-v, v]$. Thus by normality a_τ, b_τ are indeed tempered random variables. Moreover,

$$a_\tau, b_\tau \longrightarrow_\theta u_\infty \quad \text{as} \quad \tau \to \infty, \tag{2.30}$$

which also follows by normality.

For each $\tau \geq t_u$, let $\bar{a}_\tau, \bar{b}_\tau$ be the θ-stationary processes generated by a_τ, b_τ, respectively. Then

$$(\bar{a}_\tau)_s(\omega) = a_\tau(\theta_s\omega) = \inf_{t \geq \tau} u_t(\theta_{-t}\theta_s\omega) \leq u_{\tau+s}(\theta_{-(\tau+s)}\theta_s\omega) = (\rho_\tau(u))_s(\omega)$$

and, similarly,

$$(\rho_\tau(u))_s(\omega) \leq (\bar{b}_\tau)_s(\omega), \quad \forall \tau \geq t_u, \quad \forall s \geq 0, \quad \forall \omega \in \Omega.$$

Thus

$$\bar{a}_\tau \leq \rho_\tau(u) \leq \bar{b}_\tau, \quad \forall \tau \geq t_u. \tag{2.31}$$

We now return to (2.29). Using the cocycle property, we may rewrite

$$\begin{aligned}
\check{\xi}_t^{x,u}(\omega) &= \varphi(t-\tau, \theta_{-(t-\tau)}\omega, \varphi(\tau, \theta_{-t}\omega, x(\theta_{-t}\omega), u), \rho_\tau(u)) \\
&= \varphi(t-\tau, \theta_{-(t-\tau)}\omega, x_\tau(\theta_{-(t-\tau)}\omega), \rho_\tau(u)) \\
&= \check{\xi}_{t-\tau}^{x_\tau, \rho_\tau(u)}(\omega), \quad \forall \omega \in \Omega, \quad \forall t \geq \tau \geq t_u,
\end{aligned}$$

where $x_\tau \in X_\theta^\Omega$ is defined by $x_\tau := \check{\xi}_\tau^{x,u}$. Therefore

$$\|\check{\xi}_{\tau+s}^{x,u}(\omega) - (\mathscr{K}(u_\infty))(\omega)\| = \|\check{\xi}_s^{x_\tau,\rho_\tau(u)}(\omega) - (\mathscr{K}(u_\infty))(\omega)\|$$

for every $\omega \in \Omega$, for all $s \geq 0$, for all $\tau \geq t_u$. For any such ω, s, τ, we have

$$\begin{aligned}
\|\check{\xi}_s^{x_\tau,\rho_\tau(u)}(\omega) - (\mathscr{K}(u_\infty))(\omega)\| \leq{}& \|\check{\xi}_s^{x_\tau,\rho_\tau(u)}(\omega) - \check{\xi}_s^{x_\tau,\bar{a}_\tau}(\omega)\| \\
&+ \|\check{\xi}_s^{x_\tau,\bar{a}_\tau}(\omega) - (\mathscr{K}(a_\tau))(\omega)\| \\
&+ \|(\mathscr{K}(a_\tau))(\omega) - (\mathscr{K}(u_\infty))(\omega)\|.
\end{aligned}$$

From (2.30) and the continuity of $\mathscr{K}$, there exist θ-invariant subsets $\tilde{\Omega}_a$ and $\tilde{\Omega}_b$ of full measure of Ω such that

$$\|(\mathscr{K}(a_\tau))(\omega) - (\mathscr{K}(u_\infty))(\omega)\| \longrightarrow 0, \quad \text{as} \quad \tau \to \infty, \quad \forall \omega \in \tilde{\Omega}_a,$$

and

$$\|(\mathscr{K}(b_\tau))(\omega) - (\mathscr{K}(u_\infty))(\omega)\| \longrightarrow 0, \quad \text{as} \quad \tau \to \infty, \quad \forall \omega \in \tilde{\Omega}_b.$$

Similarly, from the definition of i/s characteristic, for any integer $n \geq t_u$, there exist θ-invariant subsets $\tilde{\Omega}_{a,n}$ and $\tilde{\Omega}_{b,n}$ of full measure of Ω such that

$$\|\check{\xi}_s^{x_n,\bar{a}_n}(\omega) - (\mathscr{K}(a_n))(\omega)\| \longrightarrow 0, \quad \text{as} \quad s \to \infty, \quad \forall \omega \in \tilde{\Omega}_{a,n},$$

and

$$\|\check{\xi}_s^{x_n,\bar{b}_n}(\omega) - (\mathscr{K}(b_n))(\omega)\| \longrightarrow 0, \quad \text{as} \quad s \to \infty, \quad \forall \omega \in \tilde{\Omega}_{b,n}.$$

Now by (2.31) and monotonicity, for each integer $n \geq t_u$, there exists a θ-invariant subset of full measure $\tilde{\Omega}_{\leq,n} \subseteq \Omega$ such that

$$\check{\xi}_s^{x_n,\bar{a}_n}(\omega) \leq \check{\xi}_s^{x_n,\rho_n(u)}(\omega) \leq \check{\xi}_s^{x_n,\bar{b}_n}(\omega), \quad \forall s \geq 0, \quad \forall \omega \in \tilde{\Omega}_{\leq,n}.$$

Let[8]

$$\tilde{\Omega} := \tilde{\Omega}_a \cap \tilde{\Omega}_b \cap \left(\bigcap_{n=\lceil t_u \rceil}^{\infty} \tilde{\Omega}_{a,n} \right) \cap \left(\bigcap_{n=\lceil t_u \rceil}^{\infty} \tilde{\Omega}_{b,n} \right) \cap \left(\bigcap_{n=\lceil t_u \rceil}^{\infty} \tilde{\Omega}_{\leq,n} \right).$$

Thus $\tilde{\Omega}$ is a countable intersection of θ-invariant subsets of full measure of Ω and, hence, itself a θ-invariant subset of full measure of Ω. We shall show that convergence in (2.29) occurs for every $\omega \in \tilde{\Omega}$.

Fix arbitrarily an $\omega \in \tilde{\Omega}$ and a positive integer k. It follows from the construction of $\tilde{\Omega}$ that there exists an integer $n_k \geq t_u$ such that

$$\|(\mathscr{K}(a_\tau))(\omega) - (\mathscr{K}(u_\infty))(\omega)\| < 1/k, \quad \forall \tau \geq n_k,$$

and

$$\|(\mathscr{K}(b_\tau))(\omega) - (\mathscr{K}(u_\infty))(\omega)\| < 1/k, \quad \forall \tau \geq n_k.$$

Now we can use the convergence in the definition of i/s characteristic to choose an $s_k \geq 0$ such that

$$\|\check{\xi}_s^{x_{n_k},\bar{a}_{n_k}}(\omega) - (\mathscr{K}(a_{n_k}))(\omega)\| < 1/k, \quad \forall s \geq s_k,$$

and

$$\|\check{\xi}_s^{x_{n_k},\bar{b}_{n_k}}(\omega) - (\mathscr{K}(b_{n_k}))(\omega)\| < 1/k, \quad \forall s \geq s_k.$$

[8]For any $x \in \mathbb{R}$, we write $\lceil x \rceil$ to denote the smallest integer larger than or equal to x.

Again from the construction of $\tilde{\Omega}$, we have

$$\check{\xi}_s^{x_{n_k},\bar{a}_{n_k}}(\omega) \le \check{\xi}_s^{x_{n_k},\rho_{n_k}(u)}(\omega) \le \check{\xi}_s^{x_{n_k},\bar{b}_{n_k}}(\omega), \quad \forall s \ge 0.$$

Thus

$$\|\check{\xi}_s^{x_{n_k},\rho_{n_k}(u)}(\omega) - \check{\xi}_s^{x_{n_k},\bar{a}_{n_k}}(\omega)\| \le C\,\|\check{\xi}_s^{x_{n_k},\bar{b}_{n_k}}(\omega) - \check{\xi}_s^{x_{n_k},\bar{a}_{n_k}}(\omega)\|, \quad \forall s \ge 0,$$

where $C \ge 0$ is the normality constant for U_+. Now

$$\begin{aligned}
\|\check{\xi}_s^{x_{n_k},\bar{b}_{n_k}}(\omega) - \check{\xi}_s^{x_{n_k},\bar{a}_{n_k}}(\omega)\| &\le \|\check{\xi}_s^{x_{n_k},\bar{b}_{n_k}}(\omega) - (\mathscr{K}(b_{n_k}))(\omega)\| \\
&\quad + \|(\mathscr{K}(b_{n_k}))(\omega) - (\mathscr{K}(u_\infty))(\omega)\| \\
&\quad + \|(\mathscr{K}(u_\infty))(\omega) - (\mathscr{K}(a_{n_k}))(\omega)\| \\
&\quad + \|(\mathscr{K}(a_{n_k}))(\omega) - \check{\xi}_s^{x_{n_k},\bar{a}_{n_k}}(\omega)\| \\
&\le 4/k, \quad \forall s \ge s_k.
\end{aligned}$$

We conclude that

$$\begin{aligned}
\|\check{\xi}_t^{x,u}(\omega) - (\mathscr{K}(u_\infty))(\omega)\| &= \|\check{\xi}_{t-n_k}^{x_{n_k},\rho_{n_k}(u)}(\omega) - (\mathscr{K}(u_\infty))(\omega)\| \\
&\le \|\check{\xi}_{t-n_k}^{x_{n_k},\rho_{n_k}(u)}(\omega) - \check{\xi}_{t-n_k}^{x_{n_k},\bar{a}_{n_k}}(\omega)\| \\
&\quad + \|\check{\xi}_{t-n_k}^{x_{n_k},\bar{a}_{n_k}}(\omega) - (\mathscr{K}(a_{n_k}))(\omega)\| \\
&\quad + \|(\mathscr{K}(a_{n_k}))(\omega) - (\mathscr{K}(u_\infty))(\omega)\| \\
&< 4C/k + 1/k + 1/k \\
&= (4C + 2)/k, \quad \forall t \ge n_k + s_k.
\end{aligned}$$

Since $\omega \in \tilde{\Omega}$ and the positive integer k were chosen arbitrarily, this completes the proof. □

2.4.2 Cascades

We now discuss a few applications of the "converging input to converging state" theorem just proved. Separate work in preparation deals with a small-gain theorem for random dynamical systems, a brief outline of which will be given at the end of the chapter.

Let (θ, ψ) be an autonomous RDS evolving on a space $Z = X_1 \times X_2$. We say that (θ, ψ) is *cascaded* if the flow ψ can be decomposed as

$$\psi(t,\omega,(x_1(\omega),x_2(\omega))) \equiv \begin{pmatrix} \varphi_1(t,\omega,x_1(\omega)) \\ \varphi_2(t,\omega,x_2(\omega),(\eta_1)^{x_1}) \end{pmatrix},$$

for some RDSO (θ,φ_1,h_1) with state space X_1 and output space Y_1, and some RDSI $(\theta,\varphi_2,\mathscr{U}_2)$ with state space X_2, input space $U_2 = Y_1$, and set of θ-inputs $\mathscr{U}_2$ containing all (forward) output trajectories of (θ,φ_1,h_1). In this case we write $\psi = \varphi_1 \ltimes \varphi_2$. Recall from item (1) in Theorem 2.2 that if the generator of a discrete RDS can be decomposed as in (2.14), then this RDS is a cascade. A similar decomposition can be done for systems generated by RDEI's whose generator satisfies the natural analogues of (2.14).

Example 2.4 (Bounded Outputs). Let $(\theta,\psi) := (\theta,\varphi_1 \ltimes \varphi_2)$ be a cascaded RDS as above. Suppose that (θ,φ_1,h_1) is an RDSO evolving on a normed space X_1, and such that (θ,φ_1) has a unique, globally attracting equilibrium $(\xi_1)_\infty \in X_{\mathscr{B}}^{\Omega}$:

$$(\breve{\xi}_1)_t^{x_1}(\omega) \longrightarrow (\xi_1)_\infty(\omega), \quad \text{as} \quad t \to \infty, \quad \tilde{\forall}\omega \in \Omega,\ \forall x_1 \in (X_1)_{\mathscr{B}}^{\Omega}.$$

Now suppose that $(\theta,\varphi_2,\mathscr{U}_2)$ is an RDSI satisfying the hypotheses of Theorem 2.3, and that the output function h_1 is *bounded*; in other words, there exists $M \geq 0$ such that

$$\|h_1(\omega,x_1)\| \leq M, \quad \forall x_1 \in X_1,\ \tilde{\forall}\omega \in \Omega.$$

We prove that (θ,ψ) has a unique equilibrium which is attracting for all tempered random initial states.

By continuity of h with respect to the state variable, we have

$$\begin{aligned}(\breve{\eta}_1)_t^{x_1}(\omega) &= h_1(\omega,(\breve{\xi}_1)_t^{x_1}(\omega)) \\ &\longrightarrow h_1(\omega,(\xi_1)_\infty(\omega)), \quad \text{as} \quad t \to \infty, \quad \tilde{\forall}\omega \in \Omega,\ \forall x_1 \in (X_1)_{\mathscr{B}}^{\Omega}.\end{aligned}$$

Since h_1 is bounded, the convergence and the limit are automatically tempered. Thus

$$\breve{\xi}_t^z(\omega) \longrightarrow \begin{pmatrix} (\xi_1)_\infty(\omega) \\ \mathscr{K}((u_2)_\infty)(\omega) \end{pmatrix} \quad \text{as} \quad t \to \infty, \quad \tilde{\forall}\omega \in \Omega,\ \forall z \in Z_\theta^{\Omega},$$

by Theorem 2.3. In particular, the convergence in the second coordinate is tempered.

For conditions guaranteeing that an RDS (θ,φ) would have a unique, globally attracting equilibrium in the sense above, see [4, Theorem 3.2]. The assumption that the output is bounded is very reasonable in biological applications, since there is often a cut off or saturation in the reading of the strength of a signal.

Before we consider the next example, we develop a stronger notion of regularity for output functions than continuity with respect to the state variable. We seek a property which preserves tempered convergence, and which we could check it holds in specific examples.

Definition 2.14 (Tempered Lipschitz). An output function $h\colon \Omega \times X \to Y$ is said to be *tempered Lipschitz* (with respect to a given MPDS θ) if there exists a tempered random variable $L \in (\mathbb{R}_{\geq 0})^{\Omega}_{\theta}$ such that

$$\|h(\omega, x_1) - h(\omega, x_2)\| \leq L(\omega)\|x_1 - x_2\|, \quad \forall x_1, x_2 \in X, \ \tilde{\forall}\omega \in \Omega.$$

We refer to L as a *Lipschitz random variable for* h.

For example, suppose that $X \subseteq \mathbb{R}^n$, and that $h\colon \Omega \times X \to \mathbb{R}^k$ is an output function such that $h(\omega, \cdot)$ is differentiable for all ω in a θ-invariant set of full measure $\tilde{\Omega} \subseteq \Omega$. If the norm of the Jacobian with respect to x,

$$\omega \longmapsto \|D_x h(\omega, \cdot)\| := \sup_{x \in X} |D_x h(\omega, x)|, \quad \omega \in \Omega,$$

is finite and tempered, then h is tempered Lipschitz.

Lemma 2.5. *Let $h\colon \Omega \times X \to Y$ be a tempered Lipschitz output function, $p \in \mathscr{S}^X_{\theta}$ be a θ-stochastic process in X, and let $p_\infty \in X^{\Omega}_{\mathscr{B}}$. Let $q\colon \mathscr{T}_{\geq 0} \times \Omega \to Y$ be the θ-stochastic process in Y defined by*

$$q_t(\omega) := h(\omega, p_t(\omega)), \quad (t, \omega) \in \mathscr{T}_{\geq 0} \times \Omega,$$

and $q_\infty \in Y^{\Omega}_{\mathscr{B}(Y)}$ be the random variable in Y defined by

$$q_\infty(\omega) := h(\omega, p_\infty(\omega)), \quad \omega \in \Omega.$$

If $p_t \to_\theta p_\infty$, then $q_t \to_\theta q_\infty$.

Proof. It follows from continuity with respect to $x \in X$ that

$$q_t(\omega) = h(\omega, p_t(\omega)) \longrightarrow h(\omega, p_\infty(\omega)) = p_\infty(\omega) \quad \text{as} \quad t \to \infty, \quad \tilde{\forall}\omega \in \Omega.$$

Now because $p_t \to_\theta p_\infty$, there exist $r \in (\mathbb{R}_{\geq 0})^{\Omega}_{\theta}$ and $t_0 \geq 0$ such that

$$\|p_t(\omega) - p_\infty(\omega)\| \leq r(\omega), \quad \forall t \geq t_0, \ \tilde{\forall}\omega \in \Omega.$$

Let L be a Lipschitz random variable for h. Then

$$\begin{aligned} \|q_t(\omega) - q_\infty(\omega)\| &= \|h(\omega, p_t(\omega)) - h(\omega, p_\infty(\omega))\| \\ &\leq L(\omega)\|p_t(\omega) - p_\infty(\omega)\| \\ &\leq L(\omega)r(\omega), \qquad \forall t \geq t_0, \ \tilde{\forall}\omega \in \Omega. \end{aligned}$$

By item (3) in Lemma 2.4, Lr is tempered, completing the proof. □

Now suppose that $(\theta, \psi, \mathscr{U})$ is an RDSI evolving on a state space $Z = X_1 \times X_2$. In this case we say that $(\theta, \psi, \mathscr{U})$ is *cascaded* if the flow ψ can be decomposed as

$$\psi(t, \omega, (x_1(\omega), x_2(\omega)), u) \equiv \begin{pmatrix} \varphi_1(t, \omega, x_1(\omega), u) \\ \varphi_2(t, \omega, x_2(\omega), (\eta_1)^{x_1,u}) \end{pmatrix},$$

for some RDSIO $(\theta, \varphi_1, \mathscr{U}_1, h_1)$ with state space X_1, set of θ-inputs $\mathscr{U}_1 = \mathscr{U}$ and output space Y_1, and some RDSI $(\theta, \varphi_2, \mathscr{U}_2)$ with state space X_2, input space $U_2 = Y_1$, and set of θ-inputs $\mathscr{U}_2$ containing all (forward) output trajectories of $(\theta, \varphi_1, \mathscr{U}_1, h_1)$. In this case we also write $\psi = \varphi_1 \ltimes \varphi_2$. Item (1) in Theorem 2.2 can be generalized to contemplate this kind of cascades for discrete systems, as well as systems generated by random differential equations.

Example 2.5 (Tempered Lipschitz Outputs). Suppose that $(\theta, \varphi_1, \mathscr{U}_1)$ and $(\theta, \varphi_2, \mathscr{U}_2)$ in the decomposition above satisfy both the hypotheses of Theorem 2.3. If the output function h_1 is Lipschitz continuous, then $(\theta, \psi, \mathscr{U})$ also has the "converging input to converging state" property; that is, if $u \in \mathscr{U}$ is such that $\breve{u}_t \to_\theta u_\infty$ for some $u_\infty \in U_\theta^\Omega$, then there exists a $\xi_\infty \in Z_\theta^\Omega$ such that

$$\breve{\xi}_t^{z,u} \longrightarrow_\theta \xi_\infty, \quad \forall z \in Z_\theta^\Omega, \tag{2.32}$$

as well.

To see this, let $\mathscr{K}_1\colon (U_1)_\theta^\Omega \to (X_1)_\theta^\Omega$ and $\mathscr{K}_2\colon (U_2)_\theta^\Omega \to (X_2)_\theta^\Omega$ be the i/s characteristics of $(\theta, \varphi_1, \mathscr{U}_1)$ and $(\theta, \varphi_2, \mathscr{U}_2)$, respectively. Fix

$$z = (x_1, x_2) \in Z_\theta^\Omega = (X_1)_\theta^\Omega \times (X_2)_\theta^\Omega$$

arbitrarily. From Theorem 2.3, we have

$$(\breve{\xi}_1)_t^{x_1,u} \longrightarrow_\theta \mathscr{K}_1(u_\infty).$$

Since h_1 is tempered Lipschitz, it follows from Lemma 2.5 that

$$(\breve{\eta}_1)_t^{x_1,u} \longrightarrow_\theta (u_2)_\infty,$$

where

$$(u_2)_\infty := h_1(\cdot, \mathscr{K}_1(u_\infty)(\cdot)).$$

It follows, again from Theorem 2.3, that

$$(\breve{\xi}_2)_t^{x_2,(\eta_1)^{x_1,u}} \longrightarrow_\theta \mathscr{K}_2((u_2)_\infty).$$

Hence

$$\breve{\xi}_t^{z,u} = \begin{pmatrix} (\breve{\xi}_1)_t^{x_1,u} \\ (\breve{\xi}_2)_t^{x_2,(\eta_1)^{x_1,u}} \end{pmatrix} \longrightarrow_\theta \begin{pmatrix} \mathscr{K}_1(u_\infty) \\ \mathscr{K}_2((u_2)_\infty) \end{pmatrix}.$$

Since $z \in Z_\theta^\Omega$ was picked arbitrarily, this establishes (2.32).

The procedure above can be generalized to cascades of three or more systems to show that the "converging input to converging state" property will hold provided that it holds for its individual components—and the intermediate outputs are tempered Lipschitz. In Example 2.3, suppose we assume, in addition, that the off-diagonal entries of A and all entries of B are nonnegative θ-almost everywhere. Then the RDSI generated by the RDEI in the example is monotone and thus satisfies the hypotheses of Theorem 2.3. Tempered Lipschitz output functions are not difficult to come by, as we pointed out above. This yields a class of cascaded systems having the "converging input to converging state" property.

A couple more remarks about this example are in order. A cascade of monotone systems need not itself be monotone. So the construction above provides us with a way of checking the "converging input to converging state" property for systems which do not directly satisfy the hypotheses of Theorem 2.3. But even if it would be possible to check it directly that $(\theta, \varphi, \mathscr{U})$ already satisfies the hypotheses of Theorem 2.3, it might be easier to check them for each component—for instance, if $(\theta, \varphi, \mathscr{U})$ can be decomposed as a cascade of linear systems linked by (possibly nonlinear) tempered Lipschitz output functions.

We have illustrated in Examples 2.4 and 2.5 how one may obtain global convergence results for systems decomposable into cascades, as discussed in the Introduction. Further work in preparation deals with "closed loop" systems, and how "converging input to converging state" property can be used to prove small-gain theorems for such systems. Below we provide a brief outline of the idea.

2.4.3 *Small-Gain Theorem*

A small-gain theorem for the closed-loop of monotone RDSIO's with anti-monotone outputs follows along the lines of the deterministic case [1, 10]. Assuming the input and output spaces coincide, one defines an "input to output characteristic" $\mathscr{K}^Y : U_\theta^\Omega \to U_\theta^\Omega$ by composing the i/s characteristic (assuming, of course the underlying RDSI has one) with the output function h in the natural way:

$$\big(\mathscr{K}^Y(u)\big)(\omega) := h(\omega, \big(\mathscr{K}(u)\big)(\omega)), \quad u \in U_\theta^\Omega, \quad \omega \in \Omega.$$

If the iterates $(\mathscr{K}^Y)^{(k)}(u) := (\mathscr{K}^Y \circ \cdots \circ \mathscr{K}^Y)(u)$ (k times) of $\mathscr{K}^Y$ converge to a unique equilibrium u_∞ ("small-gain condition"), then every eventually precompact solution of the closed-loop system converges to $\mathscr{K}(u_\infty)$, the state characteristic corresponding to the input u_∞.

A proof as in [1, 10] goes by appealing to the random CICS property for monotone RDSI's above, after establishing a contraction property on the "limsup" and "liminf" (defined analogously as in these references) of external signals. Mild technical assumptions on the state and input/output spaces guarantee that said limsup's and liminf's are well-defined and measurable. Reasonable ("polynomial temperedness") growth conditions on the outputs guarantee that the input to output characteristic is well-defined as a map $U_\theta^\Omega \to U_\theta^\Omega$ (preserves temperedness). Separate work in preparation will provide all the details and several examples.

Acknowledgements Work supported in part by grants NIH 1R01GM086881 and 1R01GM100473, and AFOSR FA9550-11-1-0247.

References

1. D. Angeli, E.D. Sontag, Monotone control systems. IEEE Trans. Automat. Contr. **48**(10), 1684–1698 (2003)
2. M. Arcak, E.D. Sontag, Diagonal stability for a class of cyclic systems and applications. Automatica **42**, 1531–1537 (2006)
3. L. Arnold, *Random Dynamical Systems* (Springer, Berlin, 2010)
4. F. Cao, J. Jiang, On the global attractivity of monotone random dynamical systems. Proc. Am. Math. Soc. **138**(3), 891–898 (2010)
5. I. Chueshov, *Monotone Random Systems – Theory and Applications*. Lecture Notes in Mathematics, vol. 1997 (Springer, Berlin, 2002)
6. H. Crauel, P.E. Kloeden, M. Yang, Random attractors of stochastic reaction-diffusion equations on variable domains. Stoch. Dyn. **11**, 301–314 (2011)
7. H. Deng, M. Krstic, Stochastic nonlinear stabilization, Part I: a backstepping design. Syst. Control Lett. **32**, 143–150 (1997)
8. H. Deng, M. Krstic, R. Williams, Stabilization of stochastic nonlinear systems driven by noise of unknown covariance. IEEE Trans. Automat. Contr. **46**, 1237–1253 (2001)
9. D. Del Vecchio, A.J. Ninfa, E.D. Sontag, Modular cell biology: retroactivity and insulation. Nat. Mol. Syst. Biol. **4**, 161 (2008)
10. G.A. Enciso, E.D. Sontag, Global attractivity, I/O monotone small-gain theorems, and biological delay systems. Discrete Contin. Dyn. Syst. **14**(3), 549–578 (2006)
11. P. Florchinger, Feedback stabilization of affine in the control stochastic differential systems by the control Lyapunov function method. SIAM J. Control Optim. **35**, 500–511 (1997)
12. G. B. Folland, *Real Analysis: Modern Techniques and Their Applications*, 2nd edn. (Wiley, New York, 1999)
13. T. Gedeon, *Cyclic Feedback Systems*. Memoirs of the AMS, vol. 134, no. 637 (AMS, Providence, 1998)
14. A. Goldbeter, *Biochemical Oscillations and Cellular Rhythms* (Cambridge University Press, Cambridge, 1996)
15. L.H. Hartwell, J.J. Hopfield, S. Leibler, A.W. Murray, From molecular to modular cell biology. Nature **402**(6761 Suppl), 47–52 (1999)
16. S. Hu, N.S. Papageorgiou, *Handbook of Multivalued Analysis, Vol. I: Theory* (Kluwer, Dordrecht, 1997)
17. A. Isidori, *Nonlinear Control Systems II* (Springer, London, 1999)

18. A. Jentzen, P.E. Kloeden, *Taylor Approximations of Stochastic Partial Differential Equations* CBMS Lecture Series (SIAM, Philadelphia, 2011)
19. H. Khalil, *Nonlinear Systems* (Prentice Hall, Englewood Cliffs, 2002)
20. M. Krstić, I. Kanellakopoulos, P.V. Kokotović, *Nonlinear and Adaptive Control Design* (Wiley, New York, 1995)
21. S. Lang, *Real Analysis*, 2nd edn. (Addison-Wesley, Reading, 1983)
22. D.A. Lauffenburger, Cell signaling pathways as control modules: complexity for simplicity? Proc. Natl. Acad. Sci. USA **97**(10), 5031–5033 (2000)
23. G. Lindgren, *Stationary Stochastic Processes—Theory and Applications* (Chapman and Hall, London, 2012)
24. J. Mallet-Paret, H.L. Smith, The Poincaré-Bendixson theorem for monotone cyclic feedback systems. J. Dyn. Differ. Equ. **2**, 367–421 (1990)
25. J.D. Murray, *Mathematical Biology, I, II: An Introduction* (Springer, New York, 2002)
26. Z. Pan, T. Bassar, Backstepping controller design for nonlinear stochastic systems under a risk-sensitive cost criterion. SIAM J. Control Optim. **37**, 957–995 (1999)
27. H.L. Smith, Oscillations and multiple steady states in a cyclic gene model with repression. J. Math. Biol. **25**, 169–190 (1987)
28. H.L. Smith, *Monotone Dynamical Systems: An Introduction to the Theory of Competitive and Cooperative Systems*. Mathematical Surveys and Monographs, vol. 41 (AMS, Providence, 1995)
29. E.D. Sontag, Smooth stabilization implies coprime factorization. IEEE Trans. Automat. Contr. **34**(4), 435–443 (1989)
30. E.D. Sontag, *Mathematical Control Theory. Deterministic Finite-Dimensional Systems*, 2nd edn. Texts in Applied Mathematics, vol. 6 (Springer, New York, 1998)
31. J. Tsinias, The concept of 'exponential ISS' for stochastic systems and applications to feedback stabilization. Syst. Control Lett. **36**, 221–229 (1999)
32. P. Walters, *An Introduction to Ergodic Theory* (Springer, Berlin, 2000)

Chapter 3
Canard Theory and Excitability

Martin Wechselberger, John Mitry, and John Rinzel

Abstract An important feature of many physiological systems is that they evolve on multiple scales. From a mathematical point of view, these systems are modeled as singular perturbation problems. It is the interplay of the dynamics on different temporal and spatial scales that creates complicated patterns and rhythms. Many important physiological functions are linked to time-dependent changes in the forcing which leads to nonautonomous behaviour of the cells under consideration. Transient dynamics observed in models of excitability are a prime example.
Recent developments in canard theory have provided a new direction for understanding these transient dynamics. The key observation is that canards are still well defined in nonautonomous multiple scales dynamical systems, while equilibria of an autonomous system do, in general, not persist in the corresponding driven, nonautonomous system. Thus canards have the potential to significantly shape the nature of solutions in nonautonomous multiple scales systems. In the context of neuronal excitability, we identify canards of folded saddle type as firing threshold manifolds. It is remarkable that dynamic information such as the temporal evolution of an external drive is encoded in the location of an invariant manifold—the canard.

Keywords Canards • Geometric singular perturbation theory • Excitability • Neural dynamics • Firing threshold manifold • Separatrix • Transient attractor

---

M. Wechselberger (✉) · J. Mitry
School of Mathematics and Statistics, University of Sydney, Australia
e-mail: wm@maths.usyd.edu.au; J.Mitry@maths.usyd.edu.au

J. Rinzel
Courant Institute, New York University, USA
e-mail: rinzel@cns.nyu.edu

P.E. Kloeden and C. Pötzsche (eds.), *Nonautonomous Dynamical Systems in the Life Sciences*, Lecture Notes in Mathematics 2102, DOI 10.1007/978-3-319-03080-7_3,

3.1 Motivation

Physiological rhythms and patterns are central to life. Prominent examples are the beating of the heart, the activity patterns of neurons, and the release of the hormones that regulate growth and metabolism. Although many cells in the body display *intrinsic*, spontaneous rhythmicity, many physiological functions derive from the *interaction* of these cells, with each other and with external inputs, to generate these essential rhythms. Thus it is important to analyse both the origin of the intrinsic complex nonlinear processes and the effects of stimuli on these physiological rhythms.

Cell signalling is the result of a complex interaction of feedback loops that control and modify the cell behaviour via ionic flows and currents, proteins and receptor systems. The specific feedback loops differ from one cell to another and, from a physiological point of view, the signalling seems to be extremely cell specific. The respective mathematical cell models, however, have an amazingly similar structure. This suggests *unifying mathematical mechanisms* for cell signalling and its failure.

An important feature of most physiological systems is that they evolve on *multiple scales*. For example, the rhythm of the heart beat consists of a long interval of quasi steady-state followed by a short interval of rapid variation, which is the beat itself [34]. The same feature is observed for activity patterns of neurons [34,55] and for calcium signalling in cells [34]. It is the interplay of the dynamics on different temporal or spatial scales that creates complicated rhythms and patterns.

Multiple scales problems of physiological systems are usually modelled by *singularly perturbed systems* [28, 34, 55]. The geometric theory of multiple scales dynamical systems—known as *Fenichel theory* [17, 32, 33, 49]—has provided powerful tools for studying singular perturbation problems. In conjunction with the innovative *blow-up technique* [15, 39, 57], geometric singular perturbation theory delivers rigorous results on global dynamics such as periodic and quasi-periodic relaxation oscillations in multiple time-scale problems [58]. When combined with results on Henon-like maps, this approach has the potential to explain chaotic dynamics in relaxation oscillators as observed in the periodically forced van der Pol relaxation oscillator [24].

This development within dynamical systems theory provides an excellent framework for addressing questions on how complex rhythms and patterns can be detected and controlled. The fact that equivalent stimulation can elicit qualitatively different spiking patterns in different neurons demonstrates that intrinsic coding properties differ significantly from one neuron to the next. Hodgkin recognized this and identified three basic types of neurons distinguished by their coding properties [29]. Pioneered by Rinzel and Ermentrout [31,51,52], bifurcation theory explains repetitive *(tonic)* firing patterns for adequate steady inputs (e.g. current step protocols) in *integrator (type I)* and *resonator (type II)* multiple time-scales neuronal models.

In contrast, the dynamic behaviour of *differentiator (type III)* neurons cannot be explained by standard (autonomous) dynamical systems theory. This third type of excitable neuron encodes a *dynamic change* in the input and hence they are well suited for temporal processing like phase-locking and coincidence detection [42, 53]. Auditory brain stem neurons are an important example of such neurons involved with precise timing computations. The nonautonomous (dynamic) nature of the signal is essential to determine the response of a type III neuron. A major aim of this chapter is to highlight the profound differences in the behaviour of all neuron types (I–III) when we apply a step current protocol compared to a smooth dynamic current protocol, either excitatory or inhibitory.

In a dynamical system that exhibits time-dependence in its forcing or parameters, one still expects convergence of the phase-space flow to some lower dimensional object; but this object, termed a *pullback attractor* [36, 37, 50], is now itself time-dependent. Identifying dynamic objects in phase-space that act effectively as separatrices is a major mathematical challenge. Such separatrices may influence the observed dynamics only on a certain (finite) time scale.

Recently, a *canard mechanism* was identified that leads to transient dynamics in multiple time-scales systems [26, 41, 44, 64]. Canards are exceptional solutions in singularly perturbed systems which occur on boundaries of regions corresponding to *different dynamic behaviours*. The theory on canards and their impact on *transient* dynamics of multiple scales dynamical systems is the main focus of this chapter. What makes canards so special for (driven) nonautonomous multiple scales dynamical systems? The key observation is that *canard points* (also known as *folded singularities*) are still well defined in nonautonomous multiple scales dynamical systems, while equilibria of an autonomous system will, in general, not persist in the corresponding driven, nonautonomous system. Thus canards have the potential to significantly shape the nature of solutions in nonautonomous multiple scales systems. We highlight this important point of view in Sect. 3.3.2.1.

Another class of complex oscillatory behaviour observed in neuroscience is mixed-mode oscillations (MMOs). These oscillations correspond to switching between small-amplitude oscillations and relaxation oscillations—patterns that have been frequently observed in experiments [1, 12, 25, 35, 47]. Recently, canard theory combined with an appropriate global return mechanism was used based on the multiple time-scale structure of the underlying models to explain these complicated dynamics [2, 3, 6, 22, 43, 57, 61, 63]. This is now one widely accepted explanation for MMOs; see, e.g., [5, 8, 14, 16, 27, 38, 54, 56, 60] and the current review [10].

The outline of the chapter is as follows: In Sect. 3.2 we review geometric singular perturbation theory in arbitrary dimensions with a particular emphasis on canard theory. In Sect. 3.3 we review excitable systems. We focus on external drives that are either piecewise constant or vary smoothly. The former models instantaneous (fast) changes while the later models smooth (slow) changes. We then outline the relationship between the theory of singularly perturbed systems and nonautonomous (multiple scales) systems. In particular, we show how canard theory can be used to explain excitability for smooth dynamic forcing protocols by identifying a canard of folded saddle type as the firing threshold manifold of an excitable neuron. The geometric theory is applied to neuronal and biophysical models. Finally, we conclude in Sect. 3.4.

Remark 3.1. Section 3.2 provides a comprehensive review of geometric singular perturbation theory and assumes a solid background on dynamical systems theory such as found in [23]. While the basic ideas of geometric singular perturbation theory are well known to the mathematical biology/neuroscience community, the theory presented in this section might seem at certain points too technical and/or too rigorous for this peer group. We suggest that these readers skip (parts of) the section and explore the necessary theory after reading through Sect. 3.3 on excitability. Nevertheless, we hope that many readers will appreciate the rigor and generality of the presented material.

3.2 Geometric Singular Perturbation Theory

Our focus is on a system of differential equations that has an explicit time scale splitting of the form

$$\begin{aligned} w' &= \epsilon\, g(w, v, \epsilon) \\ v' &= f(w, v, \epsilon)\,, \end{aligned} \tag{3.1}$$

where $(w, v) \in \mathbb{R}^k \times \mathbb{R}^m$ are state space variables and $k, m \geq 1$. The variables $v = (v_1, \ldots, v_m)$ are denoted *fast*, the variables $w = (w_1, \ldots, w_k)$ are denoted *slow*, the prime denotes the time derivative d/dt and $\epsilon \ll 1$ is a small positive parameter encoding the time scale separation between the slow and fast variables. The functions $f : \mathbb{R}^k \times \mathbb{R}^m \times \mathbb{R} \to \mathbb{R}^m$ and $g : \mathbb{R}^k \times \mathbb{R}^m \times \mathbb{R} \to \mathbb{R}^k$ are assumed to be C^∞ smooth. By switching from the fast time scale t to the slow time scale $\tau = \epsilon t$, system (3.1) transforms to

$$\begin{aligned} \dot{w} &= g(w, v, \epsilon) \\ \epsilon\, \dot{v} &= f(w, v, \epsilon)\,. \end{aligned} \tag{3.2}$$

where the overdot denotes the time derivative $d/d\tau$. System (3.1) respectively (3.2) are topologically equivalent and solutions often consist of a mix of slow and fast segments reflecting the dominance of one time scale or the other. We refer to (3.1) respectively (3.2) as a *singularly perturbed system*. As $\epsilon \to 0$, the trajectories of (3.1) converge during fast segments to solutions of the m-dimensional *layer (or fast) problem*

$$\begin{aligned} w' &= 0 \\ v' &= f(w, v, 0) \end{aligned} \tag{3.3}$$

while during slow segments, trajectories of (3.2) converge to solutions of

$$\begin{aligned} \dot{w} &= g(w, v, 0) \\ 0 &= f(w, v, 0) \end{aligned} \tag{3.4}$$

which is a k-dimensional differential-algebraic problem called the *reduced (or slow) problem*. Geometric singular perturbation theory [17, 32] uses these lower-dimensional sub-systems (3.3) and (3.4) to predict the dynamics of the full $(k + m)$-dimensional system (3.1) or (3.2) for $\epsilon > 0$.

3.2.1 The Layer Problem

First, we focus on the layer problem (3.3). Note that the slow variables w are parameters in this limiting system.

Definition 3.1. The set

$$S := \{(w, v) \in \mathbb{R}^k \times \mathbb{R}^m \mid f(w, v, 0) = 0\} \tag{3.5}$$

is the set of equilibria of (3.3). In general, this set S defines a k-dimensional manifold, i.e. the Jacobian $D_{(w,v)} f$ evaluated along S has full rank, and we refer to it as the *critical manifold*.

Remark 3.2. The set S could be the union of finitely many k-dimensional manifolds. All definitions regarding the critical manifold hold also for such a set.

Since we assume that f is smooth, this implies that the critical manifold is a differentiable manifold. The basic classification of singularly perturbed systems is given by the properties of the critical manifold S of the layer problem (3.3).

Definition 3.2. A subset $S_h \subseteq S$ is called *normally hyperbolic* if all $(w, v) \in S_h$ are hyperbolic equilibria of the layer problem, that is, the Jacobian with respect to the fast variables v, denoted $D_v f$, has no eigenvalues with zero real part.

- We call a normally hyperbolic subset $S_a \subseteq S$ *attracting* if all eigenvalues of $D_v f$ have negative real parts for $(w, v) \in S_a$; the layer problem describes the flow towards this set.
- $S_r \subseteq S$ is called *repelling* if all eigenvalues of $D_v f$ have positive real parts for $(w, v) \in S_r$; the layer problem describes the flow away from this set.
- If $S_s \subseteq S$ is normally hyperbolic and neither attracting nor repelling we say it is of *saddle type*.

For a normally hyperbolic manifold $S_h \subseteq S$, we have a uniform splitting of eigenvalues of $D_v f$ along S_h into two groups, i.e. for each $p \in S_h$ the Jacobian $D_v f$ has m_u eigenvalues with positive real part and m_s eigenvalues with negative real part where $m_u + m_s = m$. This enables us to define local stable and unstable manifolds of the critical manifold S_h:

Definition 3.3. The local stable and unstable manifolds of the critical manifold S_h denoted by $W^s_{loc}(S_h)$ and $W^u_{loc}(S_h)$, respectively, are the unions

$$W^s_{loc}(S_h) = \bigcup_{p \in S_h} W^s_{loc}(p)\,, \quad W^u_{loc}(S_h) = \bigcup_{p \in S_h} W^u_{loc}(p)\,. \tag{3.6}$$

The manifolds $W^s_{loc}(p)$ and $W^u_{loc}(p)$ form a family of *fast fibers* (called a *fast fibration* or *foliation*) for $W^s_{loc}(S_h)$ and $W^u_{loc}(S_h)$, respectively, with *base points* $p \in S_h$. The dimension of $W^s_{loc}(S_h)$ is $k + m_s$ and the dimension of $W^u_{loc}(S_{h,\epsilon})$ is $k + m_u$.

The geometric theory of singular perturbation problems with normally hyperbolic manifolds is referred to as *Fenichel Theory* [17, 32]. This theory guarantees the persistence of a normally hyperbolic manifold close to $S_h \subseteq S$ and corresponding local stable and unstable manifolds close to $W^s_{loc}(S_h)$ and $W^u_{loc}(S_h)$ as follows:

Theorem 3.1 (Fenichel's Theorem 1, cf. [17,32]). *Given system* (3.1) *with* $f, g \in C^\infty$. *Suppose* $S_h \subseteq S$ *is a compact normally hyperbolic manifold, possibly with boundary. Then for* $\epsilon > 0$ *sufficiently small the following holds:*

(i) For any $r < \infty$, *there exists a* C^r *smooth manifold* $S_{h,\epsilon}$, *locally invariant under the flow* (3.1), *that is* C^r $O(\epsilon)$ *close to* S_h.
(ii) For any $r < \infty$, *there exist* C^r *smooth stable and unstable manifolds*

$$W^s_{loc}(S_{h,\epsilon}) = \bigcup_{p_\epsilon \in S_{h,\epsilon}} W^s_{loc}(p_\epsilon)\,, \quad W^u_{loc}(S_{h,\epsilon}) = \bigcup_{p_\epsilon \in S_{h,\epsilon}} W^u_{loc}(p_\epsilon)\,, \tag{3.7}$$

locally invariant under the flow (3.1), *that are* C^r $O(\epsilon)$ *close to* $W^s_{loc}(S_h)$ *and* $W^u_{loc}(S_{h,\epsilon})$, *respectively.*

Remark 3.3. $S_{h,\epsilon}$ is, in general, not unique but all representations of $S_{h,\epsilon}$ lie exponentially close in ε from each other, i.e. all r-jets are uniquely determined.

Remark 3.4. We assume that a compact, simply connected, k-dimensional smooth manifold with boundary implies that its boundary is a $(k-1)$-dimensional smooth manifold. A compact manifold with boundary is called *overflowing invariant*, if the vector field inside the manifold is tangent to the manifold and along the boundary it points everywhere outward. The proof of Fenichel's theorem is based on this definition.

3.2.1.1 Folded Critical Manifolds

Normal hyperbolicity fails at points on S where $D_v f$ has (at least) one eigenvalue with zero real part, i.e. a bifurcation occurs in the layer problem under the variation of the parameter set w. Generically, such points are *folds* in the sense of singularity theory [59].

Definition 3.4. The critical manifold S (3.5) of the singularly perturbed system (3.2) is (locally) folded if there exists a set F that forms a $(k-1)$-dimensional manifold in the k-dimensional critical manifold S defined by

$$\begin{aligned} F := \{(w,v) \in \mathbb{R}^k \times \mathbb{R}^m; \mid f(w,v,0) = 0\,, \mathrm{rk}(D_v f)(w,v,0) = m-1\,, \\ l \cdot [(D^2_{vv} f)(w,v,0)\,(r,r)] \neq 0,\; l \cdot [(D_w f)(w,v,0)] \neq 0\} \end{aligned} \tag{3.8}$$

with corresponding left and right null vectors l and r of the Jacobian $D_v f$. The set F denotes the *fold points* of the critical manifold.

A fold corresponds to a saddle-node bifurcation in the layer problem which is one of the generic codimension-one bifurcations in a dynamical system.

3.2.2 *The Reduced Problem*

The reduced problem (3.4) is a differential algebraic problem and describes the evolution of the slow variables w constrained to the critical manifold S. As a consequence, S defines an interface between the two sub-systems (3.3) and (3.4).

Definition 3.5. Given the reduced problem (3.4). A vector field on the critical manifold S (3.5) is a C^1-mapping $g : S \to \mathbb{R}^k$ such that $g(w,v) \in T_{(w,v)}S$ for all $(w,v) \in S$.

In other words, the reduced vector field (3.4) has to be in the tangent bundle TS of the critical manifold S. The total (time) derivative of $f(w,v,0) = 0$, i.e. $D_v f \cdot \dot{v} + D_w f \cdot \dot{w} = 0$ provides exactly the definition for a tangent vector $(\dot{w}, \dot{v})$ of an integral curve $(w(\tau), v(\tau)) \in \mathbb{R}^{k+m}$ to be constrained to the tangent bundle TS. This leads to the following representation of the reduced problem (3.4):

$$\begin{aligned} \dot{w} &= g(w,v,0) \\ -D_v f \cdot \dot{v} &= (D_w f \cdot g)\,(w,v,0) \end{aligned} \tag{3.9}$$

where $(w,v) \in S$. Let adj $(D_v f)$ denote the *adjoint* of the matrix $D_v f$ which is the transpose of the co-factor matrix of $D_v f$, i.e. $\mathrm{adj}\,(D_v f) \cdot D_v f = D_v f \cdot \mathrm{adj}\,(D_v f) = \det(D_v f)\, I$.

Remark 3.5. In the case $m = 1$, $D_v f = \det D_v f = \frac{\partial f}{\partial v} = f_v$ is a scalar and $\mathrm{adj}\,(D_v f) := 1$. Note that the adjoint of a square matrix is well defined for both regular and singular matrices. This is in contrast to the definition of the inverse of a square matrix which is only defined in the regular case.

We apply adj $(D_v f)$ to both sides of the second equation in (3.9) to obtain

$$\begin{aligned} \dot{w} &= g(w, v, 0) \\ -\det(D_v f)\,\dot{v} &= \text{adj}\,(D_v f)\cdot D_w f\cdot g\,(w, v, 0) \end{aligned} \tag{3.10}$$

where $(w, v) \in S$. System (3.10) provides a representation of the original reduced problem (3.4) in any (local) coordinate chart on the manifold S.

Remark 3.6. A coordinate chart on an n-dimensional smooth manifold S is a pair $(U;\phi)$, where U is an open subset of S and $\phi : U \to \tilde{U}$ is a diffeomorphism from U to an open subset $\tilde{U} = \phi(U) \subset \mathbb{R}^k$. A well-known (and often used) example is the graph of a smooth function $F : U \to \mathbb{R}^n$ which is a subset of $\mathbb{R}^n \times \mathbb{R}^k$ defined by $\{(x;y) \in \mathbb{R}^n \times \mathbb{R}^k \,:\, x \in U\,, y = F(x)\}$.

Suppose that the critical manifold S is normally hyperbolic, i.e. $D_v f$ has full rank for all $(w, v) \in S$. The implicit function theorem implies that S is given as a graph $v = h(w)$. In other words, S can be represented in a single chart given by the slow variable base $w \in \mathbb{R}^k$. The reduced problem (3.10) on S_h is then given in this coordinate chart by

$$\dot{w} = g(w, h(w), 0)\,. \tag{3.11}$$

Fenichel theory [17, 32] guarantees the persistence of a slow flow on $S_{h,\epsilon}$ close to the reduced flow of S_h in the following way:

Theorem 3.2 (Fenichel's Theorem 2, cf. [17,32]). *Given system* (3.1) *with $f, g \in C^\infty$. Suppose $S_h \subseteq S$ is a compact normally hyperbolic manifold, possibly with boundary. Then for $\epsilon > 0$ sufficiently small, Theorem 3.1(i), holds and the following:*

(iii) The slow flow on $S_{h,\epsilon}$ converges to the reduced flow on S_h as $\epsilon \to 0$.

Since S_h is a graph $v = h(w)$ it follows that $S_{h,\epsilon}$ is also a graph $v_\epsilon = h(w, \epsilon)$ for sufficiently small $\epsilon \ll 1$. Thus the slow flow on $S_{h,\epsilon}$ fulfills

$$\dot{w} = g(w, h(w, \epsilon), \epsilon)\,, \tag{3.12}$$

and we are dealing with a regular perturbation problem on $S_{h,\epsilon}$ which is a remarkable result. Consequently, we have

Corollary 3.1. *Hyperbolic equilibria of the reduced problem* (3.11) *persist as hyperbolic equilibria of the full problem* (3.2) *for sufficiently small $\epsilon \ll 1$.*

For $\epsilon > 0$, the base points $p_\epsilon \in S_{h,\epsilon}$ of the fast fibers $W^s_{loc}(p_\epsilon)$, respectively $W^u_{loc}(p_\epsilon)$, evolve according to (3.12). Hence, the individual fast fibers $W^s_{loc}(p_\epsilon)$, respectively $W^u_{loc}(p_\epsilon)$, are not invariant, but the families of fibers (3.7) are invariant in the following sense:

Theorem 3.3 (Fenichel's Theorem 3, cf. [17]). *Given system* (3.1) *with* $f, g \in C^\infty$. *Suppose* $S_h \subseteq S$ *is a compact normally hyperbolic manifold, possibly with boundary. Then for* $\epsilon > 0$ *sufficiently small, Theorem 3.1(ii) holds and the following:*

(iv) The foliation $\{W^s_{loc}(p_\epsilon) | \, p_\epsilon \in S_{h,\epsilon}\}$ *is (positively) invariant, i.e.*

$$W^s_{loc}(p_\epsilon) \cdot t \subset W^s_{loc}(p_\epsilon \cdot t)$$

for all $t \geq 0$ *such that* $p_\epsilon \cdot t \in S_{h,\epsilon}$, *where* $\cdot t$ *denotes the solution operator of system* (3.1).

(v) The foliation $\{W^u_{loc}(p_\epsilon) | \, p_\epsilon \in S_{h,\epsilon}\}$ *is (negatively) invariant, i.e.*

$$W^u_{loc}(p_\epsilon) \cdot t \subset W^u_{loc}(p_\epsilon \cdot t)$$

for all $t \leq 0$ *such that* $p_\epsilon \cdot t \in S_{h,\epsilon}$, *where* $\cdot t$ *denotes the solution operator of system* (3.1).

This theorem implies that the exponential decay of a trajectory in the stable manifold $W^s(S_{h,\epsilon})$ towards its corresponding base point $p_\epsilon \in S_{h,\epsilon}$ is inherited from the unperturbed case. The same is true in backward time for a trajectory in the unstable manifold $W^u(S_{h,\epsilon})$ and summarized in the following:

Theorem 3.4 (Fenichel's Theorem 4, cf. [17,32]). *Let* $\alpha_s < 0$ *be an upper bound* $Re\,\lambda_i < \alpha_s < 0$, $i = 1, \ldots, m_s$, *for the stable eigenvalues of the critical manifold* S_h. *There exists a constant* $\kappa_s > 0$, *so that if* $p_\epsilon \in S_{h,\epsilon}$ *and* $q_\epsilon \in W^s_{loc}(p_\epsilon)$ *then*

$$\|q_\epsilon \cdot t - p_\epsilon \cdot t\| \leq \kappa_s \exp(\alpha_s t)$$

for all $t \geq 0$ *such that* $p_\epsilon \cdot t \in S_{h,\epsilon}$.

Similarly, let $\alpha_u > 0$ *be a lower bound* $Re\,\lambda_j > \alpha_u > 0$, $j = 1, \ldots, m_u$, *for the unstable eigenvalues of the critical manifold* S_h. *There exists a constant* $\kappa_u > 0$, *so that if* $p_\epsilon \in S_{h,\epsilon}$ *and* $q_\epsilon \in W^u_{loc}(p_\epsilon)$ *then*

$$\|q_\epsilon \cdot t - p_\epsilon \cdot t\| \leq \kappa_u \exp(\alpha_u t)$$

for all $t \leq 0$ *such that* $p_\epsilon \cdot t \in S_{h,\epsilon}$.

If we assume that $S_h = S_a$ is an attracting normally hyperbolic manifold then Fenichel theory implies that the dynamics of system (3.2) are completely described (after some initial transient time) by the dynamics on the k-dimensional slow manifold $S_{a,\epsilon}$ which to leading order can be completely determined by the reduced flow on S_a. This result justifies certain model reduction techniques often found in the mathematical biology literature on biochemical reactions.

Example 3.1. A classic biophysical example of a normally hyperbolic problem is given by *Michaelis-Menten* enzyme kinetics (see, e.g., [34] for details):

$$S + E \underset{k_{-1}}{\overset{k_1}{\rightleftharpoons}} C \overset{k_2}{\rightharpoonup} P + E\,, \tag{3.13}$$

which models an enzymatic reaction with substrate S, enzyme E, an intermediate complex C and product P. Using the law of mass action gives the following system of differential equations

$$\begin{aligned} \tfrac{d[S]}{dt} &= k_{-1}[C] - k_1[S][E], \\ \tfrac{d[C]}{dt} &= k_1[S][E] - (k_{-1} + k_2)[C], \\ \tfrac{d[E]}{dt} &= (k_{-1} + k_2)[C] - k_1[S][E], \\ \tfrac{d[P]}{dt} &= k_2[C], \end{aligned} \tag{3.14}$$

where $[X]$ denotes the concentration of $X = S,\ C,\ E,\ P$ with initial concentrations

$$[S](0) = S_0, \quad [C](0) = 0, \quad [E](0) = E_0, \quad [P](0) = 0.$$

Notice that $[P]$ can be found by direct integration, and there is a conserved quantity since $d[C]/dt + d[E]/dt = 0$, so that $[C] + [E] = E_0$. Hence it suffices to study the first two equations of system (3.14) with $[E] = E_0 - [C]$. Using dimensional analysis gives the corresponding two-dimensional dimensionless system,

$$\begin{aligned} \frac{ds}{d\tau} &= \quad \dot{s} = \alpha_1 c - s(1-c) = g(s,c) \\ \epsilon\frac{dc}{d\tau} &= \epsilon\dot{c} = s(1-c) - (\alpha_1 + \alpha_2)c = f(s,c)\,, \end{aligned} \tag{3.15}$$

with (dimensionless) substrate and complex concentration $s = [S]/S_0$ and $c = [C]/E_0$, initial conditions $s(0) = 1$ and $c(0) = 0$, time $\tau = E_0 k_1 t$ and parameters $\alpha_1 = k_{-1}/(S_0 k_1) > 0$, $\alpha_2 = k_2/(S_0 k_1) > 0$, $\varepsilon = E_0/S_0 \ll 1$. Here, the initial enzyme concentration E_0 is considered significantly smaller than the initial substrate concentration S_0 which is a realistic condition for enzyme reactions. Thus, the obtained dimensionless system is a singularly perturbed system with s slow and c fast.

The critical manifold is given by $f(s,c) = 0$. The Jacobian of the layer problem is the derivative $f_c = -(s + \alpha_1 + \alpha_2) < 0$ for all $s \geq 0$. Hence, the critical manifold is an attracting normally hyperbolic manifold S_a for the biophysically relevant domain of $s \geq 0$ and is given as a graph

$$c = h(s) = \frac{s}{s + \alpha_1 + \alpha_2}\,.$$

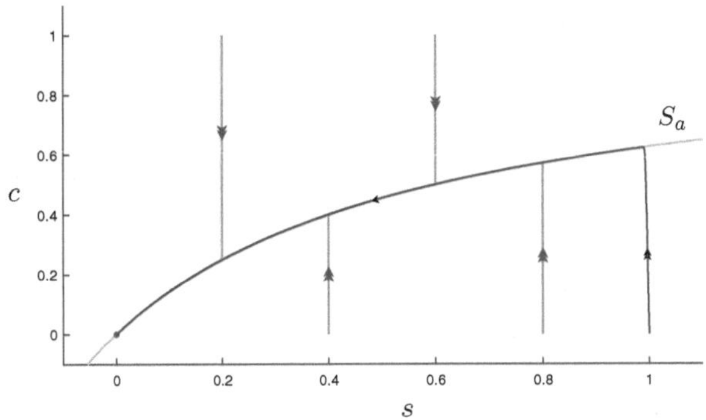

Fig. 3.1 Michaelis-Menten kinetics: from the initial condition $(s, c) = (1, 0)$, the complex c builds quickly up (along fast fibers) until it reaches the normally hyperbolic manifold S_a. Then the slow uptake of the substrate s starts (slow flow along S_a towards the rest state $(s, c) = (0, 0)$)

The reduced problem is then given in the single coordinate chart $s \in \mathbb{R}$ by

$$\dot{s} = g(s, h(s)) = -\frac{\alpha_2 s}{s + \alpha_1 + \alpha_2} \leq 0\,, \quad \forall s \geq 0\,. \tag{3.16}$$

This differential equation has a hyperbolic equilibrium at $s = 0$ which is stable. Since the initial condition $(s(0), c(0)) = (1, 0)$ is not on the critical manifold S_a, we expect an initial fast transient behavior towards the slow manifold $S_{a,\epsilon}$ close to the stable fast fiber at $s = 1$. Then the slow dynamics will take over and the substrate concentration will slowly decay towards zero along the slow manifold $S_{a,\epsilon}$ as predicted by the reduced flow. Figure 3.1 confirms the predictions of Fenichel theory. The reduced problem (3.16) is indeed a good approximation of the substrate concentration dynamics after a transient initial time. The rate of uptake of the substrate s described by (3.16) is often referred to as the *Michaelis-Menten law.*

3.2.2.1 Reduced Problem on Folded Critical Manifolds

Similar to the normally hyperbolic case, a (local) graph representation of the critical manifold S is used to analyse the k-dimensional reduced problem (3.10) in the case of a folded critical manifold. From the definition (3.8) of the folded critical manifold follows that there exists (at least) one slow variable w_j, $j \in \{1, \ldots, k\}$ with $l \cdot [(D_{w_j} f)(w, v, 0)] \neq 0$. Without loss of generality, let w_1 be this slow variable. One is then able to replace one column in $D_v f$ (we assume, without loss of generality, that this column is $D_{v_1} f$) by the column of $D_{w_1} f$ such that $\text{rk}\, D_{(w_1, v_2, \ldots, v_m)} f = m$ along S (including F). In the case $k = 1$ respectively $k \geq 2$, the implicit function theorem then implies that S is (locally) a graph $y = h(v_1)$ respectively $y = h(w_2, \ldots, w_k, v_1)$ where $y = (w_1, v_2, \ldots, v_m)$. In the case $k = 1$, incorporating this graph representation of S leads to the projection of the reduced problem (3.10) onto the coordinate chart $v_1 \in \mathbb{R}$,

$$-\det(D_v f)\,\dot{v}_1 = \text{adj}\,(D_v f)_1 \cdot D_w f \cdot g\,(v_1, 0)\,, \tag{3.17}$$

respectively in the case $k \geq 2$ it leads to the projection of the reduced problem (3.10) onto the coordinate chart $(w_2, \ldots, w_k, v_1) \in \mathbb{R}^k$:

$$\begin{aligned} \dot{w}_j &= g_j(w_2, \ldots, w_k, v_1, 0)\,, \quad j = 2, \ldots k \\ -\det(D_v f)\,\dot{v}_1 &= \text{adj}\,(D_v f)_1 \cdot D_w f \cdot g\,(w_2, \ldots, w_k, v_1, 0)\,, \end{aligned} \tag{3.18}$$

where adj $(D_v f)_1$ denotes the first row of the adjoint matrix adj $(D_v f)$.

Remark 3.7. This row vector adj $(D_v f)_1$ represents the left null-vector l of the matrix $D_v f$. As mentioned before, the scalar adj $(D_v f)_1 \cdot D_{w_1} f \neq 0$ and, hence, the row vector adj $(D_v f)_1 \cdot D_w f$ is non-singular.

Looking at the reduced problem (3.17), respectively (3.18), we observe $\det(D_v f) = 0$ along the fold F, i.e. (3.17), respectively (3.18), is singular along F.

Definition 3.6. *Regular fold points* $p \in F$ of the reduced flow (3.17) respectively (3.18) satisfy the *transversality condition (normal switching condition)*

$$\text{adj}\,(D_v f)_1 \cdot D_w f \cdot g \neq 0\,. \tag{3.19}$$

The condition $l \cdot [(D^2_{vv} f)(w, v, 0)\,(r, r)] \neq 0$ along F implies that $\det(D_v f)$ has different signs on adjacent subsets (branches) of the critical manifold S bounded by F. Hence, in the neighborhood of regular fold points $p \in F$ the flow is directed either towards or away from the fold F. Solutions of the reduced problem will reach the fold F in finite (forward or backward) time where they cease to exist.

We can circumvent the problem of the singular nature of the reduced problem along the fold F by introducing a new time τ_1 defined by $d\tau = -\det(D_v f)\,d\tau_1$, (this is a space dependent time rescaling and, hence, the differential form is needed), and rescaling time in system (3.17) respectively (3.18) which then gives the *desingularized problem*

$$\dot{v}_1 = \text{adj}\,(D_v f)_1 \cdot D_w f \cdot g\,(v_1, 0)\,, \tag{3.20}$$

respectively

$$\begin{aligned} \dot{w}_j &= -\det(D_v f) \cdot g_j(w_2, \ldots, w_k, v_1, 0)\,, \quad j = 2, \ldots k \\ \dot{v}_1 &= \text{adj}\,(D_v f)_1 \cdot D_w f \cdot g\,(w_2, \ldots, w_k, v_1, 0) \end{aligned} \tag{3.21}$$

where the overdot denotes now $d/d\tau_1$. From the time rescaling it follows that the direction of the flow in (3.20) respectively (3.21) has to be reversed on branches where $\det(D_v f) > 0$ to obtain the corresponding reduced flow (3.17), respectively (3.18). Otherwise, the flows of (3.17) and (3.20), respectively (3.18) and (3.21), are equivalent. Obviously, the analysis of the desingularized problem (3.20), respectively (3.21), is preferable.

3.2.3 *Folded Singularities and Singular Canards*

Our aim is to understand the properties of the reduced problem (3.17), respectively (3.18), based on properties of the desingularized problem (3.20), respectively (3.21). Keeping that in mind, we define the following:

Definition 3.7. We distinguish between two possible types of singularities of the desingularized problem (3.20), respectively (3.21):

- *Ordinary singularities* which are defined by $g = 0$.
- *Folded singularities* which are defined by

$$\det(D_v f) = 0\,, \quad \mathrm{adj}\,(D_v f)_1 \cdot D_w f \cdot g = 0\,. \tag{3.22}$$

Ordinary singularities correspond to equilibria of the reduced problem (3.17), respectively (3.18). Generically, they are positioned away from the fold F, i.e. $\det(D_v f) \neq 0$, and they are isolated singularities. In other words, these singularities correspond to equilibria in both the reduced and desingularized system.

Folded singularities are positioned on the fold F. There is a crucial difference between the case $k = 1$ and $k \geq 2$ and we will study these two cases separately.

3.2.3.1 The Case $k = 1$

Recall from Remark 3.7 that the scalar $\mathrm{adj}\,(D_v f)_1 \cdot D_w f \neq 0$. Hence, the folded singularity condition (3.22) can only be fulfilled for $g = 0$. This folded singularity is generically a hyperbolic equilibrium for the desingularized problem (3.20), but it *does not* correspond to an equilibrium of the reduced problem (3.17). In fact, the reduced problem has finite non zero speed at the folded singularity (due to a cancellation of a simple zero). This allows solutions of the reduced problem to cross (in forward or backward time) from one branch of S via the fold F to the other branch of S.

Definition 3.8. Given a singularly perturbed system (3.2) with a (locally) folded critical manifold $S = S_a \cup F \cup S_{s/r}$ where S_a denotes an attracting branch and $S_{s/r}$ denotes a repelling branch (case $m = 1$) respectively a saddle type branch (case $m \geq 2$). A trajectory of the reduced problem (3.17) that has the ability to cross in finite time from the S_a branch of the critical manifold to the $S_{r/s}$ branch via a folded singularity is called a *singular canard*.

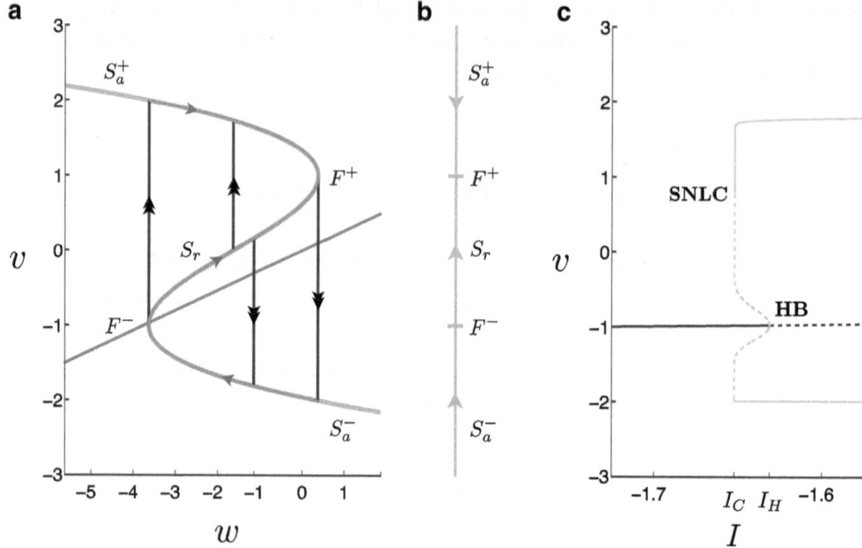

Fig. 3.2 The FitzHugh-Nagumo (FHN) model (3.23) with $a = -\sqrt{3}$, $b = \sqrt{3}$, $c = 4/15$, $\epsilon = 8/100$: (**a**) singular canard cycles and relaxation oscillation cycles for $I = I_f^- = -7/4$ obtained through continuous concatenations of slow orbit segments (*gray*) and fast fibers (*black*); (**b**) the corresponding reduced flow projected onto coordinate chart $v \in \mathbb{R}$ indicates the crossing of a singular canard from S_a^- to S_r; (**c**) Bifurcation diagram includes a singular subcritical Andronov-Hopf bifurcation (HB) at $I = I_H \approx -1.632$, a branch of canard cycles and relaxation oscillation cycles and a saddle-node of limit cycles (SNLC) bifurcation at $I = I_C \approx -1.653$

Example 3.2. The *FitzHugh-Nagumo (FHN)* model [20,45] is a qualitative (dimensionless) description of action potential generation in a class of conductance based, Hodgkin-Huxley-type models [30], given by

$$\begin{aligned} w' &= \epsilon g(w,v) = \epsilon(v - cw) \\ v' &= f(w,v) = v(v-a)(b-v) - w + I\,, \end{aligned} \tag{3.23}$$

where we assume $b > 0 > a$. For $I = 0$, this system may have one, two or three equilibria depending on (a,b,c). We restrict the parameter set to $4/(a-b)^2 > c > 0$ which guarantees only one equilibrium. Note, for sufficiently small $c > 0$ there will be only one equilibrium in system (3.23) for any choice of I.

The critical manifold S of system (3.23) is not normally hyperbolic since $f_v = -3v^2 + 2(a+b)v - ab$ vanishes for $v^{\pm} = (a + b \pm \sqrt{a^2 - ab + b^2})/3$. At these values, $f_{vv}(v^{\pm}) = \mp 2\sqrt{a^2 - ab + a^2} \neq 0$. Furthermore $f_w = -1 \neq 0$ which shows that the FHN model has a cubic-shaped critical manifold $S = S_a^- \cup F^- \cup S_r \cup F^+ \cup S_a^+$ with outer attracting branches $S_a^{\pm}$ and repelling middle branch S_r; see Fig. 3.2a.

The critical manifold S is given as a graph $w = h(v) = v(v-a)(b-v) + I$. Thus we project the reduced problem on the single coordinate chart $v \in \mathbb{R}$,

$$-(-3v^2 + 2(a+b)v - ab)\dot{v} = -(v - c(v(v-a)(b-v) + I)) . \quad (3.24)$$

The corresponding desingularized problem is given by

$$\dot{v} = -(v - c(v(v-a)(b-v) + I)) . \quad (3.25)$$

We have to reverse direction of the desingularized flow on S_r to obtain the corresponding reduced flow. Otherwise, the desingularized flow is equivalent to the reduced flow. There exist parameter values $I = I_f^{\pm}$ such that the right hand side of (3.25) evaluated at $v = v^{\pm}$ vanishes. These particular parameter values define folded singularities of the reduced problem (3.24). Figure 3.2a,b shows the case $I = I_f^-$ where the folded singularity exists at the lower fold F^-. We observe a singular canard crossing from S_a^- to S_r.

This enables us to construct a whole family of singular limit cycles known as singular *canard cycles* that are formed through continuous concatenations of slow orbit segments including canard segments (gray segments, one arrowhead) and fast fibers (black segments, two arrowheads). We distinguish two types of canard cycles, known as *canards without head* and *canards with head* [2,62]. Both are illustrated in Fig. 3.2a: a canard without head is a continuous concatenation of a singular canard segment from S_a^- to S_r (grey) and a fast fiber segment connecting S_r with S_a^- (black). Obviously, a jump back along a fast fiber segment from any base point on S_r works. This gives the family of canards without head.

Similarly, a canard with head is a continuous concatenation of a singular canard segment from S_a^- to S_r (grey), a fast fiber segment connecting S_r to S_a^+ (black), a slow segment on S_a^+ connecting to the upper fold F^+ (grey) and, finally, a fast fiber segment connecting F^+ to S_a^- (black). Again, a jump forward along a fast fiber segment from any base point on S_r works and we obtain a whole family of canards with head. All these singular canard cycles have $O(1)$ amplitude and have a frequency on the order of the slow time scale. The canard cycles are bounded by a singular relaxation cycle, a continuous concatenation of a slow segment on S_a^- connecting to F^- (grey), a fast fiber segment connecting F^- to S_a^+ (black), a slow segment on S_a^+ connecting to F^+ (grey) and, finally, a fast fiber segment connecting F^+ to S_a^- (black).

3.2.3.2 The Case $k \geq 2$

Here, the folded singularity condition (3.22) can be fulfilled for $g \neq 0$. Such generic folded singularities *do not* correspond to equilibria of the reduced problem (3.18). The set of these folded singularities, denoted M_f, forms a submanifold of codimension one in the $(k-1)$-dimensional set of fold points F.

Remark 3.8. In the case $k = 2$, the set M_f consists of isolated folded singularities. This makes the following description of associated geometric objects sometimes simpler or even trivial. The reader should keep that in mind since we do not distinguish between $k = 2$ and $k > 2$ throughout this section and Sect. 3.2.4.2.

Generically, the set M_f viewed as a set of equilibria of the desingularized system (3.21) has $(k - 2)$ zero eigenvalues and two eigenvalues $\lambda_{1/2}$ with nonzero real part. Thus for $k \geq 3$, M_f represents a normally hyperbolic manifold of equilibria in system (3.21). The classification of folded singularities is based on these two nonzero eigenvalues $\lambda_{1/2}$ and follows that of singularities in two-dimensional vector fields.

Definition 3.9. *Classification of generic folded singularities* (3.22):

- In the case that $\lambda_{1/2}$ are real, let us denote the eigenvalue ratio by

$$\mu := \lambda_1/\lambda_2$$

  where we assume without loss of generality that $|\lambda_1| \leq |\lambda_2|$. Then the corresponding singularity is either a *folded saddle* if $\mu < 0$, or a *folded node* if $0 < \mu \leq 1$.
- In the case that $\lambda_{1/2}$ are complex conjugates and $\mathrm{Re}\,\lambda_{1/2} \neq 0$ then the corresponding singularity is a *folded focus*.

For a generic folded singularity, the algebraic multiplicity of the corresponding singularities on both sides of the last equation in the reduced problem (3.18) is the same (i.e. one). This leads in the case of a folded saddle or a folded node to a nonzero but finite speed of the reduced flow through a folded singularity. Hence, folded saddles and folded nodes create possibilities for the reduced flow to cross to different (normally hyperbolic) branches of the critical manifold S via such folded singularities. This is the hallmark of singular canards in systems with two or more slow variables. Definition 3.8 of singular canards applies here as well. We restate it here for convenience:

Definition 3.10. Given a singularly perturbed system (3.2) with a folded critical manifold $S = S_a \cup F \cup S_{s/r}$ where S_a denotes an attracting branch and $S_{s/r}$ denotes a repelling branch (case $m = 1$) respectively a saddle type branch (case $m \geq 2$). A trajectory of the reduced problem (3.18) that has the ability to cross in finite time from the S_a branch of the critical manifold to the $S_{r/s}$ branch via a folded singularity is called a *singular canard*.

Remark 3.9. In the case of a folded focus there are no singular canards. Only the flow direction changes along the fold F at the folded focus. All solutions starting near a folded focus reach the set of fold-points F/M_f in finite forward or backward time where they cease to exist due to finite time blow-up.

In the folded saddle case, $\mu < 0$, there exists a $(k - 1)$-dimensional centre-stable manifold W_{cs} and a $(k - 1)$-dimensional centre-unstable manifold W_{cu} along the $(k - 2)$-dimensional normally hyperbolic manifold $W_c = W_{cs} \cap W_{cu} = M_f$. Both manifolds, W_{cs} and W_{cu}, are uniquely foliated by one-dimensional fast fibers

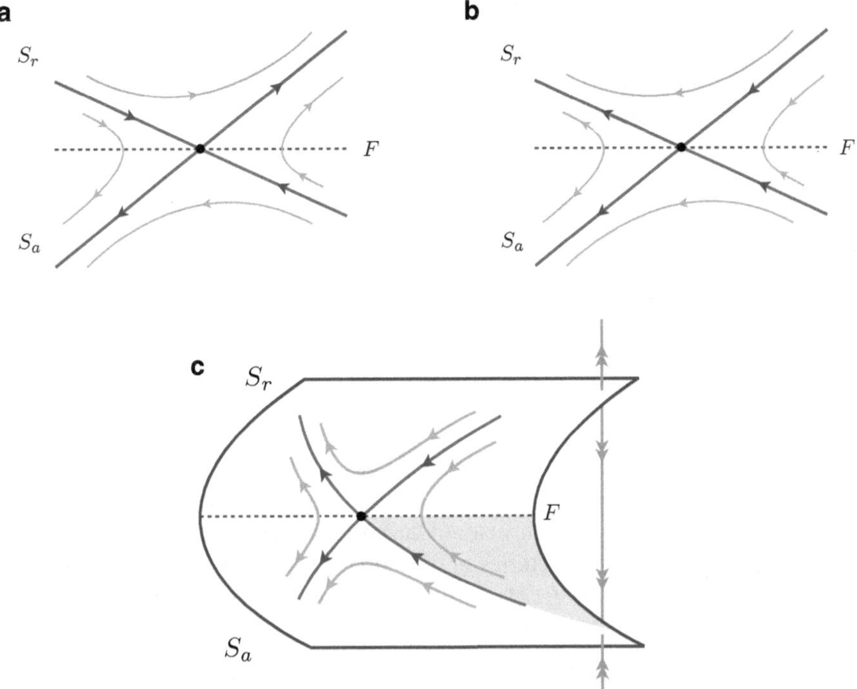

Fig. 3.3 A folded saddle singularity: (**a**) desingularized flow; (**b**) corresponding reduced flow and (**c**) reduced flow on the critical manifold S. There are two singular canards that cross the fold F at the folded singularity (*black dot*), one from S_a to S_r and the other (called faux canard) from S_r to S_a. The *shaded region* indicates a region of solutions on S_a that will reach the fold F in forward time

W_s respectively W_u over the base M_f where the fibers are tangent to the stable respectively unstable eigenvector of the corresponding folded singularity $p_f \in M_f$, i.e. the corresponding base point.

Recall that the reduced flow is obtained from the desingularized flow by changing the direction of the flow on $S_{r/s}$. Thus, trajectories that start in a stable fiber $W_s \subset W_{cs} \subset S_a$ approach M_f in finite time and cross tangent to the stable eigenvector of the corresponding folded singularity on M_f to the unstable branch $W_{cu} \subset S_{r/s}$. These are singular canards of folded saddle type.

All other trajectories of the reduced flow starting in S_a (close to F) reach either the set of fold-points F/M_f in finite forward or backward time where they cease to exist due to finite time blow-up or they do not reach the set F/M_f at all. Figure 3.3 shows the folded saddle case for $k = 2$.

Remark 3.10. Trajectories starting on an unstable fiber $W_u \subset W_{cu} \subset S_{r/s}$ approach M_f in finite time and cross tangent to the unstable eigenvector of the corresponding folded singularity on M_f to the stable branch $W_{cs} \subset S_a$. Such solutions are called *singular faux canards*.

In the folded node case $\mu > 0$, assuming $\lambda_{1/2} < 0$ are both negative, the whole phase space S is equivalent to W_{cs}. Let us define $W_{ss} \subset W_{cs}$ as the $(k-1)$-dimensional subset of unique fast fibers corresponding to the span of the strong stable eigenvectors along the base M_f. Again, the reduced flow is obtained from the desingularized flow by changing the direction of the flow on S_r.

Definition 3.11. The set W_{ss} together with the $(k-1)$-dimensional set of fold points F bounds a sector in S_a, called the *singular funnel*, with the property that every trajectory starting in the singular funnel reaches the set of folded node singularities M_f in finite time and subsequently crosses the set F transversely to the other branch $S_{r/s}$ in the direction that is tangent to the weak stable eigenvector of the corresponding folded node singularity on M_f.

Thus, every trajectory within a singular funnel is a singular canard. Trajectories that start on the boundary set $W_{ss} \subset S_a$ reach also the set M_f in finite time but cross tangent to the strong stable eigenvector of the corresponding folded node singularity (by definition). All other trajectories of the reduced flow starting in S_a (close to F) reach the set of fold-points F/M_f in finite forward or backward time where they cease to exist due to finite time blow-up.

Remark 3.11. In the folded node case $\mu > 0$ with $\lambda_{1/2} > 0$, we are dealing with a whole family of faux canards.

3.2.4 Maximal Canards

Next, we are concerned with the persistence of singular canards as canards of the full system (3.1). We first provide a geometric definition of canards for $\epsilon > 0$. Recall that the branches S_a and $S_{r/s}$ are normally hyperbolic away from the fold F. Thus, Fenichel theory implies the existence of (non-unique but exponentially close) invariant slow manifolds $S_{a,\epsilon}$ and $S_{r/s,\epsilon}$ away from F. Fix a representative for each of these manifolds $S_{a,\epsilon}$ respectively $S_{r/s,\epsilon}$

Definition 3.12. A *maximal canard* corresponds to the intersection of the manifolds $S_{a,\epsilon}$ and $S_{r/s,\epsilon}$ extended by the flow of (3.1) into the neighborhood of the set $M_f \subset F$.

Such a maximal canard defines a family of canards nearby which are exponentially close to the maximal canard, i.e. a family of solutions of (3.1) that follow an attracting branch $S_{a,\epsilon}$ of the slow manifold towards the neighbourhood of the set $M_f \subset F$, pass close to $M_f \subset F$ and then follow, rather surprisingly, a repelling/saddle branch $S_{r/s,\epsilon}$ of the slow manifold for a considerable amount of slow time. The existence of this family of canards is a consequence of the non-uniqueness of $S_{a,\epsilon}$ and $S_{r/s,\epsilon}$. However, in the singular limit $\epsilon \to 0$, such a family of canards is represented by a unique singular canard.

Remark 3.12. The key to understanding the local dynamics near the set of folded singularities by means of geometric singular perturbation theory is the *blow-up* technique. The "blow-up" desingularizes degenerate singularities such as the set of folded singularities or the fold itself. With this procedure, one gains enough hyperbolicity on the blown-up locus B to apply standard tools from dynamical system theory. For a detailed description of the blow-up technique and its application to singularly perturbed systems we refer the interested reader to [15,39,40,57,58,61,63].

Remark 3.13. A folded critical manifold S implies a single zero eigenvalue of the m-dimensional layer problem. Hence in system (3.1), there exist locally invariant manifolds W_{cs} (centre-stable) and W_{cu} (centre-unstable) near the fold F where $W_{cu} \cup W_{cs}$ spans the whole phase space and $W_c = W_{cu} \cap W_{cs}$ corresponds to a $(k+1)$-dimensional centre-manifold. A centre manifold reduction of system (3.1) onto this $(k+1)$-dimensional subspace W_c captures the local dynamics near the fold F. Note that the reduced problem (3.17) respectively (3.18) reflects already such a center manifold reduction (on the linear level) through the projection onto the nullvector $l = \text{adj}\,(D_v f)_1$ corresponding to the zero eigenvalue of the Jacobian $D_v f$. In the following, we present results that are based on such a reduction. The interested reader is referred to, e.g. [6,63,65], for details.

3.2.4.1 Case $k = 1$: Singular Hopf Bifurcation and Canard Explosion

Recall from the FHN model that a folded singularity and associated singular canards exist only for a specific parameter value $I = I_f$. In the case $k = 1$, this shows that a folded singularity is degenerate, i.e. a codimension-one phenomenon. Furthermore, the condition for the folded singularity coincides with the equilibrium condition $g = 0$. This indicates a bifurcation of the equilibrium state in the full system under the variation of I. This can be easily seen when looking at a planar slow-fast system

$$\begin{aligned} w' &= \epsilon g(w,v) \\ v' &= f(w,v,I)\,. \end{aligned} \tag{3.26}$$

The trace and the determinant of the Jacobian are given by

$$\text{tr}\,J = f_v + \epsilon g_w\,, \qquad \det J = \epsilon(f_v g_w - f_w g_v)\,. \tag{3.27}$$

Close to the fold F, a bifurcation of equilibria defined by $f = g = 0$ happens for $0 < \epsilon \ll 1$ when $\text{tr}\,J = 0$. This implies $f_v = -\epsilon g_w = O(\epsilon)$ and, in the singular limit, this gives the fold condition $f_v = 0$. The existence of singular canards is given if the equilibrium $g = 0$ of the desingularized problem (3.20) is stable. This implies that $f_w g_v < 0$ evaluated at $g = 0$ and, hence, $\det J = O(\epsilon) > 0$. So, we are expecting a *singular Andronov-Hopf bifurcation* for $I = I_H$ that creates small $O(\sqrt{\epsilon})$ amplitude limit cycles with nonzero frequencies of order $O(\sqrt{\epsilon})$ [39].

Hence, the singular nature of the Andronov-Hopf bifurcation is encoded in both, amplitude and frequency. Figure 3.2c shows an example of a singular subcritical Andronov-Hopf bifurcation.

Note in Fig. 3.2c that the $O(\sqrt{\epsilon})$ branch of the Andronov-Hopf bifurcation suddenly changes dramatically near $I = I_c$. This almost vertical branch marks the unfolding of the canard cycles within an exponentially small parameter interval of the bifurcation parameter I. This is often referred to as a *canard explosion* [2,15,39]. The following summarizes these observations:

Theorem 3.5 (cf. [39]). *Given a planar slow-fast system*

$$\begin{aligned} w' &= \epsilon g(w, v) \\ v' &= f(w, v, I), \end{aligned} \tag{3.28}$$

with a (locally) folded critical manifold $S = S_a \cup F \cup S_r$. Assume there exists a folded singularity for $I = I_f$ that also allows for the existence of singular canards. Then a singular Andronov-Hopf bifurcation and a canard explosion occur at

$$I_H = I_f + H_1\,\epsilon + O(\epsilon^{3/2}) \qquad \textit{and} \tag{3.29}$$

$$I_c = I_f + (H_1 + K_1)\,\epsilon + O(\epsilon^{3/2}). \tag{3.30}$$

The coefficients H_1 and K_1 can be calculated explicitly and, hence, the type of Andronov-Hopf bifurcation (super- or subcritical).

In the singular limit, we have $I_H = I_c = I_f$. By definition, we associate one maximal canard with the canard explosion. In Fig. 3.2a, this maximal canard is represented by the singular canard that moves along the middle branch S_r right up to the upper fold F^+. It delineates between jump back canards that form small amplitude canard cycles—*canards without head*—and jump away canards that form large amplitude canard cycles—*canards with head.*

In Fig. 3.2c, the branch of canard cycles then connects to the branch of stable relaxation oscillation cycles with large amplitude. Note, there is also a saddle-node of limit cycles bifurcation of the canard cycles where the stability property changes. Since canards are exponentially sensitive to parameter variations, they are hard to detect. In reality, this makes canard cycles rather exceptional.

3.2.4.2 Case $k \geq 2$: Folded Saddle and Folded Node Canards

Here, folded singularities are generic, i.e. they persist under small parameter variations. This makes these canards robust creatures, i.e. their impact on the dynamics of a singularly perturbed system is observable. In the following, we present persistence results of canards.

Theorem 3.6 (cf. [57, 63]). *In the folded saddle case ($\mu < 0$) of a singularly perturbed system* (3.1)*, the $(k-1)$-dimensional set W_{cs} of singular canards perturb to a $(k-1)$-dimensional set of maximal canards for sufficiently small $\epsilon \ll 1$.*

Thus, there is a one-to-one correspondence between singular and maximal canards in the case of folded saddles. Note that these canards form a separatrix set for solutions that either reach the fold F locally near the set M_f or not. This separatrix set of folded saddles will play an important role in the analysis of neural excitability (see Sect. 3.3).

Theorem 3.7 (cf. [6,57,61,63]). *In the folded node case $0 < \mu \leq 1$ of a singularly perturbed system* (3.1)*, we have the following results:*

(i) The $(k-1)$-dimensional set W_{ss} of singular strong canards perturb to a $(k-1)$-dimensional set of maximal strong canards called primary strong canards for sufficiently small $\epsilon \ll 1$.
(ii) If $1/\mu \notin \mathbb{N}$ then the $(k-1)$-dimensional set of singular weak canards perturb to a $(k-1)$-dimensional set of maximal weak canards called primary weak canards for sufficiently small $\epsilon \ll 1$.
(iii) If $2l+1 < \mu^{-1} < 2l+3$, $l \in \mathbb{N}$ and $\mu^{-1} \neq 2l+2$, then there exist l additional sets of maximal canards, all $(k-1)$-dimensional, called secondary canards for sufficiently small $\epsilon \ll 1$. These l sets of secondary canards are $O(\epsilon^{(1-\mu)/2})$ close to the set of primary strong canards in an $O(1)$ distance from the fold F.

Note the difference to the folded saddle case. In the folded node case, only a finite number of maximal canards persists under small perturbations $0 < \epsilon \ll 1$ out of the continuum of singular canards given in the singular limit $\epsilon = 0$. Furthermore, these maximal canards create some counter-intuitive geometric properties of the invariant manifolds $S_{a,\epsilon}$ and $S_{r/s,\epsilon}$ near the set of folded singularities M_f. In particular, the $(k-1)$-dimensional set of primary weak canards forms locally an "axis of rotation" for the k-dimensional sets $S_{a,\epsilon}$ and $S_{r/s,\epsilon}$ and hence also for the set of primary strong canards and the set of secondary canards; this follows from [61], case $k=2$. These rotations happen in an $O(\sqrt{\epsilon})$ neighbourhood of F. The rotational properties of maximal canards are summarized in the following result:

Theorem 3.8 (cf. [6,57,61,63]). *In the folded node case of a singularly perturbed system* (3.1) *with $2l+1 < \mu^{-1} < 2l+3$, $l \in \mathbb{N}$ and $\mu^{-1} \neq 2l+2$,*

(i) the set of primary strong canards twists once around the set of primary weak canards in an $O(\sqrt{\epsilon})$ neighbourhood of F,
(ii) the j-th set of secondary canards, $1 \leq j \leq l$, twists $(2j+1)$-times around the set of primary weak canards in an $O(\sqrt{\epsilon})$ neighbourhood of F,

where a twist corresponds to a half rotation. Thus each set of maximal canards has a distinct rotation number.

As a geometric consequence, the funnel region of the set of folded nodes M_f in S_a is split by the secondary canards into $(l+1)$ sub-sectors I_j, $j = 1, \dots, l+1$, with distinct rotational properties. I_1 is the sub-sector bounded by the primary strong canard and the first secondary canard, I_2 is the sub-sector bounded by the first and second secondary canard, I_l is the sub-sector bounded by the $(l-1)$-th and the l-th secondary canard and finally, I_{l+1} is bounded by the l-th secondary canard and the set of fold points F. Trajectories with initial conditions in the interior of I_j, $1 \le j < l+1$, make $(2j + 1/2)$ twists around the set of primary weak canards, while trajectories with initial conditions in the interior of I_{l+1} make at least $[2(l+1) - 1/2]$ twists around the set of primary weak canards. All these solutions are forced to follow the *funnel* created by the manifolds $S_{a,\sqrt{\epsilon}}$ and $S_{r/s,\sqrt{\epsilon}}$. After solutions leave the funnel in an $O(\sqrt{\epsilon})$ neighbourhood of F they get repelled by the manifold $S_{r/s,\sqrt{\epsilon}}$ and will follow close to a fast fiber of system (3.1). Hence, folded node type canards form separatrix sets in the phase space for different rotational properties near folded critical manifolds. Canard induced *mixed mode oscillations (MMOs)* are a prominent example of a complex rhythm that can be traced to folded node singularities. We refer the interested reader to, e.g., [5, 6, 10, 43, 61].

3.3 Excitable Systems

The notion of excitability was first introduced in an attempt to understand firing behaviors of neurons. Neural action potentials are responsible for transmitting information through the nervous system. Most neurons are excitable, i.e. they are typically silent but can fire an action potential or produce a firing pattern in response to certain forms of stimulation. While the biophysical basis of action potential generation *per se* is well established, the coding properties of single neurons are less well understood. A first answer to the question of the neuron's computational properties was given by Hodgkin [29] who identified three basic types (classes) of excitable axons distinguished by their different responses to injected steps of currents of various amplitudes.

Type I (class I) axons are able to *integrate* the input strength of an injected current step, i.e. the corresponding *frequency-current (f-I) curve* is continuous.

Type II (class II) axons have a discontinuous f-I curve because of their inability to maintain spiking below a certain frequency. The frequency band of a type II neuron is very limited and, hence, relatively insensitive to the strength of the injected current. It appears that type II neurons *resonate* with a preferred frequency input.

Type III (class III) axons will only fire a single or a few action potentials at the onset of the injected current step, but are not able to fire repetitive action potentials like type I and type II neurons (besides for extremely strong injected currents). Type III neurons are able to *differentiate*, i.e. they are able to encode the occurrence of a "change" in the stimulus. Such *phasic* firing (versus *tonic* or repetitive firing) identifies these type III neurons as *slope detectors*. Obviously, the f-I curve is not defined for type III neurons.

Rinzel and Ermentrout [52] pioneered a mathematical framework based on bifurcation theory that distinguishes type I and type II neural models. In Sect. 3.3.1, we will briefly review this approach but with a slight twist. We will emphasise the inherent multiple time-scales structure found in many neuronal models and apply geometric singular perturbation theory together with bifurcation theory to define the different types of excitability.

In Sect. 3.3.2 we will go a step further and ask more general questions about excitability. In particular, we want to focus on dynamic inputs beyond (current) step protocols (that are usually applied in laboratory settings). For example, synapses produce excitatory or inhibitory inputs and these synaptic inputs may be activated (resp. inactivated) fast or slow. We will model sufficiently smooth dynamic inputs and apply these inputs to the 2D slow-fast excitable system models introduced in Sect. 3.3.1. The geometric key to the understanding of excitability will be to identify threshold manifolds (aka separatrices). This is very much in the spirit of FitzHugh's work on excitability [18–20] (see also Izhikevich [31], Chap. 7), but it extends FitzHugh's ideas to the dynamic, nonautonomous case.

3.3.1 Slow-Fast Excitable Systems with Step Protocols

We focus on a class of 2D excitable models given by

$$\begin{aligned} w' &= \epsilon g(w, v, \epsilon) \\ v' &= f(w, v, \epsilon, I) = f_1(w, v, \epsilon) + I \end{aligned} \tag{3.31}$$

where $I \in [I_0, I_1] \subset \mathbb{R}$ is an external (constant) drive of the excitable system, and the following assumptions hold (for many two-dimensional neuronal models):

Assumption 1. *The critical manifold S of system* (3.31) *is cubic shaped, i.e.*

$$S = S_a^- \cup F^- \cup S_r \cup F^+ \cup S_a^+ ,$$

with attracting outer branches $S_a^{\pm}$, repelling middle branch S_r, and folds $F^{\pm}$.

Assumption 2. *The (unforced) system* (3.31) *with* $I = 0$ *has one, two or three equilibria. In the corresponding reduced problem, one equilibrium is located on the lower attracting branch S_a^- and it is stable. Each of the other two equilibria, if they exist, are located on the middle branch S_r.*

Example 3.3. The *Morris-Lecar (ML)* model [46] was originally developed to study the electrical activity of barnacle muscle fiber. Later it was popularised as a model for neural excitability; see e.g. Izhikevich [31] where a large collection of minimal conductance based 2D ML-type models is introduced. We use the ML-type model from Rinzel and Ermentrout [52] in the following form (as in Prescott et al [48]):

$$\begin{aligned} w' &= \phi[w_\infty(V) - w]/\tau_w(V) \\ CV' &= -I_{ion}(w, V) + I_{stim}\,, \end{aligned} \tag{3.32}$$

with functions

$$\begin{aligned} I_{ion}(w, V) &= -(g_f m_\infty(V)(V - E_f) + g_s w(V - E_s) + g_l(V - E_l)) \\ m_\infty(V) &= [1 + \tanh((V - V_1)/V_2)]/2 \\ w_\infty(V) &= [1 + \tanh((V - V_3)/V_4)]/2 \\ \tau_w(V) &= 1/\cosh((V - V_3)/(2V_4))\,. \end{aligned} \tag{3.33}$$

V models the voltage membrane potential and $I_{ion}(V) = I_{fast} + I_{slow} + I_{leak}$ represents the ionic currents of the model which consist of a *fast* non-inactivating current $I_{fast} = g_f m_\infty(V)(V - E_f)$, a *delayed* rectifier type current $I_{slow} = g_s w(V - E_s)$, and a leak current $I_{leak} = g_l(V - V_l)$. The parameter I_{stim} represents the injected current step. The activation variable w of the I_{slow} current provides the *slow* voltage-dependent negative feedback required for excitability. Its dynamics are described by the sigmoidal activation function $w_\infty(V)$ and the bell-shaped voltage dependent time-scale $\tau_w(V)/\phi$. The activation of the fast I_{fast} current is assumed instantaneous and, hence, its activation variable is set to $m = m_\infty(V)$.

A representative parameter set of this ML model is given by $g_f = 20\,\text{mS/cm}^2$, $g_s = 20\,\text{mS/cm}^2$, $g_l = 2\,\text{mS/cm}^2$ (maximal conductances of ion channels), $E_f = 50\,\text{mV}$, $E_s = -100\,\text{mV}$, $E_l = -70\,\text{mV}$ (Nernst potentials), capacitance $C = 2\,\mu\text{F/cm}^2$, time scale factor $\phi = 0.1\,\text{ms}^{-1}$ and auxiliary voltage parameters $V_1 = -1.2\,\text{mV}$, $V_2 = 18\,\text{mV}$, $V_3 = 0\,\text{mV}$, $V_4 = 10\,\text{mV}$.

To identify a slow-fast timescale structure explicitly in (3.32) we have to non-dimensionalise the model. This is done by introducing dimensionless variables $v = V/k_v$ and $t_1 = t/k_t$ with typical reference scales for voltage $k_v = 100\,\text{mV}$ and time $k_t = C/g_{max} = 0.1\,\text{ms}$ where $g_{max} = 20\,\text{mS}$ is a reference conductance scale. This leads to the dimensionless ML model,

$$\begin{aligned} w' &= \epsilon[w_\infty(v) - w]/\tau_w(v) = \epsilon g(w, v) \\ v' &= -\bar{I}_{ion}(w, v) + \bar{I}_{stim} = f(w, v, \bar{I}_{stim}) = f_1(w, v) + \bar{I}_{stim}\,, \end{aligned} \tag{3.34}$$

with functions

$$\begin{aligned} \bar{I}_{ion}(w, v) &= -(\bar{g}_f m_\infty(v)(v - \bar{E}_f) + \bar{g}_s w(v - \bar{E}_s) + \bar{g}_l(v - \bar{E}_l)) \\ m_\infty(v) &= [1 + \tanh((k_v v - V_1)/V_2)]/2 \\ w_\infty(v) &= [1 + \tanh((k_v v - V_3)/V_4)]/2 \\ \tau_w(v) &= 1/\cosh((k_v v - V_3)/(2V_4))\,. \end{aligned} \tag{3.35}$$

where $\bar{g}_x = g_x/g_{max}$, $\bar{E}_x = E_x/k_v$ $(x = f, s, l)$, $\bar{I}_{stim} = I_{stim}/(k_v g_{max})$ and

$$\epsilon := (C/g_{max})\phi = 0.01$$

is the singular perturbation parameter that measures the time-scale separation between the fast v dynamics and the slow w dynamics. This timescale separation can be enhanced by decreasing the capacitance C, slowing the w-dynamics via ϕ or increasing the maximum conductance of the ion channels. Hence, system (3.34) can be viewed as a singularly perturbed system.

Using the parameter values from above, it can be shown that the critical manifold is cubic shaped (Assumption 1) and that it has three equilibria for $I = 0$, one on the lower attracting branch and the other two are on the repelling middle branch (Assumption 2). By changing the system parameters, this model can be transformed into all three excitable neuron types; see [48] for more details.

Example 3.4. We introduce a dimensionless hybrid of the Morris-Lecar and the FitzHugh-Nagumo (ML-FHN) model that combines important features of both:

$$\begin{aligned} w' &= \epsilon g(w, v) = \epsilon(w_\infty(v) - w) \\ v' &= f(w, v, I) = v(v - a)(b - v) - w + I = F(v) - w + I\,, \end{aligned} \tag{3.36}$$

with

$$w_\infty(v) = [1 + \tanh((v - v_3)/v_4)]/2$$

with dimensionless parameters $b > 0 > a$, v_3, $v_4 > 0$, I is the primary bifurcation parameter and $\epsilon \ll 1$ as the singular perturbation parameter. Again, this singularly perturbed system has a cubic-shaped critical manifold (Assumption 1). Furthermore, the sigmoidal shaped activation function $w_\infty(v)$ allows us to explore more easily the cases of different numbers of equilibria as described by Assumption 2. We focus on this ML-FHN model (3.36) to explore the notion of excitability. We fix the parameter $a = -0.5$, $b = 1$, and vary (I, v_3, v_4).

3.3.1.1 The Geometry of Excitability

A classical physiology definition of excitability is that a large enough brief stimulus ("supra-threshold" pulse) triggers an action potential (large regenerative excursion). This implies the existence of a "threshold" that the stimulus must pass to evoke an action potential with a fairly constant amplitude. On the other hand, a graded response with intermediate amplitudes was already observed in the Hodgkin-Huxley model of the squid giant axon [29] as well as the FHN model [19, 20] which contradicts the traditional view that the action potential is an *all-or-none* event with a fixed amplitude. We will focus on a geometrical definition of excitability to avoid this ambiguity.

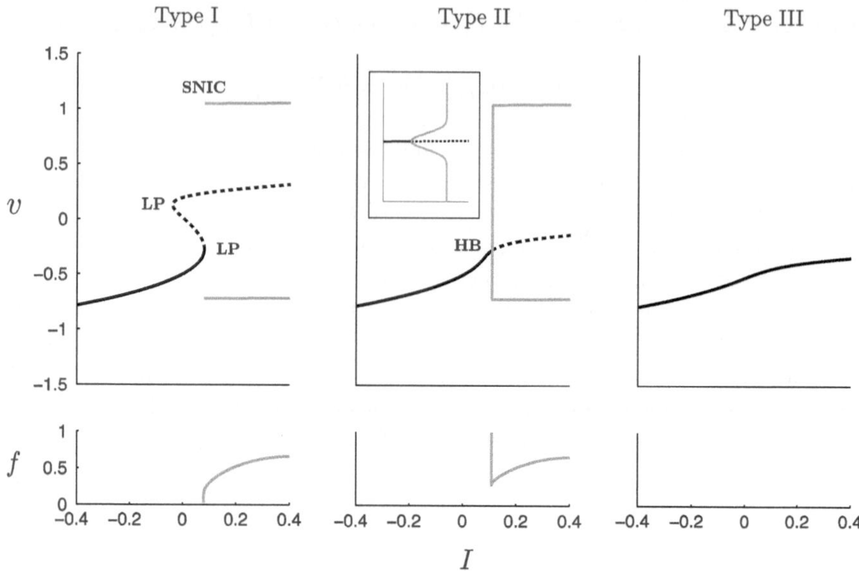

Fig. 3.4 Bifurcation diagrams for type I–III neurons of the ML-FHN model (3.36) together with $f - I$ curves: (Type I) $v_3 = 0.3$, $v_4 = 0.1$; we observe a SNIC bifurcation for $I = I_{bif} = 0.079$ where the frequency approaches zero; (Type II) $v_3 = -0.1$, $v_4 = 0.1$; we observe a singular HB bifurcation for $I = I_{bif} = 0.109$; note the small frequency band for the relaxation oscillation branch; (Type III) $v_3 = -0.3$, $v_4 = 0.1$; there are no bifurcation for $I = [-0.4, 0.4]$

As mentioned in the introduction, Hodgkin [29] identified three distinct types (classes) of excitability by applying a current step protocol to neurons:

- Type I neurons: depending on the strength of the injected current, action potentials can be generated with arbitrary low frequency; see Fig. 3.4.
- Type II neurons: Action potentials are generated in a certain frequency band that is relatively insensitive to changes in the strength of the injected current; see Fig. 3.4.
- Type III neurons: A single action potential is generated in response to a pulse of injected current. Repetitive spiking is not possible or can be only generated for extremely strong injected current.

Type I and type II neurons are able to fire trains of action potentials (tonic firing) if depolarized sufficiently strong which distinguishes them from type III neurons. This distinction points to a bifurcation in type I and type II neurons where the cell changes from an excitable to an oscillatory state. The main bifurcation parameter is given by I, the magnitude of the current step protocol. This leads to the following classical definition of excitability via bifurcation analysis under the variation of the applied current I [52]:

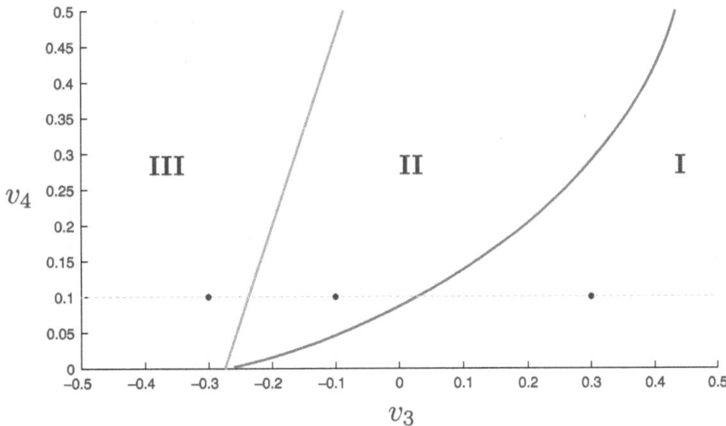

Fig. 3.5 Two parameter bifurcation diagram (v_3, v_4) of the ML-FHN model (3.36) with boundaries of type I, type II and type III neurons: the three dots indicate the parameter values used in Fig. 3.4

- Type I : The stable equilibrium (resting state) disappears via a *saddle-node on invariant circle (SNIC)* bifurcation; see Fig. 3.4.
- Type II: The stable equilibrium (resting state) loses stability via an *Andronov-Hopf bifurcation*; see Fig. 3.4.
- Type III: The equilibrium (resting state) remains stable for $I \in [I_0, I_1]$; see Fig. 3.4.

In the ML-FHN model (3.36) we are able to identify all three types of neurons by varying the parameters (v_3, v_4) which change the position (v_3) and maximum slope (v_4) of the sigmoidal function $w_\infty(v)$. Figure 3.5 shows the different regions in the parameter-space (v_3, v_4) that correspond to the different excitability classes. The boundaries were found numerically using the software package AUTO [13]. The boundary between type II and type III is a continuation of the Andronov-Hopf bifurcation at a fixed $I = I_1$. Hence, its position depends on the definition of the interval $I \in [I_0, I_1]$ where the type III neuron must stay excitable. The boundary between type I and type II is a continuation of a cusp-bifurcation where the two folds coalesce. This boundary is not exact but defines a small strip where the transition happens. Note that fixing v_4 (slope) and varying v_3 (position) provides us with a simple way to change the model from type I to type II and to type III. Figure 3.4 was obtained in that way. Throughout the rest of the chapter, we will fix $v_4 = 0.1$ and use v_3 as our second bifurcation parameter.

Since the ML-FHN model (3.36) is a singularly perturbed system, we are able to provide the corresponding definition of excitability based on geometric singular perturbation theory:

- Type I: The stable equilibrium on the lower attracting branch S_a^- disappears via a *singular Bogdanov-Takens bifurcation* at the lower fold F^-; see Fig. 3.6.

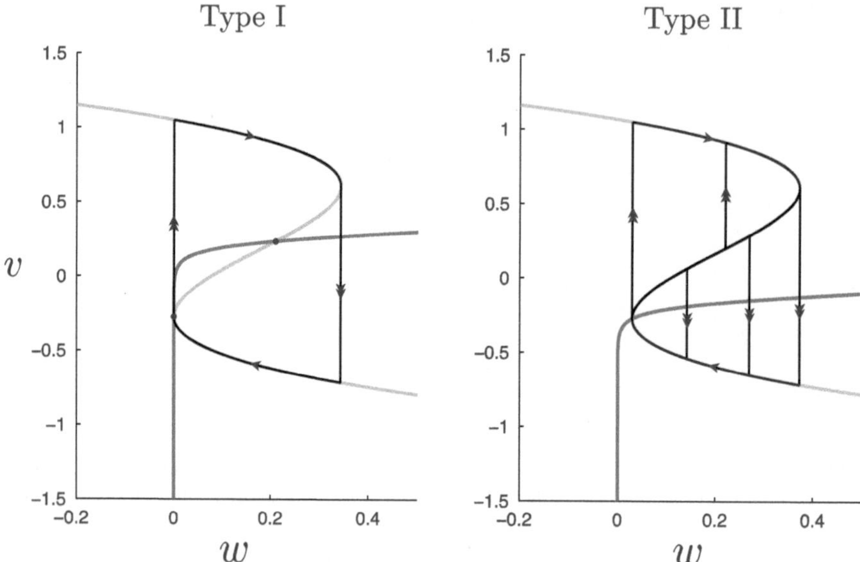

Fig. 3.6 ML-FHN model (3.36), singular limit bifurcations and their singular limit orbits: (type I) saddle-node homoclinic (SNIC) for $v_3 = 0.3$ and $I = I_{bif} = I_{bif}(v_3) \approx 0.079$; (type II) canard cycles for $v_3 = -0.1$ and $I = I_{bif} = I_{bif}(v_3) \approx 0.109$

- Type II: The stable equilibrium on the lower attracting branch S_a^- bifurcates via a *singular Andronov-Hopf bifurcation* at the lower fold F^-; see Fig. 3.6.
- Type III: The stable equilibrium on the lower attracting branch S_a^- remains stable for $I \in [I_0, I_1]$.

To identify the different types of excitability one has to look at the nullclines of the ML-FHN model (3.36). As can be seen in Fig. 3.7, for $I = I_{bif}$ a type I neuron has a saddle-node bifurcation of equilibria at the lower fold F^-. This allows for the construction of a singular homoclinic orbit as follows (see Fig. 3.6): we start at the saddle-node equilibrium at the lower fold F^- and concatenate a fast fiber of the layer problem that connects to the upper stable branch S_a^+. Then we follow the reduced (slow) flow towards the upper fold F^+ where we concatenate a fast fiber at F^+ that connects back towards the lower attracting branch S_a^-. Finally, we follow the reduced (slow) flow on S_a^- towards the lower fold F^- and hence end up at the saddle-node equilibrium. This homoclinic orbit is the singular limit representation of the SNIC shown in Fig. 3.4. The unfolding of this singular limit object is quite intricate [9], is closely related to a local slow-fast Bogdanov-Takens bifurcation [7] at the lower fold F^- and goes beyond the aim of this chapter.

In the case of a type II neuron, the stable equilibrium on the lower branch S_a^- crosses the lower fold F^- at $I = I_{bif} = I_{bif}(v_3)$ (note, it is a different value than for the type I case) and moves onto the unstable middle branch S_r; see Fig. 3.7 (note, a precise definition of $I = I_{thr}$ will be given in Sect. 3.3.2). This

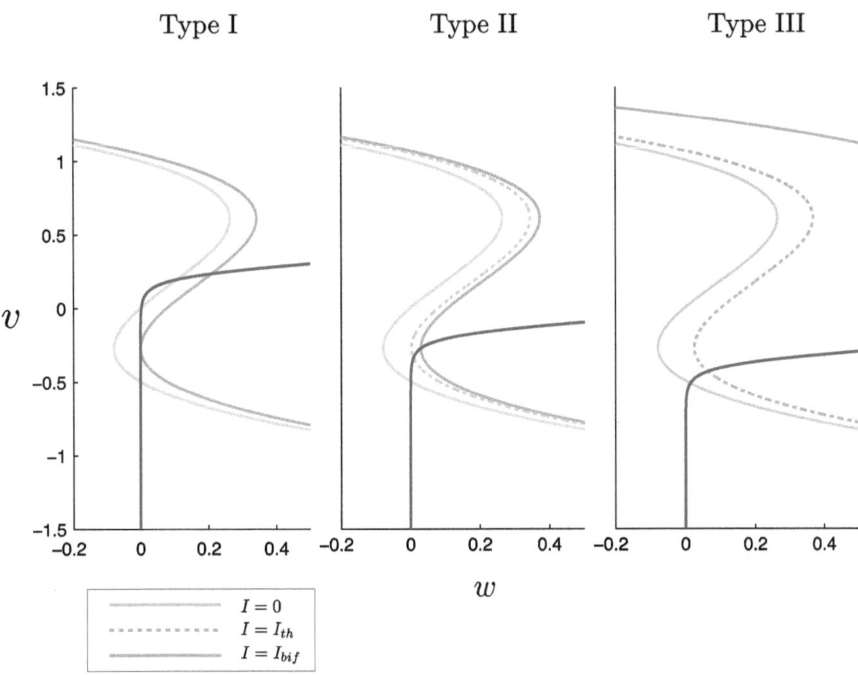

Fig. 3.7 ML-FHN model (3.36): nullclines under variation of I which leads to the (singular limit) definition of I_{thr} and I_{bif} for type I–III: (type I) $I = I_{bif}$ at a saddle node bifurcation; (type II) $I = I_{bif}$ at singular HB bifurcation; (type III) no bifurcation; (type I–III) $I = I_{thr}$ when the w-coordinate of the equilibrium on S_a^- for $I = 0$ equals the w-coordinate of the lower fold F^- for $I = I_{thr}$. In the type I case, $I_{bif} = I_{thr}$. The bifurcation values I_{bif} respectively threshold values I_{thr} are not the same for the different types

is the same mechanism as shown for the FHN model in Fig. 3.2. Hence, one can construct singular canard cycles that are formed through concatenations of slow canard segments and fast fibers as shown in Fig. 3.6. Note that these singular canard cycles have $O(1)$ amplitude and have a frequency $O(1)$ on the order of the slow time scale. These singular canard cycles will unfold to actual canard cycles as we turn on the singular perturbation parameter. The unfolding of these canard cycles, the canard explosion, happens within an exponentially small parameter interval of the bifurcation parameter near $I = I_C$. This canard explosion is preceded by a singular supercritical Andronov-Hopf bifurcation at $I = I_H$ that creates small stable $O(\sqrt{\epsilon})$ amplitude limit cycles with nonzero intermediate frequencies of order $O(\sqrt{\epsilon})$ [39] and succeeded by relaxation oscillations with frequencies of order $O(1)$; see Fig. 3.4. Hence the singular nature of the Andronov-Hopf bifurcation is encoded in both, amplitude and frequency. Note that the classic definition of type II excitability refers to the slow frequency band of the relaxation oscillations which does not vary much.

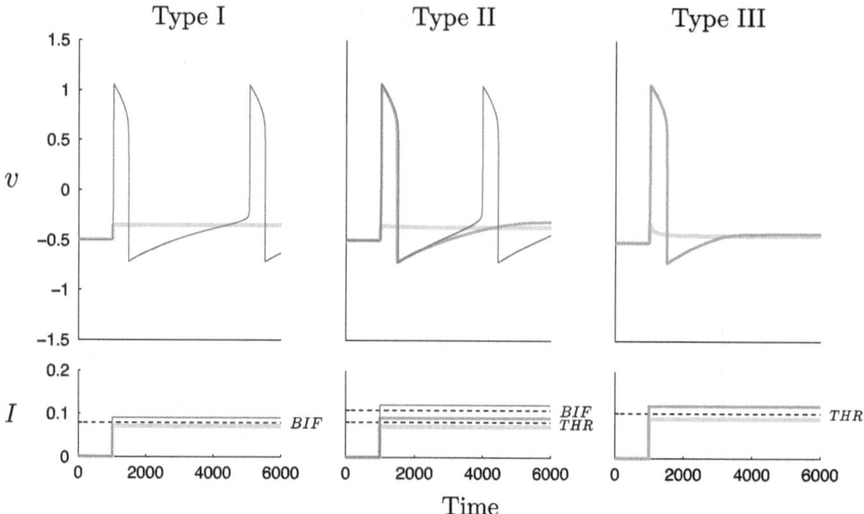

Fig. 3.8 ML-FHN model (3.36) with current step protocols for type I–III and time traces for different step currents I: (type I) for $I < I_{bif} = I_{thr}$ no spiking while for $I > I_{bif}$ there is periodic (tonic) spiking; (type II) for $I < I_{thr}$ no spiking, for $I_{thr} < I < I_{bif}$ a transient spike while for $I > I_{bif}$ there is periodic (tonic) spiking; (type III) for $I < I_{thr}$ no spiking while for $I_{thr} < I$ there is a transient spike

In the case of type III neurons, there is no bifurcation (see Fig. 3.4) and, hence, a type III neuron is excitable for all $I \in [I_0, I_1]$, i.e. a type III neuron does not spike repetitively. On the other hand, as one observes in Fig. 3.8, the type III neuron is indeed excitable—it is able to elicit a single spike for a sufficiently strong injected current step $I > I_{thr}$.

3.3.1.2 Transient Responses

Let us consider possible transient responses of type I and type II neurons for $I < I_{bif}$, the minimum injected current step I_{bif} required for periodic tonic spiking. Type II neurons are also able to elicit a single spike for a sufficiently strong injected current step $I_{thr} < I < I_{bif}$; see Fig. 3.8. On the other hand, type I neurons are not able to elicit a single spike below the minimum injected current step I_{bif} required for periodic tonic spiking; see Fig. 3.8. Obviously, this transient behavior for type II neurons cannot be explained by the bifurcation structure identified in Fig. 3.7 since this transient behavior is found for $I_{thr} < I < I_{bif}$. It points to the ability of type II and III neurons to elicit single transient spikes under a current step protocol, while type I neurons are not able to produce this transient behaviour.

Figure 3.9 provides an explanation for the firing threshold $I = I_{thr}$ in the case of a type III neuron. The rest state, $I = 0$ case in Fig. 3.9, on the lower attracting branch (the resting membrane potential of a neuron) is given by the intersection of the two nullclines, the critical manifold S and the sigmoidal $w = w_\infty(v)$. When

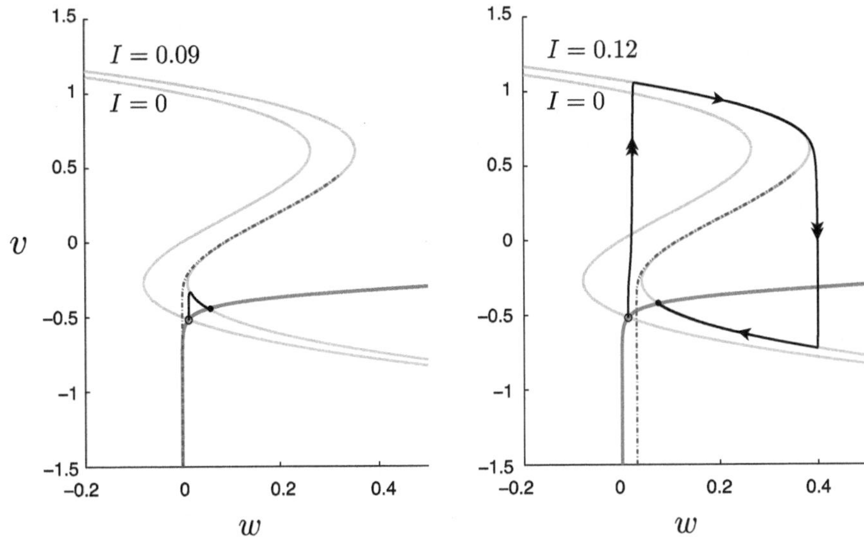

Fig. 3.9 Explanation of transient spiking property of type III neuron shown in Fig. 3.8: *open circle* indicates resting state for $I = 0$ while *filled circle* indicates the resting state for $I > 0$; the firing threshold manifold (*dashed curve*) for $I > 0$ is the extension of the middle repelling branch $S_{r,\epsilon}$ in backward time; (*left panel*) no spike since initial rest state is to the right of the firing threshold manifold; (*right panel*) transient spike since initial rest state is to the left of the firing threshold manifold

a current step I is injected, the critical manifold S shifts to the right. The old rest state is suddenly off S and it will follow the fast dynamics to find a stable attractor. If it follows a fast fiber of the lower stable branch $S^-_{a,\epsilon}$ then the cell is not able to fire (right panel), but if it follows a fast fiber of the upper stable branch $S^+_{a,\epsilon}$ then the cell will fire an action potential before it returns to the lower $S^-_{a,\epsilon}$ and the new resting state (left panel). The firing threshold manifold [11, 18, 42, 48] is shown as a dashed curve. It is the extension of the unstable middle branch $S_{r,\epsilon}$ in backward time. In the singular limit, this firing threshold manifold is given by the concatenation of the branch S_r and the layer fiber attached to the lower fold F^-. By looking at Fig. 3.7 and the position of the equilibrium state for $I = 0$ relative to the nullclines for $I > 0$ it becomes now apparent why type II and type III neurons can fire transient spikes while type I neurons cannot.

This also points to a well known phenomenon in neuronal dynamics known as *post-inhibitory rebound (PIR)* [4, 21], where excitable neurons are able to fire an action potential when they are released after having received an inhibitory current input for a sufficient amount of time. Again, only type II and III neurons are able to create a post-inhibitory rebound while type I neurons cannot. Simply note that Fig. 3.8 could also be interpreted as a PIR current step protocol where cells have been held sufficiently long at $I = 0$ before they are released back to the original state $I > 0$.

It seems that the transient firing behavior observed for type II and type III neurons is a function of the external current amplitude only. This is actually a misconception because it also depends crucially on the dynamics of the external input. In a current step protocol, we are dealing with an instantaneous (fast) change in the external input, i.e. a *fast input modulation.* In the following, we will show that these transient behaviors can be explained in a more general context by applying a dynamic, nonautonomous approach to the problem under study.

3.3.2 *Slow-Fast Excitable Systems with Dynamic Protocols*

We focus on the ML-FHN model (3.36) with an external drive $I(t)$:

$$\begin{aligned} w' &= \epsilon g(w,v) = \epsilon(w_\infty(v) - w) \\ v' &= f(w,v,t) = F(v) - w + I(t)\,, \end{aligned} \tag{3.37}$$

where we assume that $I(t)$ is a sufficiently smooth function. This excludes the case of the current step protocol used in the previous section. We replace this protocol by a mollified version such as given by a smooth ramp or by a smooth pulse which resemble qualitatively certain classes of neuronal synaptic or network inputs.

System (3.37) is a singularly perturbed nonautonomous system. Is it possible to apply geometric singular perturbation theory to the nonautonomous case as well? In the following, we briefly highlight connections between geometric singular perturbation theory and nonautonomous attractor theory (see also Chap. 1 of this book).

3.3.2.1 Nonautonomous Systems and Canard Theory

Given a nonautonomous singularly perturbed system

$$\begin{aligned} w' &= \epsilon g(w,v,\epsilon,t) \\ v' &= f(w,v,\epsilon,t) \end{aligned} \tag{3.38}$$

where $w = (w_1,\ldots,w_{k-1}) \in \mathbb{R}^{k-1}$ and $v = (v_1,\ldots,v_m) \in \mathbb{R}^m$ are slow and fast phase space variables, $t \in \mathbb{R}$ is the fast time scale and the prime denotes the time derivative d/dt. It is well known that such a nonautonomous system can be viewed as an extended autonomous system by increasing the phase space dimension by one, i.e.

$$\begin{aligned} w' &= \epsilon g(w,v,\epsilon,s) \\ v' &= f(w,v,\epsilon,s) \\ s' &= 1 \end{aligned} \tag{3.39}$$

where $s \in \mathbb{R}$ is an additional (fast) dummy phase-space variable. Note, this system has no critical manifold S. Hence, the previously introduced geometric singular perturbation theory is only of limited use here. To be more precise, the fast dynamics of the (v, s)-variables are dominant throughout the phase space. Thus we can interpret this system as a regularly perturbed nonautonomous problem [36, 37, 50].

This apparent shortfall with respect to geometric singular perturbation theory diminishes immediately if we assume that the nonautonomous nature of the problem evolves slowly, i.e. $g(v, w, \epsilon, \tau = \epsilon t)$ and $f(v, w, \epsilon, \tau = \epsilon t)$ where $\epsilon \ll 1$ indicates the scale separation between the fast time scale t and the slow time scale τ which leads to

$$\begin{aligned} s' &= \epsilon \\ w' &= \epsilon g(w, v, \epsilon, s) \\ v' &= f(w, v, \epsilon, s)\,. \end{aligned} \tag{3.40}$$

This system represents a special case of a *singularly perturbed system* (3.1) where $(w, s) \in \mathbb{R}^k$ are slow variables and $v \in \mathbb{R}^m$ are fast variables. The critical manifold is given by $f = 0$ and we can apply the theory given in Sect. 3.2. In particular, folded critical manifolds provide singularly perturbed systems with the opportunity to switch from the slow time scale to the fast time scale or from one attracting sheet of a critical manifold to another. As we have seen before, most models of excitability have cubic shaped critical manifolds (i.e. they have two folds) and, hence, have the ability to switch between different states (e.g. silent and active).

Furthermore, while system (3.40) possesses, in general, no equilibria, it may possess folded singularities. As described in Sect. 3.2.4, canards of folded saddle and folded node type have the potential to act as "effective separatrices" between different local attractor states in a dynamically driven multiple scales system. A dynamic drive itself (e.g., in the case of a periodic signal that regularly rises and falls) has the potential to create folded singularities and to form and change these effective separatrices. Hence, the specific nature of the dynamic drive determines which local attractor states can be reached through global mechanisms. This point of view has profound consequences in the analysis of excitable systems as we will show next. In particular, we will identify canards of folded saddle type as firing threshold manifolds.

3.3.2.2 Slow External Drive Protocols

We analyse a 2D singularly perturbed system with slow external drive $I(\epsilon t)$ given by

$$\begin{aligned} w' &= \epsilon g(w, v) = \epsilon(w_\infty(v) - w) \\ v' &= f(w, v, t) = F(v) - w + I(\epsilon t)\,, \end{aligned} \tag{3.41}$$

where the autonomous part of this model, i.e. system (3.41) with $I(\epsilon t) \equiv 0$, fulfills Assumption 1 and 2. Obviously, the ML-FHN model (3.37) with slow external drive $I(\epsilon t)$ fulfills this requirement. We recast the 2D nonautonomous singularly perturbed problem as a 3D autonomous singularly perturbed problem,

$$\begin{aligned} s' &= \epsilon \\ w' &= \epsilon g(w, v) = \epsilon (w_\infty(v) - w) \\ v' &= f(w, v, s) = F(v) - w + I(s)\,, \end{aligned} \tag{3.42}$$

where $(s, w) \in \mathbb{R}^2$ are the slow variables and $v \in \mathbb{R}$ is the fast variable.

Assumption 3. *The external slow drive $I(s)$ is a C^∞ function which is constant outside a finite interval $[s^-, s^+]$, i.e. $I(s) = I_0$ for $s < s^-$ and $I(s) = I_1$ for $s > s^+$. $I(s)$ is bounded, i.e. $I_{min} \leq I(s) \leq I_{max}$, $\forall s \in \mathbb{R}$, such that type I and II neurons are in an excitable state for the maximal constant drive $I = I_{max} < I_{bif}$.*

Remark 3.14. By Assumption 3, the function $I'(s) = I_s$ is compactly supported. This is not necessary for the following analysis but makes it more convenient. We could relax the smoothness assumption on $I(s)$. The constant states could also be relaxed to asymptotic states.

Example 3.5 (Ramp). This is a mollified version of the current step protocol and is given by

$$I(s) = \frac{I_1}{2}\left(1 + \tanh\left(\frac{2(s - s_0)}{s_1}\right)\right), \quad \forall s \in [s^-, s^+]\,, \tag{3.43}$$

$I(s) = 0 = I_{min}$ for $s < s^-$ and $I(s) = I_1 = I_{max}$ for $s > s^+$ for a sufficiently large choice of $[s^-, s^+]$ centered around s_0. The ramp has a maximal slope of I_1/s_1 when $I(s_0) = I_1/2$.

Example 3.6 (Pulse). We model a symmetric pulse given by

$$I(s) = \frac{I_1}{\cosh\left(\frac{2(s-s_0)}{s_1}\right)}, \quad \forall s \in [s^-, s^+]\,, \tag{3.44}$$

$I(s) = 0 = I_{min}$ for $s < s^-$ and for $s > s^+$ for a sufficiently large choice of $[s^-, s^+]$ centered around s_0 with $I(s_0) = I_1 = I_{max}$. The pulse has its maximal slope of I_1/s_1 when $I(s_0 + \frac{s_1}{4}\ln(3 - 2\sqrt{2})) = I_1/\sqrt{2}$.

Figure 3.10 shows ramp and pulse protocol examples. Note that the maximal drive I_1 is the same in both cases, only the maximal slope of the ramp respectively the pulse varies (slightly). In both cases, a single spike is elicited if the slope of rising exceeds a certain threshold value. This clearly indicates that this type II neuron is a *slope detector* (for $I < I_{bif}$). The same can be observed for type III neurons.

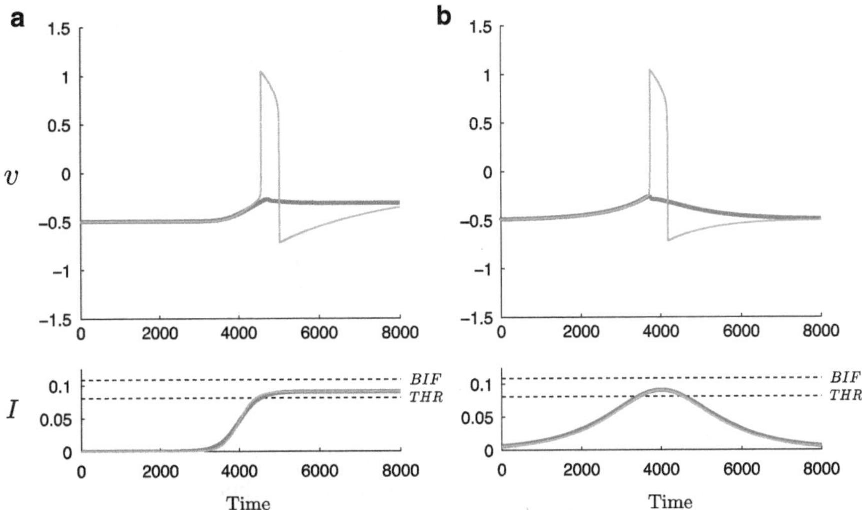

Fig. 3.10 Type II neuron ($v_3 = -0.1$) for $I_{th} < I_1 = 0.09 < I_{bif}$: (**a**) ramp protocol (3.43) with $s_1 = 0.5$ (*dark grey*, no spike) and $s_1 = 0.4$ (*bright grey*, transient spike); note that the difference in the two ramp protocols is barely visible; (**b**) pulse protocol (3.44) with $s_1 = 2.3$ (*dark grey*, no spike) and $s_1 = 2.2$ (*bright grey*, transient spike). Note that the difference in the two ramp protocols is barely visible

In the following, we will use geometric singular perturbation theory to explain this phenomenon in detail.

3.3.2.3 Geometric Singular Perturbation Analysis

The critical manifold S of system (3.42) is given as a graph

$$w = W(s, v) = F(v) + I(s) . \tag{3.45}$$

By Assumption 1, this manifold is cubic shaped, i.e. S has two folds $F^{\pm}$ for $v = v^{\pm}$ where $F_v(v^{\pm}) = 0$ and $F_{vv}(v^{\pm}) \neq 0$ for all $s \in \mathbb{R}$. Note that $F_v = dF/dv$, $F_{vv} = d^2F/dv^2$. The geometry of the critical manifold S together with the stability properties of the three branches of S, outer branches $S_a^{\pm}$ are stable and middle branch S_r is unstable, imply that $F_{vv}(v^+) < 0$ while $F_{vv}(v^-) > 0$. Figure 3.11 shows the critical manifold in the case of a ramp respectively pulse protocol.

Assumptions 2 and 3 are concerned with properties of the reduced problem of system (3.42). Since the critical manifold S is a graph $w = W(s, v)$, we are able to project the reduced problem onto a single coordinate chart $(s, v) \in \mathbb{R}^2$ (compare with Sect. 3.2.2.1):

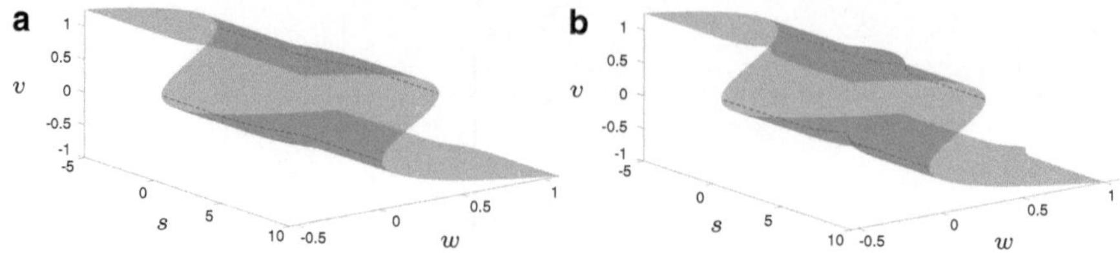

Fig. 3.11 Critical manifold (3.45) for (**a**) ramp and (**b**) pulse protocol; note that the ramp and pulse profiles are visible in the geometry of the critical manifold

$$\begin{aligned} \dot{s} &= 1 \\ -F_v \dot{v} &= F(v) - w_\infty(v) + I(s) + I_s \,, \end{aligned} \tag{3.46}$$

where $I_s = dI/ds$. This system is singular along the folds $F^\pm$ where F_v vanishes. We rescale time by $d\tau = -F_v \, d\tau_1$ in system (3.46) to obtain the desingularized system

$$\begin{aligned} \dot{s} &= -F_v \\ \dot{v} &= F(v) - w_\infty(v) + I(s) + I_s \,, \end{aligned} \tag{3.47}$$

where the overdot denotes now $d/d\tau_1$. From the time rescaling it follows that the direction of the flow in (3.47) has to be reversed on the middle branch S_r where $F_v > 0$ to obtain the corresponding reduced flow (3.46). Otherwise, the reduced flow (3.46) and the desingularized flow (3.47) are equivalent.

By Assumption 3, the drive $I(s)$ is constant for $s \in (-\infty, s^-) \cup (s^+, \infty)$. From Assumption 2 it follows that

$$\dot{v} = F(v) - w_\infty(v) + I(s) + I_s = F(v) - w_\infty(v) + \bar{I} < 0 \,, \forall s \in (-\infty, s^-) \cup (s^+, \infty) \,,$$

where $\bar{I} = I_0$ or $\bar{I} = I_1$. This shows that the reduced flow cannot reach the lower fold F^- from S_a^- for a constant drive $\bar{I}$ confirming $I < I_{bif}$ for type I and type II neurons. For an action potential to occur we need necessarily that $\dot{v} > 0$ somewhere along the lower fold F^- within the dynamic range of $I(s)$ where $I_s > 0$ (on the rising phase of $I(s)$). This implies that $\dot{v}$ must vanish in system (3.47) along the lower fold F^-, i.e.

$$\tilde{I}(s) := F(v^-) - w_\infty(v^-) + I(s) + I_s = 0 \,, \tag{3.48}$$

which is Definition 3.7 of a folded singularity. The type of these folded singularities is obtained by calculating the Jacobian of system (3.47),

$$J = \begin{pmatrix} 0 & -F_{vv} \\ I_s + I_{ss} & F_v - w_{\infty,v} \end{pmatrix} . \tag{3.49}$$

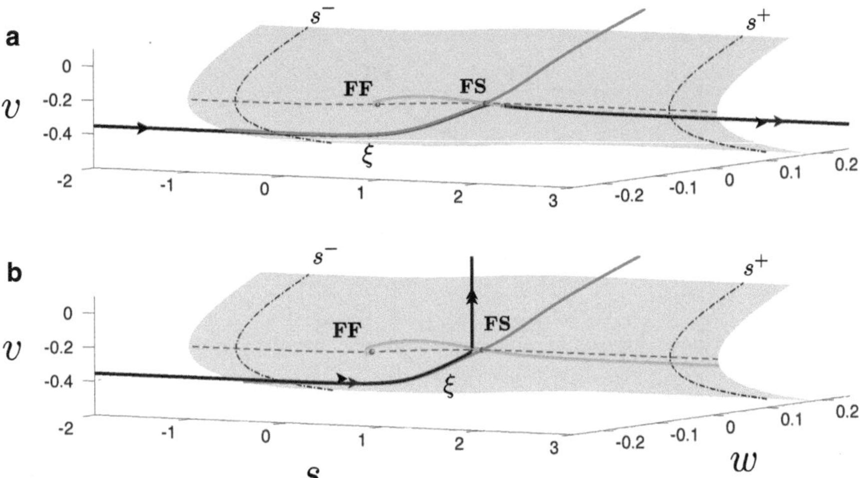

Fig. 3.12 Reduced flow on the critical manifold S near the lower fold F^- corresponding to Fig. 3.10a, ramp protocol: there are two folded singularities, a folded saddle (FS) and a folded focus (FF), ξ denotes the folded saddle canard that crosses from the lower stable branch S_a^- via the FS singularity onto the repelling middle branch S_r. The canard ξ forms the firing threshold manifold. Note, the other canard of the FS singularity (the faux canard) crosses from S_r to S_a^-. The segment on S_a^- forms a boundary that prevents trajectories to the right of ξ to spike

Note that $w_{\infty,v} > 0$. Hence, the trace of the Jacobian evaluated along F^- is

$$\mathrm{tr}(J) = -w_{\infty,v} < 0\,.$$

The determinant of the Jacobian evaluated along F^- is given by

$$\det(J) = F_{vv}(I_s + I_{ss})\,.$$

Recall, we have $F_{vv} > 0$ along F^-. The function $\tilde{I}(s)$ defined in (3.48) is constant and negative for $s \in (-\infty, s^-) \cap (s^+, \infty)$. This implies, in general, an even number of folded singularities (if they exist). The derivative $\tilde{I}_s$ must be positive at the first (odd) folded singularity while negative at the second (even) singularity. Hence, $\det(J) > 0$ for an odd folded singularity and $\det(J) < 0$ for an even folded singularity. This implies that an odd folded singularity is either of folded node or folded focus type while an even folded singularity is of folded saddle type. From the structure of the Jacobian (3.49) it follows that the eigenvectors corresponding to negative eigenvalues have a positive slope while eigenvectors corresponding to positive eigenvalues have a negative slope.

Figures 3.12 and 3.13 show an example of system (3.42) for a type II neuron with ramp protocol (3.43), where $F(v)$ is given by the ML-FHN model (3.36). Figure 3.12 is a three-dimensional representation of the critical manifold S near the lower fold F^- and Fig. 3.13 is the corresponding reduced flow on S projected

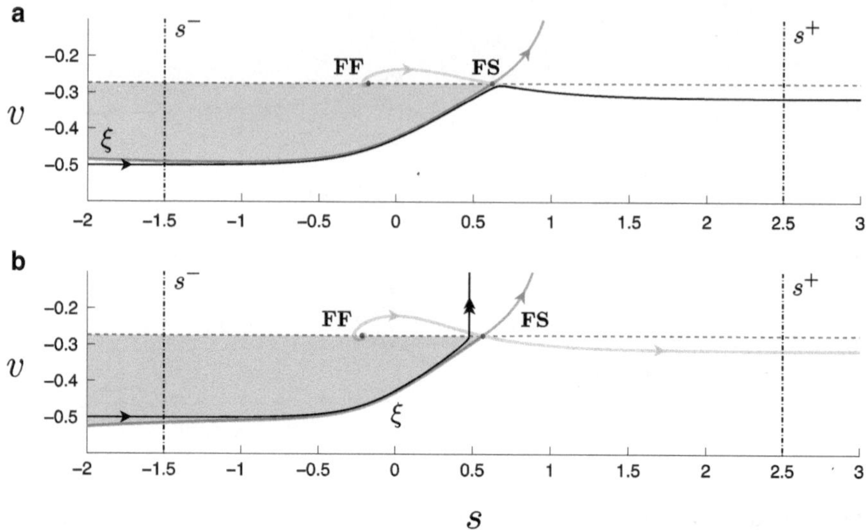

Fig. 3.13 Reduced flow shown in Fig. 3.12 projected onto coordinate chart (s, v) corresponding to Fig. 3.10a, ramp protocol: there are two folded singularities, a folded saddle (FS) and a folded focus (FF), ξ denotes the folded saddle canard that crosses from the lower stable branch S_a^- via the FS singularity onto the repelling middle branch S_r. The canard ξ forms the firing threshold manifold. Note, the other canard of the FS singularity (the faux canard) crosses from S_r to S_a^-. The segment on S_a^- forms a boundary that prevents trajectories to the right of ξ to spike

onto the coordinate chart (s, v). The initial state on the stable branch S_a of the critical manifold is (s^-, v_{rest}^-) where $v = v_{rest}^-$ (the horizontal trajectory for $s \leq s^-$) corresponds to the resting membrane potential of the neuron for $s \leq s^-$, i.e. for $I = 0$. On the lower fold F^- (dashed horizontal line), we observe two folded singularities, a folded focus (FF) respectively a folded saddle (FS). Note, $\dot{v} > 0$ along the segment of F^- bounded by the two folded singularities. To reach this segment of the lower fold F^- and, hence, to be able to elicit a spike, the initial state (s^-, v_{rest}^-) must be in the *'domain of attraction'* of this segment (shown as a shaded region). This domain is bounded by the folded saddle canard ξ and a segment of the lower fold F^-. Thus the folded saddle canard ξ forms the firing threshold manifold on S_a^-.

As can be also seen in Fig. 3.13, the position of the canard ξ changes as the (maximal) slope of the drive $I(s)$ changes. Clearly, folded singularities and their canards encode the complete temporal information of the drive $I(s)$, i.e. amplitude, slope, curvature, etc. Figure 3.13a predicts no spike while Fig. 3.13b predicts a spike. These correspond to the two cases shown in Figs. 3.10a and 3.12 for the ramp protocol.

Similarly, Fig. 3.14a predicts no spike while Fig. 3.14b predicts a spike. These correspond to the two cases shown in Fig. 3.10b for the pulse protocol. Therefore, we can view this type II excitable neuron shown in Fig. 3.10 as a slope detector.

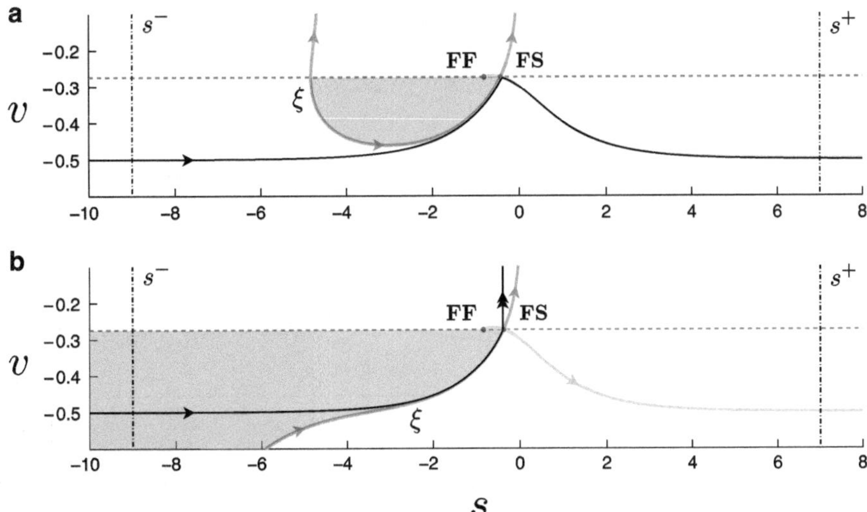

Fig. 3.14 Reduced flow projected onto coordinate chart (s, v) corresponding to Fig. 3.10b, pulse protocol: ξ denotes the folded saddle canard that forms the firing threshold manifold; see Fig. 3.13 caption for details

It is remarkable that dynamic, nonautonomous information such as the evolution profile of the external drive $I(s)$ is encoded in the location of an invariant manifold of a singular perturbation problem, the canard. Here, in particular, we observe that only changing the slope is sufficient to elicit a spike. In general, our analysis provides a *slow input modulation* condition for transient phenomena based on canard theory.

3.3.2.4 Firing Threshold Amplitude I_{thr}

The previous analysis showed that the existence of a folded saddle singularity is a necessary but not a sufficient condition for a neuron model to be able to fire an action potential. At the heart of the issue lies the relative position of the folded saddle canard ξ that forms the firing threshold manifold in these models to the initial condition. Numerically, we found that any ramp with a maximal drive $I_1 < I_{thr}$ is not able to elicit a spike independent of the slope of the ramp. Although a folded saddle singularity might exist, the domain of attraction for firing a spike bounded by the folded saddle canard ξ never encloses the initial condition given by the resting membrane potential.

Even if we formally take the limit $s_1 \rightarrow 0$ (at $s = s_0$) which transforms the smooth ramp into a discontinuous step protocol, we are not able to elicit a spike. By looking at Fig. 3.7, it becomes immediately clear why the model neuron cannot spike. The shift of the critical manifold is not sufficient to pass the lower fold as

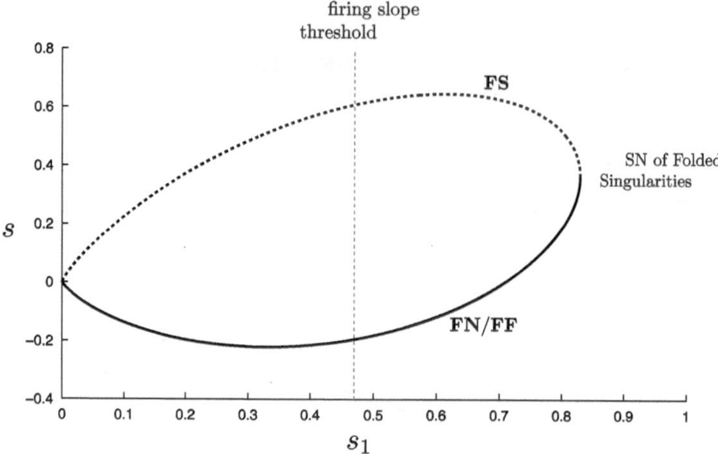

Fig. 3.15 Bifurcation diagram of folded singularities under variation of the slope s_1 for type II neuron with ramp protocol shown in Fig. 3.13: $I_1 = 0.09$ is fixed. For $s_1 \approx 0.83$ we observe a folded saddle-node (FSN) type I bifurcation where two folded singularities annihilate each other. The vertical line at $s_1 \approx 0.47$ indicates the firing slope threshold, i.e for sufficiently steep slope $s_1 < 0.47$, the type II neuron will transiently spike; compare with Fig. 3.10a

discussed at the end of Sect. 3.3.1.1. Although the canard ξ represents the firing threshold manifold for slow dynamic changes, it can be continued towards the fast time-scale limit and it will converge to the firing threshold shown in Fig. 3.9 (dashed curve). Hence $I = I_{thr}$ represents the fast time-scale limit of the minimum current amplitude needed to elicit a spike. The closer the amplitude $I > I_{thr}$ to this limit I_{thr} the steeper the slope of the profile has to be. The same holds for type III neurons. This relates the concepts of fast and slow input modulations.

3.3.2.5 Bifurcation of Canards

The existence of folded singularities and their associated canards is necessary for the transient spiking phenomenon observed. They are slope detectors. Figure 3.15 shows a folded singularity bifurcation diagram for a type II neuron under the variation of the slope s_1 of the ramp (the amplitude is fixed). Folded singularities bifurcate at a saddle-node bifurcation of a folded saddle with a folded node. (Subsequently, the branch of folded nodes becomes a branch of folded foci.) This bifurcation is known as a *folded saddle-node (FSN) of type I* [40, 57]. Here, the type refers to the bifurcation, not to the type of neuron. This points to the importance of this bifurcation for the excitability of neurons.

In contrast, a *FSN type II* bifurcation [40, 57] indicates a transcritical bifurcation of a folded and an ordinary singularity bifurcation. This type of bifurcation usually happens in type II neurons close to $I = I_{bif}$. It corresponds to the unfolding of a singular Hopf bifurcation in singularly perturbed systems with two or more slow variables. The interested reader is referred to [10] and references therein.

3.4 Conclusion

There is a growing synergy between neurophysiology and dynamical systems. The abstraction and generalization of the mathematical approach can lead to the identification of, and deep insights into, common mathematical structures of rhythmicity and excitability across contexts. The demand for an adequate high-level description of cell function raises a number of challenges at the forefront of present-day research in the field of dynamical systems. We face fundamental challenges in trying to understand the relationships between intrinsic dynamics, stimuli, coupling, and patterns of synchrony in network models. The ability of neuronal networks to create spatio-temporal patterns, spontaneously or driven, and the ways in which neuromodulators reshape or totally change these patterns is of eminent interest for understanding neuronal dynamics.

The application of concepts and techniques from dynamical systems theory to neuronal dynamics continues to mature, especially to stationary rhythms and steady state attractors. Meanwhile, there is increasing awareness that transient dynamics play an important physiological role. Excitability of neurons [18, 29, 52] and networks are prime examples of transient dynamics, especially as responses to brief or non-stationary time-varying inputs. Recent developments in canard theory [26, 63, 64] have provided a new direction for understanding these transient dynamics that are modelled as nonautonomous multiple time-scale systems. It is well known that a nonautonomous system can be viewed as an extended autonomous system by increasing the phase space dimension by one. The key observation is that folded singularities are still well defined, while equilibria of the unforced system will not persist in the extended system. Thus canards have the potential to significantly shape the nature of solutions in nonautonomous multiple time-scales systems. We would like to stress this important point of view.

The take-home message lies in the realisation that folded singularities and associated canards create local transient "attractor" states in multiple scales problems. This is due to the fact that trajectories in the domain of attraction of folded singularities will reach and pass these folded singularities in finite slow time; folded singularities are not equilibrium states. In the context of neuronal excitability and as shown in [64], we identify *canards of folded saddle type* as *firing threshold manifolds*. We have demonstrated the role of such structures in comparing the dynamics of spike generation for neuron models in the different behavioral regimes of type I, II and III excitability. For type II and III we have revealed and characterized stimulus features that lead to spike generation for transient stimuli, most notably that a stimulus must rise fast enough for excitation.

Dynamic forcing has the potential to create folded singularities and to form these effective separatrices or to change the global return mechanism. Hence, the specific nature of the dynamic forcing determines which local attractor states can be reached through global mechanisms. This point of view has profound consequences in the analysis of excitable physiological systems such as in auditory brain stem neurons [42], modeling propofol anesthesia [41, 44] and cell calcium dynamics [26]. From a mathematical point of view, the time is "*ripe*" for forging (more) connections between nonautonomous attractor theory [36,37] and geometric singular perturbation theory [17, 32].

Acknowledgements This work was supported in part by NIH grant DC008543-01 (JR) and by the ARC grant FT120100309 (MW). MW would like to thank the organisers of the *Inzell workshop* for their hospitality and financial support.

References

1. R. Amir, M. Michaelis, M. Devor, Burst discharge in primary sensory neurons: triggered by subthreshold oscillations, maintained be depolarizing afterpotentials. J. Neurosci. **22**, 1187–1198 (2002)
2. E. Benoît, J. Callot, F. Diener, M. Diener, Chasse au canard. Collectanea Math. **31–32**, 37–119 (1981)
3. E. Benoît, Systémes lents-rapides dans $\mathbb{R}^3$ et leur canards. Asterisque **109–110**, 159–191 (1983)
4. A. Borisyuk, J. Rinzel, Understanding neuronal dynamics by geometric dissection of minimal models, in *Models and Methods in Neurophysics, Proc. Les Houches Summer School 2003, (Session LXXX)*, ed. by C. Chow, B. Gutkin, D. Hansel, C. Meunier, J. Dalibard (Elsevier, 2005), pp. 19–72. ISBN:978-0-444-51792-0
5. M. Brøns, T. Kaper, H. Rotstein, Focus issue: mixed mode oscillations: experiment, computation, and analysis. Chaos **18**, 015101 (2008), 1–4
6. M. Brøns, M. Krupa, M. Wechselberger, Mixed mode oscillations due to the generalized canard phenomenon. Fields Inst. Comm. **49**, 39–63 (2006)
7. P. De Maesschalck, F. Dumortier, Slow-fast Bogdanov-Takens bifurcations. J. Differ. Equat. **250**, 1000–1025 (2011)
8. P. De Maesschalck, F. Dumortier, M. Wechselberger, Special issue on bifurcation delay. Discrete Cont. Dyn. Sys. S **2**(4), 723–1023 (2009)
9. P. De Maesschalck, M. Wechselberger, Unfolding of a singularly perturbed system modelling type I excitability, preprint (2013)
10. M. Desroches, J. Guckenheimer, B. Krauskopf, C. Kuehn, H. Osinga, M. Wechselberger, Mixed-mode oscillations with multiple time-scales. SIAM Rev. **54**, 211–288 (2012)
11. M. Desroches, M. Krupa, S. Rodrigues, Inflection, canards and excitability threshold in neuronal models. J. Math. Biol. **67**(4), 989–1017 (2013)
12. C. Dickson, J. Magistretti, M. Shalinsky, E. Fransen, M. Hasselmo, A. Alonso, Properties and role of I(h) in the pacing of subthreshold oscillations in entorhinal cortex layer II neurons. J. Neurophysiol. **83**, 2562–2579 (2000)
13. E. Doedel, A. Champneys, T. Fairgrieve, Y. Kuznetsov, B. Sandstede, X. Wang, AUTO 97: continuation and bifurcation software for ordinary differential equations (with HomCont)
14. J. Drover, J. Rubin, J. Su, B. Ermentrout, Analysis of a canard mechanism by which excitatory synaptic coupling can synchronize neurons at low firing frequencies. SIAM J. Appl. Math. **65**, 65–92 (2004)
15. F. Dumortier, R. Roussarie, Canard cycles and center manifolds. In: Memoirs of the American Mathematical Society, **577** (1996)

16. B. Ermentrout, M. Wechselberger, Canards, clusters and synchronization in a weakly coupled interneuron model. SIAM J. Appl. Dyn. Syst. **8**, 253–278 (2009)
17. N. Fenichel, Geometric singular perturbation theory. J Differ. Equat. **31**, 53–98 (1979)
18. R. FitzHugh, Mathematical models of threshold phenomena in the nerve membrane. Bull. Math. Biophys. **7**, 252–278 (1955)
19. R. FitzHugh, Thresholds and plateaus in the Hodgkin-Huxley nerve equations. J. Gen. Physiol. **43**, 867–896 (1960)
20. R. FitzHugh, Impulses and physiological states in theoretical models of nerve membrane. Biophys. J. **1**, 455–466 (1961)
21. R. FitzHugh, Anodal excitation in the Hodgkin-Huxley nerve model. Biophys J. **16**, 209–226 (1976)
22. J. Guckenheimer, Singular Hopf bifurcation in systems with two slow variables. SIAM J. Appl. Dyn. Syst. **7**, 1355–1377 (2008)
23. J. Guckenheimer, P. Holmes, *Nonlinear Oscillations, Dynamical Systems, and Bifurcations of Vector Fields* (Springer, New York, 1983)
24. J. Guckenheimer, M. Wechselberger, L.-S. Young, Chaotic attractors of relaxation oscillators. Nonlinearity **19**, 709–720 (2006)
25. Y. Gutfreund, Y. Yarom, I. Segev, Subthreshold oscillations and resonant frequency in guinea-pig cortical neurons: physiology and modelling. J. Physiol. **483**, 621–640 (1995)
26. E. Harvey, V. Kirk, H. Osinga, J. Sneyd, M. Wechselberger, Understanding anomalous delays in a model of intracellular calcium dynamics. Chaos **20**, 045104 (2010)
27. E. Harvey, V. Kirk, J. Sneyd, M. Wechselberger, Multiple timescales, mixed-mode oscillations and canards in models of intracellular calcium dynamics. J. Nonlinear Sci. **21**, 639–683 (2011)
28. G. Hek, Geometric singular perturbation theory in biological practice. J. Math. Biol. **60**, 347–386 (2010)
29. A.L. Hodgkin, The local electric changes associated with repetitive action in a non-medullated axon. J. Physiol. **107**, 165–181 (1948)
30. A.L. Hodgkin, A.F. Huxley, A quantitative description of membrane current and its application to conduction and excitation in nerve. J. Physiol. **117**, 500–544 (1952)
31. E. Izhikevich, Dynamical systems in neuroscience: the geometry of excitability and bursting, *Computational Neuroscience* (MIT Press, Cambridge, MA, 2007)
32. C.K.R.T. Jones, Geometric singular perturbation theory, in dynamical systems. Springer Lect. Notes Math. **1609**, 44–120 (1995)
33. T. Kaper, An introduction to geometric methods and dynamical systems theory for singular perturbation problems. Proc. Symp. Appl. Math. **56**, 85–131 (1999)
34. J. Keener, J. Sneyd, *Mathematical Physiology* (Springer, New York, 1998)
35. S. Khosrovani, R. van der Giessen, C. de Zeeuw, M. de Jeu, In vivo mouse inferior olive neurons exhibit heterogeneous subthreshold oscillations and spiking patterns. PNAS **104**, 15911–15916 (2007)
36. P.E. Kloeden, C. Pötzsche, *Nonautonomous Dynamical Systems in the Life Sciences*, Chap. 1 (Springer, Heidelberg, 2013)
37. P.E. Kloeden, M. Rasmussen, *Nonautonomous Dynamical Systems* (American Mathematical Society, Providence, 2011)
38. M. Krupa, N. Popovic, N. Kopell, H. Rotstein, Mixed-mode oscillations in a three time-scale model for the dopaminergic neuron. Chaos **18**, 015106 (2008)
39. M. Krupa, P. Szmolyan, Relaxation oscillations and canard explosion. J. Differ. Equat. **174**, 312–368 (2001)
40. M. Krupa, M. Wechselberger, Local analysis near a folded saddle-node singularity. J. Differ. Equat. **248**, 2841–2888 (2010)
41. M. McCarthy, N. Kopell, The effect of propofol anesthesia on rebound spiking. SIAM J. Appl. Dyn. Syst. **11**, 1674–1697 (2012)
42. X. Meng, G. Huguet, J. Rinzel, Type III excitability, slope sensitivity and coincidence detection. Discrete Cont. Dyn. Syst. A **32**, 2729–2757 (2012)

43. A. Milik, P. Szmolyan, H. Löffelmann, E. Gröller, The geometry of mixed-mode oscillations in the 3d-autocatalator. Int. J. Bifurcat. Chaos **8**, 505–519 (1998)
44. J. Mitry, M. McCarthy, N. Kopell, M. Wechselberger, Excitable neurons, firing threshold manifold and canards. J. Math. Neurosci. **3**, 12 (2013)
45. J.S. Nagumo, S. Arimoto, S. Yoshizawa, An active pulse transmission line simulating nerve axon. Proc. IRE **50**, 2061–2070 (1962)
46. C. Morris, H. Lecar, Voltage oscillations in the barnacle giant muscle fiber. Biophys. J. **35**, 193–213 (1981)
47. C. Del Negro, C. Wilson, R. Butera, H. Rigatto, J. Smith, Periodicity, mixed-mode oscillations, and quasiperiodicity in a rhythm-generating neural network. Biophys. J. **82**, 206–14 (2002)
48. S. Prescott, Y. de Koninck, T. Sejnowski, Biophysical basis for three distinct dynamical mechanisms of action potential initiation. PLoS Comput. Biol. **4**(10), e1000198 (2008)
49. R. O'Malley, *Singular Perturbation Methods for Ordinary Differential Equations* (Springer, New York, 1991)
50. M. Rasmussen, Attractivity and bifurcation for nonautonomous dynamical systems, *Lecture Notes in Mathematics*, vol. 1907 (Springer, Heidelberg, 2007)
51. J. Rinzel, Excitation dynamics: insights from simplified membrane models. Fed. Proc. **44**, 2944–2946 (1985)
52. J. Rinzel, G. Ermentrout, Analysis of neural excitability and oscillations, in *Methods in Neuronal Modelling: From Synapses To Networks, 2nd edn.*, ed. by C. Koch, I. Segev (MIT Press, Cambridge, MA, 1998), pp. 251–291
53. J. Rothman, P. Manis, The roles potassium currents play in regulating the electric activity of ventral cochlear neucleus neurons. J. Neurophysiol. **89**, 3097–3113 (2003)
54. H. Rotstein, M. Wechselberger, N. Kopell, Canard induced mixed-mode oscillations in a medial enorhinal cortex layer II stellate cell model. SIAM J. Appl. Dyn. Syst. **7**, 1582–1611 (2008)
55. J. Rubin, D. Terman, Geometric singular perturbation analysis of neuronal dynamics, in *Handbook of Dynamical Systems*, vol. 2, ed. by B. Fiedler (Elsevier Science B.V., Amsterdam, 2002)
56. J. Rubin, M. Wechselberger, Giant Squid - Hidden Canard: the 3D geometry of the Hodgkin Huxley model. Biol. Cyb. **97**, 5–32 (2007)
57. P. Szmolyan, M. Wechselberger, Canards in $\mathbb{R}^3$. J. Differ. Equat. **177**, 419–453 (2001)
58. P. Szmolyan, M. Wechselberger, Relaxation oscillations in $\mathbb{R}^3$. J. Differ. Equat. **200**, 69–104 (2004)
59. F. Takens, Constrained equations; a study of implicit differential equations and their discontinuous solutions, in Structural stability, the theory of catastrophes, and applications in the sciences. *Lecture Notes in Mathematics*, vol. 525 (Springer, Berlin/New York, 1976)
60. T. Vo, R. Bertram, J. Tabak, M. Wechselberger, Mixed-mode oscillations as a mechanism for pseudo-plateau bursting. J. Comp. Neurosci. **28**, 443–458 (2010)
61. M. Wechselberger, Existence and bifurcation of canards in $\mathbb{R}^3$ in the case of a folded node. SIAM J. Appl. Dyn. Syst. **4**, 101–139 (2005)
62. M. Wechselberger, Canards. Scholarpedia **2**(4), 1356 (2007)
63. M. Wechselberger, À propos de canards (Apropos canards). Trans. Am. Math. Soc. **364**, 3289–3309 (2012)
64. S. Wieczorek, P. Ashwin, C. Luke, P. Cox, Excitability in ramped systems: the compost-bomb instability. Proc. R. Soc. A **467**, 1243–1269 (2011)
65. W. Zhang, V. Kirk, J. Sneyd, M. Wechselberger, Changes in the criticality of Hopf bifurcations due to certain model reduction techniques in systems with multiple timescales. J. Math. Neurosci. **1**, 9 (2011)

Part II
Applications

Chapter 4
Stimulus-Response Reliability of Biological Networks

Kevin K. Lin

Abstract If a network of cells is repeatedly driven by the same sustained, complex signal, will it give the same response each time? A system whose response is reproducible across repeated trials is said to be *reliable*. Reliability is of interest in, e.g., computational neuroscience because the degree to which a neuronal network is reliable constrains its ability to encode information via precise temporal patterns of spikes. This chapter reviews a body of work aimed at discovering network conditions and dynamical mechanisms that can affect the reliability of a network. A number of results are surveyed here, including a general condition for reliability and studies of specific mechanisms for reliable and unreliable behavior in concrete models. This work relies on qualitative arguments using random dynamical systems theory, in combination with systematic numerical simulations.

Keywords Reliability • Spike-time precision • Coupled oscillators • Random dynamical systems • Neuronal networks • Lyapunov exponents • SRB measures

4.1 Introduction

If a network of neurons is repeatedly presented with the same complex signal, will its response be the same each time? A network for which the answer is affirmative is said to be *reliable*. This property is of interest in computational neuroscience because neurons communicate information via brief electrical impulses, or *spikes*, and the degree to which a system is reliable constrains its ability to transmit information via precise temporal patterns of spikes. Thus, whether a given system is capable of reliable response can affect the mode and rate with which it transmits and processes information.

K.K. Lin
Department of Mathematics, University of Arizona, Tucson, AZ, USA
e-mail: klin@math.arizona.edu

P.E. Kloeden and C. Pötzsche (eds.), *Nonautonomous Dynamical Systems in the Life Sciences*, Lecture Notes in Mathematics 2102, DOI 10.1007/978-3-319-03080-7_4,

The reliability of *single neurons* has been well studied both experimentally and theoretically. In particular, in vitro experiments have found that single, synaptically isolated neurons are reliable under a broad range of conditions, i.e., the spike times of an isolated neuron in response to repeated injections of a fixed, fluctuating current signal tend to be repeatable across multiple trials [6, 16, 33]. Theoretical studies have also found that models of isolated neurons tend to be reliable [15, 16, 20, 38, 39, 41, 53]. Less is known at the network or systems level; see, e.g., [2, 4, 7, 17, 31, 34] for some relevant experimental findings, and [40] for a theoretical treatment using a different approach.

This chapter reviews a body of work aimed at discovering network conditions and dynamical mechanisms that can affect the reliability of a network. The ergodic theory of random dynamical systems, i.e., the measure-theoretic analog of the theories surveyed in Chaps. 1 and 2 of the present collection, plays a key role in this work: it provides a natural mathematical framework for precisely formulating the notion of reliability and providing tools that, in combination with numerical simulations, enable the analysis of concrete network models.

Much of the material and exposition here follow [27–29]. The first of these papers is concerned with mathematically-motivated questions, while the latter two concentrate on a more biological class of networks. These papers, as well as the present review, mainly focus on networks of oscillatory (i.e., tonically spiking) neurons. Networks of excitable neurons, which can behave rather differently, are the subject of a recent study [22] (see the discussion).

Relevance Outside Neuroscience. Reliability is a general dynamical property that is potentially relevant for a wide range of signal processing systems. Since biological systems, on scales ranging from single genes to entire organisms (and even populations), must respond to unpredictable environmental signals, it is possible that some of the mathematical framework and perhaps even the approaches and ideas outlined here may be of use in studying other types of biological information processing. Concepts analogous to reliability have also found use in areas far from biology, e.g., in engineered systems like coupled lasers [42] and in molecular dynamics simulations [43].

The rest of this chapter is organized as follows: in Sect. 4.2, the concept of reliability is given a precise formulation, and some relevant results from random dynamical systems theory are reviewed. Section 4.3 presents a general condition that guarantees reliability, and Sect. 4.4 examines specific mechanisms for reliable and unreliable behavior in some concrete network models.

4.2 Problem Statement and Conceptual Framework

This section describes a class of models which will be used throughout the rest of this chapter. The concept of reliability is given a precise formulation in this context, and relevant ideas from random dynamical systems theory are reviewed.

4.2.1 Model Description and a Formulation of Reliability

To illustrate our ideas, we use networks of so-called "theta neurons" (see, e.g., [12]). These are idealized models of neurons that spike periodically at a fixed frequency ω in the absence of external forcing.[1] Single theta neurons have the form

$$\dot{\theta}(t) = \omega + z\big(\theta(t)\big)I(t) \tag{4.1}$$

Here, the state of the neuron is given by an angle $\theta \in S^1$, which is here mapped onto the interval $[0, 1]$ with endpoints identified; $\omega > 0$ is the intrinsic frequency of the neuron; $I(t)$ represents the sum of all the stimuli driving the neuron; and z is the *phase response curve* (PRC) of the neuron. The angle θ represents the fraction of the cycle that the neuron has completed; the neuron is viewed as generating a spike at $\theta = 0$. If $I(t) \equiv 0$, (4.1) is just the equation for a phase oscillator. A nonzero input $I(t)$ modulates the firing rate of the neuron, and the phase response z captures the state-dependent response of the neuron to stimuli.

Phase models like (4.1) are often used in biology to model rhythmic activity (see, e.g., [12, 49], and also Chap. 5). In the context of neuroscience, the choice of PRC determines the response of the neuron model to stimuli, and a variety of PRCs are commonly used. In this chapter, $z(\theta)$ is taken to be $\frac{1}{2\pi}\big(1-\cos(2\pi\theta)\big)$, which models so-called "Type I" neurons [5, 11]. This PRC has the property that it is positive when the neuron spikes, and for $\theta \approx 0$, $z(\theta) = O(\theta^2)$. The latter represents a form of *refractory effect:* at the moment when the neuron spikes, it is insensitive to its inputs, and is unable to generate a second spike immediately. This PRC is sometimes justified formally by truncating the normal form of neuron models near a saddle-node-on-invariant-circle bifurcation[2]; for our purposes, it mainly serves as convenient phenomenological model for neuronal response.

The class of models used in this chapter are networks of theta neurons. The network equations have the form

$$\dot{\theta}_i = \omega_i + z(\theta_i)\left(\sum_{j\neq i} a_{ji}\, g(\theta_j) + I_i(t)\right), \quad i = 1, 2, \cdots, N, \tag{4.2}$$

where the function $g : [0, 1] \to \mathbb{R}$ is an approximate delta function, i.e., it is a smooth function supported in a small interval $[-\delta, \delta]$ (here $\delta \sim 1/20$) satisfying $\int_0^1 g(\theta)\, d\theta = 1$; such pulse couplings are simple models for relatively fast synapses. The coupling matrix $A = (a_{ji})$ encodes the network structure; for simplicity we assume $a_{ii} = 0$ for all i, i.e., no self-loops. A number of different

[1] Theta neurons can also model neurons operating in an *excitable* regime. The reliability of excitable theta neuron networks is studied in [22].

[2] See, e.g., [5], but note that phase truncations can sometimes miss important dynamical effects [30], and their use in biological modeling should be carefully justified.

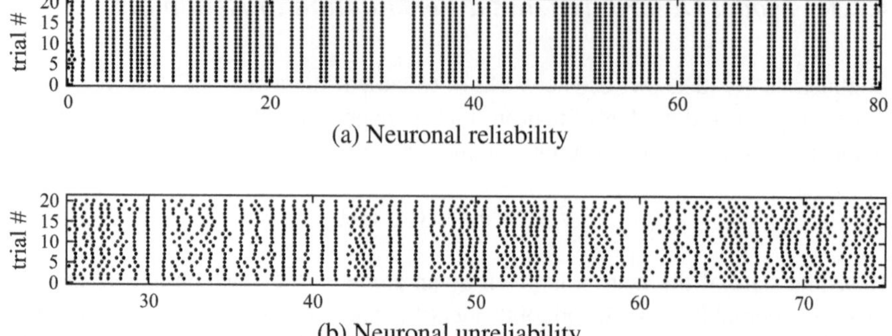

Fig. 4.1 Raster plots showing the spike times of two neurons across 20 trials in response to a fixed stimulus. (**a**) Neuronal reliability, (**b**) neuronal unreliability. The neuron shown in (**a**) is chosen at random from a network with 100 neurons, with parameters chosen so that the network response is reliable across trials (as are single neuron responses). In (**b**), the neuron comes from an unreliable network. Figure adapted from [28]

network architectures are considered in this review; these are specified along the way.

The stimuli $I_i(t)$ in (4.2) are modeled as white noise, i.e., $I_i(t)dt = \epsilon_i \, dW_t^i$ where W_t^i denotes a standard Wiener process. This is an idealization of sustained, fluctuating signals, and has the convenient mathematical consequence that (4.2) is a (possibly quite large) system of stochastic differential equations (SDEs). *A priori,* the W_t^i for $i = 1, \cdots, N$ may be independent or correlated; for simplicity let us assume they are either independent or identical, allowing some neurons to receive the same input. Note that in (4.2), the stochastic forcing terms solely represent external stimuli driving the neuron, and not sources of neuronal or synaptic noise (but see the discussion at the end of the chapter).

These network models, though highly idealized, are broad enough to generate both reliable and unreliable network response without requiring careful tuning of parameters. That is, upon repeated trials with the same realizations of the $I_i(t)$ but different initial conditions, they can generate responses that are essentially the same across trials *(reliable)* or differ substantially across trials *(unreliable).* These behaviors are illustrated in Fig. 4.1.

A Notion of Neuronal Reliability

What would it mean for a system of the form (4.2) to be reliable? Suppose we fix a single realization of the stimulus $(I_i(t))$ and drive (4.2) with the stimulus realization over a number of repeated "trials," with a new initial condition on each trial. The system (4.2) is said to be *neuronally reliable* (or simply "reliable") if

$$\lim_{t \to \infty} \mathrm{dist}(\Theta(t), \Theta'(t)) = 0 \,, \tag{4.3}$$

where $\Theta(t) = (\theta_1(t), \cdots, \theta_N(t))$ denotes the state of the entire network at time t, and Θ and Θ' are two trajectories with initial states $\Theta(0) \neq \Theta'(0)$ [28]; it is assumed that $\Theta(0)$ is sampled independently from a fixed probability density ρ, say ρ =Lebesgue, on each trial. In other words, (4.3) means that given enough time, the entire network state is reproducible across repeated trials with random initial conditions.

As will be seen in Sect. 4.2.2, this notion of reliability can be naturally studied within the framework of random dynamical systems theory, making it a convenient mathematical definition. In biological terms, if a network is neuronally reliable, any network output that is a function of the network state will also be reproducible across repeated trials, so that neuronal reliability is in a sense the strongest form of reliability one might consider. Note, however, that there are other biologically relevant notions of reliability; some of these are mentioned in Sect. 4.2.3.

4.2.2 *Relevant Mathematical Background*

I begin by reviewing some relevant mathematical ideas [1, 3]; these can be viewed as ergodic-theoretic analogs of the theories reviewed in Chap. 1 by Kloeden and Pötzsche, and Chap. 2 by de Freitas and Sontag. There is, in particular, some overlap (both in overall goals and specific results) with the latter, though the perspective and emphasis here are different. The setting is a general SDE

$$dx_t = a(x_t)\, dt + \sum_{i=1}^{k} b_i(x_t) \circ dW_t^i \,, \tag{4.4}$$

where $x_t \in M$ with M a compact Riemannian manifold, and the W_t^i are independent standard Brownian motions. Clearly, (4.2) is a special case of (4.4): $x_t = (\theta_1(t), \ldots, \theta_N(t))$, $M = \mathbb{T}^N \equiv S^1 \times S^1 \times \cdots \times S^1$.

(To make sense of the theory outlined below on a general manifold M, Stratonovich calculus is necessary. But for $M = \mathbb{T}^N$ one can use either Itô or Stratonovich, and for simplicity Itô is used in Sect. 4.3 and beyond.)

Stochastic Flows. In most physical applications involving SDEs, one fixes an initial x_0, and looks at the distribution of x_t for $t > 0$. These distributions evolve in time according to the Fokker-Planck equation, and under fairly general conditions converge to a unique stationary measure μ as $t \to \infty$. Since reliability is about a system's reaction to a single stimulus, i.e., a single realization of the driving Wiener processes $(W_t^1, \cdots, W_t^N)$, at a time, and concerns the simultaneous evolution of all or large sets of initial conditions, of relevance to us are not the distributions of x_t but *flow-maps* $F_{t_1,t_2;\omega}$, where $t_1 < t_2$ are two points in time, ω is a sample Brownian path, and $F_{t_1,t_2;\omega}(x_{t_1}) = x_{t_2}$ where x_t is the solution of (4.4) corresponding to ω. A well known theorem states that such *stochastic flows of diffeomorphisms* are well defined if the functions $a(x)$ and $b(x)$ in (4.4) are sufficiently smooth; see, e.g., [21]. More precisely, the maps $F_{t_1,t_2;\omega}$ are well defined for almost every ω, and they are invertible, smooth transformations with smooth inverses. Moreover, $F_{t_1,t_2;\omega}$ and $F_{t_3,t_4;\omega}$ are independent for $t_1 < t_2 < t_3 < t_4$. These results allow us to treat the evolution of systems described by (4.4) as compositions of random, IID, smooth maps. Many of the techniques for analyzing smooth deterministic systems have been extended to this random setting (see, e.g., Chaps. 1 and 2); the resulting body of results is collectively called "RDS theory" in this review.

The stationary measure μ, which gives the steady-state distribution averaged over all realizations ω of the driving Wiener processes, does not describe what we see when studying a system's reliability. Of relevance are the *sample measures* $\{\mu_\omega\}$, defined by

$$\mu_\omega = \lim_{t\to\infty} (F_{-t,0;\omega})_*\mu \tag{4.5}$$

where $(F_{-t,0;\omega})_*\mu$ denotes the push-forward of the stationary measure μ along the flow $F_{-t,0;\omega}$, and we think of ω as defined for all $t \in (-\infty,\infty)$ and not just for $t > 0$. (That the limit in (4.5) exists follows from a martingale convergence argument; see, e.g., [18].) One can view μ_ω as the conditional measures of μ given the past history of ω; it describes the distribution of states at $t = 0$ given that the system has experienced the input defined by ω for all $t < 0$. The family of measures $\{\mu_\omega\}$ is invariant in the sense that $(F_{0,t;\omega})_*(\mu_\omega) = \mu_{\sigma_t(\omega)}$ where $\sigma_t(\omega)$ is the time-shift of the sample path ω by t; for this reason they are also sometimes called *random invariant measures.* Sample measures are measure-theoretic analogs of *pullback attractors* (see Chaps. 1 and 2), and are the distributions of *equilibria* (Chap. 2).

If our initial distribution is given by a probability density ρ and we apply the stimulus corresponding to ω, then the distribution at time t is $(F_{0,t;\omega})_*\rho$. For t sufficiently large, and assuming ρ and μ are both sufficiently smooth, one expects in most situations that $(F_{0,t;\omega})_*\rho$ is very close to $(F_{0,t;\omega})_*\mu$, which is essentially given by $\mu_{\sigma_t(\omega)}$ for large times t. (The time-shift by t of ω is necessary because by definition, μ_ω is the conditional distribution of μ at time 0.)

Figure 4.2 shows some snapshots of $(F_{0,t;\omega})_*\rho$ for a system with $N = 2$ cells, for two different sets of parameters. As noted earlier, these distributions approximate $\mu_{\sigma_t(\omega)}$ for t sufficiently large. In these simulations, the initial distribution ρ is the stationary density of (4.2) with a small-amplitude noise, the interpretation being that the system is intrinsically noisy even in the absence of external stimuli; this distribution is then pushed forward in time using a fixed stimulus ω. Observe that these pictures evolve with time, and for large enough t, they have similar qualitative properties depending on the underlying system. This is in agreement with RDS theory, which tells us in fact that the $\mu_{\sigma_t(\omega)}$ obey a statistical law for almost all ω.

The measure μ_ω gives the distribution of all possible states that a system may attain starting in a random state in the past and receiving a given stimulus for a sufficiently long time. Its structure is therefore of natural interest in reliability

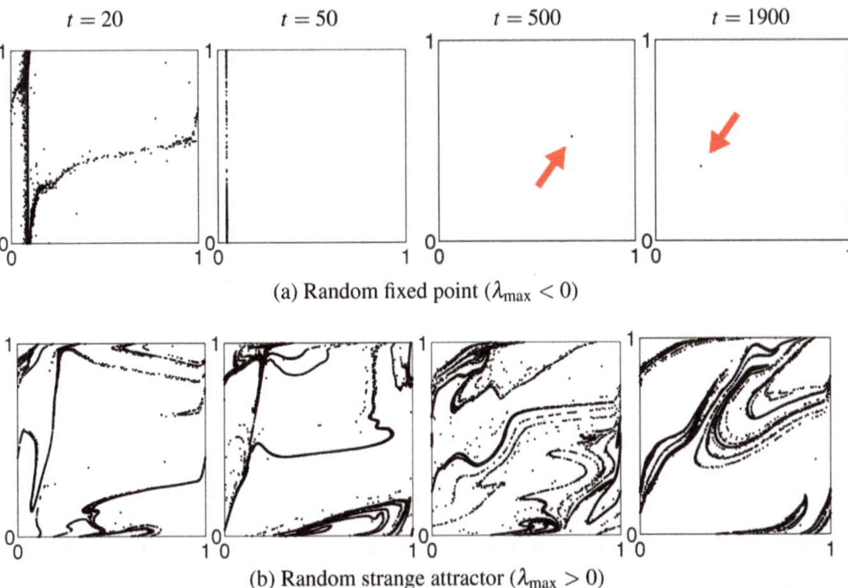

Fig. 4.2 Temporal snapshots of sample measures for (4.2) with $N = 2$ oscillators driven by a single stimulus realization. Two different sets of parameters are used in (**a**) Random fixed point ($\lambda_{\max} < 0$) and (**b**) Random strange attractor ($\lambda_{\max} > 0$). In (**a**), the sample measures converge to a random fixed point. In (**b**), the sample measures converge to a random strange attractor. These figures (adapted from [27]) illustrate Theorem 4.1

studies. Below we recall two mathematical results that pertain to the structure of μ_ω, specifically relating Lyapunov exponents to μ_ω.

Lyapunov Exponents and Sample Distributions. For a fixed stimulus realization ω, any $x \in M$, and any nonzero tangent vector $v \in T_x M$, define the *Lyapunov exponent* [50]

$$\lambda_\omega(x, v) = \lim_{t\to\infty} \frac{1}{t} \log |DF_{0,t;\omega}(x) \cdot v| \tag{4.6}$$

when the limit exists. If μ is a stationary measure of the stochastic flow, then for almost every ω and μ-a.e. x, $\lambda_\omega(x, v)$ is well defined for all v. Moreover, if the invariant measure is ergodic, then $\lambda_\omega(x, v)$ is *non-random*, i.e., there exists a set $\{\lambda_1, \cdots, \lambda_r | \lambda_i \in \mathbb{R}\}$, $1 \leq r \leq \dim(M)$, such that for a.e. ω and x and every v, $\lambda_\omega(x, v) = \lambda_i$ for some i. (We can have $r < \dim(M)$ because some of the λ_i may have multiplicity > 1.) In what follows, we assume the invariant measure is indeed ergodic, and let $\lambda_{\max} = \max_i \lambda_i$.

As in deterministic dynamics, Lyapunov exponents measure the exponential rates of separation of nearby trajectories. In particular, a positive $\lambda_{\max}$ means the flow is sensitive to small variations in initial conditions, which is generally synonymous with chaotic behavior, while $\lambda_{max} < 0$ means nearby trajectories converge in time; in the deterministic context, this is usually associated with the presence of stable fixed points.

In Theorem 4.1 below, we present two results from RDS theory that together suggest that the sign of λ_{max} is a good criterion for distinguishing between reliable and unreliable behavior:

Theorem 4.1. *In the setting of (4.4), let μ be an ergodic stationary measure.*

(1) (Random sinks) [23] *If $\lambda_{max} < 0$, then with probability 1, μ_ω is supported on a finite set of points.*
(2) (Random strange attractors) [24] *If μ has a density and $\lambda_{max} > 0$, then with probability 1, μ_ω is a random Sinai-Ruelle-Bowen (SRB) measure.*

In Part (1) above, if in addition to $\lambda_{max} < 0$, two non-degeneracy conditions (on the relative motions of two points embedded in the stochastic flow) are assumed, then almost surely μ_ω is supported on a single point, and (as required in (4.3)) any pair of trajectories will almost surely converge in time [3]. Observe that this corresponds exactly to reliability for almost every ω as defined in Sect. 4.2.1, namely the collapse of trajectories starting from almost all initial conditions to a single, distinguished trajectory. This is the situation in Fig. 4.2a. In view of this interpretation, we will equate $\lambda_{max} < 0$ with reliability in the rest of this chapter.

The conclusion of Part (2) requires clarification: in deterministic dynamical systems theory, SRB measures are natural invariant measures that describe the asymptotic dynamics of chaotic dissipative systems, in the same way that Liouville measures are the natural invariant measures for Hamiltonian systems. SRB measures are typically singular with respect to Lebesgue, and are concentrated on unstable manifolds, which are families of curves, surfaces etc., that wind around in a complicated way in the phase space [10, 51]. Part (2) of Theorem 4.1 generalizes these ideas to random dynamical systems. Here, random (i.e., ω-dependent) SRB measures live on random unstable manifolds, which are complicated families of curves, surfaces, etc. that evolve with time. In particular, in a system with random SRB measures, different initial conditions acted on by the same stimulus may lead to very different outcomes at time t; this is true for all $t > 0$, however large. Note that in principle, a random strange attractor may still be supported in e.g. a small ball at all times, for the class of oscillator networks at hand we have not observed this to occur. It is, therefore, natural to regard $\lambda_{max} > 0$ as a signature of unreliability.

In the special case where the phase space is a circle, such as in the case of a single oscillator, the fact that $\lambda \leq 0$ is an immediate consequence of Jensen's inequality. In more detail,

$$\lambda(x) = \lim_{t\to\infty} \frac{1}{t} \log F'_{0,t;\omega}(x)$$

for typical ω by definition. Integrating over initial conditions x, we obtain

$$\lambda = \int_{\mathbb{S}^1} \lim_{t\to\infty} \frac{1}{t} \log F'_{0,t;\omega}(x)\, dx = \lim_{t\to\infty} \frac{1}{t} \int_{\mathbb{S}^1} \log F'_{0,t;\omega}(x)\, dx\ .$$

The exchange of integral and limit is permissible because the required integrability conditions are satisfied in stochastic flows [18]. Jensen's inequality then gives

$$\int_{S^1} \log F'_{0,t;\omega}(x)\, dx\ \leq\ \log \int_{S^1} F'_{0,t;\omega}(x)\, dx\ =\ 0\ . \tag{4.7}$$

The equality above follows from the fact that $F_{0,t;\omega}$ is a circle diffeomorphism. Since the gap in the inequality in (4.7) is larger when $F'_{0,t;\omega}$ is farther from being a constant function, we see that $\lambda < 0$ corresponds to $F'_{0,t;\omega}$ becoming "exponentially uneven" as $t \to \infty$. This is consistent with the formation of random sinks.

The following results from general RDS theory shed some light on the situation when the system is multi-dimensional:

Proposition 4.1 (see, e.g., Chap. 5 of [18] or [23]). *In the setting of (4.4), assume μ has a density, and let $\lambda_1, \cdots, \lambda_d$ be the Lyapunov exponents of the system counted with multiplicity. Then*

(i) $\sum_i \lambda_i \leq 0$;
(ii) $\sum_i \lambda_i = 0$ *if and only if* $F_{s,t,\omega}$ *preserves* μ *for almost all* ω *and all* $s < t$;
(iii) if $\sum_i \lambda_i < 0$, *and* $\lambda_i \neq 0$ *for all* i, *then* μ_ω *is singular.*

A formula giving the dimension of μ_ω is proved in [24] under mild additional conditions.

The reliability of a single oscillator, i.e. that $\lambda < 0$, is also easily deduced from Proposition 4.1: μ has a density because the transition probabilities have densities, and no measure is preserved by all the $F_{s,t,\omega}$ because different stimuli distort the phase space differently. Proposition 4.1(i) and (ii) together imply that $\lambda < 0$. See also [35, 38, 39, 41].

For the 2-oscillator system illustrated in Fig. 4.2, assuming that μ has a density (this is straightforward to show; see Part I of [27]), Proposition 4.1(i) and (ii) together imply that $\lambda_1 + \lambda_2 < 0$. Here $\lambda_1 = \lambda_{\max}$ can be positive, zero, or negative. If it is > 0, then it will follow from Proposition 4.1(i) that $\lambda_2 < 0$, and by Proposition 4.1(iii), the μ_ω are singular. From the geometry of random SRB measures, we conclude that different initial conditions are attracted to lower dimensional sets that depend on the stimulus history. Thus even in unreliable dynamics, the responses are highly structured and far from uniformly distributed, as illustrated in Fig. 4.2b.

Note on Numerical Computation of Lyapunov Exponents. As is usually the case for concrete models, $\lambda_{\max}$ for (4.2) can only be computed numerically. As in the deterministic context, the maximum Lyapunov exponent $\lambda_{\max}$ can be computed by solving the variational equation associated with the SDE. In the examples shown here, this is done using the Milstein scheme [19].

4.2.3 *Reliability Interpretations*

RDS theory provides a useful framework for analyzing the reliability properties of specific systems. Before proceeding, however, let us discuss some issues related to the interpretation of the foregoing theory in reliability studies, as these interpretations are useful to keep in mind in what follows.

Lyapunov Exponents and Neuronal Reliability

Advantages of Lyapunov Exponents as a Theoretical and Numerical Tool. A natural question is: since reliability measures across-trial variability, why does one not simply perform a number of trials using the same input, and compute e.g. cross-trial variances? Here are some reasons for using λ_{max}—without asserting that it is better for all circumstances:

(1) λ_{max} *Is a Convenient Summary Statistic.* Consider a network of size N. Should one carry out the above procedure for a single neuron, a subset of neurons, or for all N of them? Keeping track of N neurons is potentially computationally expensive, but other possibilities (e.g., using a subset of neurons or other small sets of observables) may involve arbitrary choices and/or auxiliary parameters. A virtue of using λ_{max} is that it is a single non-random quantity, depending only on system parameters. It thus sums up the stability property of a system in a compact way, without requiring any auxiliary, tunable parameters. When plotted as a function of system parameters it enables us to view at a glance the entire landscape, and to identify emerging trends.

(2) Useful, Well-understood Mathematical Properties. A second reason is that known mathematical properties of Lyapunov exponents can be leveraged. For example, under fairly general conditions, λ_{max} varies continuously, even smoothly, with parameters. This means that if λ_{max} is found to be very negative for a system, then it is likely to remain negative for a set of nearby parameters; the size of this parameter region can sometimes be estimated with knowledge of how fast λ_{max} is changing. Knowing that a system has zero across-trial variance alone will not yield this kind of information.

(3) Computational Efficiency. Lyapunov exponents are defined in terms of infinitesimal perturbations. This means they are easily computed by simulating a single long trajectory, rather than requiring evolving an ensemble of trajectories. The latter can be quite expensive computationally.

(4) λ_{max} as a Measure of Relative Reliability. Theorem 4.1 justifies using the *sign* of the Lyapunov exponent as a way to detect reliability. Reliability, however, is often viewed in relative terms, i.e., one might view some systems as being "more reliable" or "less reliable" than others. We claim that, *all else being equal,* the *magnitude* of λ_{max} carries some useful information. There are no theorems to cite here, but ideas underlying the results discussed above tell us that other things being equal, the more negative λ_{max}, the stronger the tendencies of trajectories to coalesce, hence the more reliable the system. Conversely, for $\lambda_{max} > 0$, the larger its magnitude, the greater the instability, which often translates into greater sensitivity to initial conditions. (But there are some caveats; see below.)

In practice, this can be rather useful because for reliable systems, a variance computation will yield 0, while the magnitude of λ_{max} indicates how reliable the system is. If one were to, say, model various sources of noise by adding random terms to (4.2) that vary from trial to trial, a system with a more negative λ_{max} is likely to be have greater tolerance for such system noise.

Limitations. Having explained some of the advantages, it is important to keep in mind that Lyapunov exponents have a number of serious limitations. The first is that λ_{max} *is a long-time average.* As was pointed out in Chap. 1, Lyapunov exponents measure asymptotic stability. For one thing, this means λ_{max} may not reflect the initial response of the network upon presentation of the stimulus, which can be important as biological signal processing always occurs on finite timescales. In a bit more detail, for a system with $\lambda_{max} < 0$, one can view initial transients as an "acquisition" period, during which the system has yet to "lock on" to the signal and after which the system can respond reliably. The magnitude of λ_{max} does not give direct information about this acquisition timescale.

Second, λ_{max} *reflects only net expansion in the fastest-expanding direction.* Because of the pulsatile nature of the interactions in our networks, the action of the stochastic flow map on phase space is extremely uneven: at any one time, some degrees of freedom may be undergoing rapid change, while others evolve at a more modest pace. One therefore expects phase space expansion to occur only in certain directions at a time, and that expansion may coexist with strong phase space contraction. A positive λ_{max} only tells us that, on balance, expansion wins over contraction in *some* phase space direction. It does not give information about the relative degrees of expansion and contraction, nor does it tell us the directions in which expansion is taking place. In situations where one is interested in the response of specific "read-out" neurons in the network (see below), it may well be that phase space expansion occurs mainly in directions that do not significantly affect the reliability of these read-out neurons; λ_{max} will not be a useful indicator of relative reliability in that case. (See [22,27] for examples.)

Moreover, as discussed below, in many applications one may be interested in other notions of reliability other than neuronal reliability. While $\lambda_{max} < 0$ will typically ensure other forms of reliability, the magnitude of λ_{max} need not necessarily map onto other reliability measures in a one-to-one fashion.

Other Notions of Reliability

From the point of view of encoding information in the response of networks of biological neurons, neuronal reliability is very stringent: in a large biological network, where individual neurons and synapses may be rather noisy [13], it is clearly quite idealistic to insist that the detailed, microscopic dynamics of each neuron be reproducible across multiple trials. A less stringent way to formulate reliability is to project the dynamics onto a number $k \ll N$ of lower-dimensional signals, e.g., by applying a function Φ to the network state $\Theta(t)$, and to ask whether $\Phi(\Theta(t))$ is reproducible across trials. A neuronally reliable system is clearly guaranteed to be reliable for any choice of Φ, but a neuronally *unreliable* system may still produce reliable responses for some observables Φ. In [28], this idea was carried out in the form of "pooled-response reliability": a "pool," or subset C, of neurons is chosen, and their synaptic outputs are averaged; this average output defines a function Φ_C. A system is said to exhibit *pooled-response reliability* if the signal $\Phi_C(\Theta(t))$ is reproducible. In [28], it is found that (as one might suspect) neuronally unreliable systems often retain some degree of pooled-response reliability. Pooled-response reliability is, however, more difficult to work with, both computationally and mathematically.

There are also a number of other notions of spike-time reliability and precision in use in neuroscience that are distinct from neuronal reliability. Some of these focus on the reproducibility of *spike counts* within narrow time windows, while others focus on the variability of the spike times themselves. (See, e.g., [14] for a discussion.) A recent numerical study [22] has found that while $\lambda_{max} > 0$ does imply unreliability as indicated by other reliability measures, the quantitative behavior of these other measures are not always captured by trends in λ_{max}.

4.3 Acyclic Networks and Modular Decompositions

We now consider the neuronal reliability of *acyclic networks,* i.e., networks in which there is no feedback, and thus a well defined direction of information flow. We will show that acyclic networks are never unreliable. The proof technique will also suggest a broader class of networks that is more accessible to analysis, namely networks that admit a decomposition into modules with acyclic inter-module connections. The exposition here closely follows Part II of [27], with some details omitted.

For simplicity, we assume throughout that the stimuli are independent; it is trivial to modify the results of this section to accommodate the situation when some of them are identical to each other.

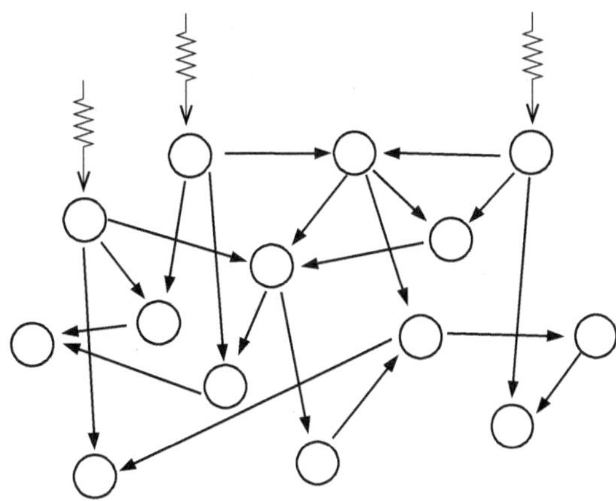

Fig. 4.3 Schematic of an acyclic network. Figure adapted from [27]

4.3.1 Skew-product Representations

We first describe the connection graph that correspond to an acyclic network. Each node of this graph, $i \in \{1, \cdots, N\}$, corresponds to an oscillator. If oscillator i provides input to oscillator j, i.e., if $a_{ij} \neq 0$, we assign a directed edge from node i to node j and write "$i \to j$". A *cycle* in such a directed graph is a sequence of nodes such that $i_1 \to i_2 \to \cdots i_k \to i_1$ for some k.

Definition 4.1. An oscillator network is *acyclic* if its connection graph has no cycles.

Given any two vertices i and j, let us write "$i \gtrsim j$" if there exists a path from i to j or if $i = j$. It is well known that if a graph is acyclic, then the relation $\gtrsim$ is a partial ordering, so that $i \gtrsim j$ and $j \gtrsim i$ if and only if $i = j$. In terms of information flow in an acyclic network, this means that for any pair of oscillators in an acyclic network, either they are "unrelated" (i.e., not comparable with respect to $\gtrsim$), or one is "upstream" from the other. Unrelated oscillators are not necessarily independent: they may receive input from the same source, for example. Acyclic networks can still be quite complex, with many branchings and recombinations; see Fig. 4.3 for an example.

Now let φ_t denote a flow on $\mathbb{T}^N$ with zero inputs, i.e., with $\epsilon_i \equiv 0$. We say φ_t *factors into a hierarchy of skew-products with one-dimensional fibers* if after relabeling the N oscillators, the following holds: for each $k = 1, \cdots, N$, there is a vector field $X^{(k)}$ on $\mathbb{T}^k$ such that if $\varphi_t^{(k)}$ is the flow generated by $X^{(k)}$, then (i) $\varphi_t^{(k)}$ describes the dynamics of the network defined by the first k oscillators and the relations among them, and (ii) $\varphi_t^{(k+1)}$ is a *skew-product* over $\varphi_t^{(k)}$, that is, the vector field $X^{(k+1)}$ on $\mathbb{T}^{k+1}$ has the form

$$X^{(k+1)}(\theta_1, \cdots, \theta_{k+1}) = (X^{(k)}(\theta_1, \cdots, \theta_k), Y_{(\theta_1, \cdots, \theta_k)}(\theta_{k+1})) \qquad (4.8)$$

where $\{Y_{(\theta_1,\cdots,\theta_k)}\}$ is a family of vector fields on S^1 parametrized by $(\theta_1,\cdots,\theta_k) \in \mathbb{T}^k$. In particular, $\varphi_t^{(N)} = \varphi_t$. In the system defined by (4.8), we refer to $\varphi_t^{(k)}$ on $\mathbb{T}^k$ as the flow on the *base*, and each copy of S^1 over $\mathbb{T}^k$ as a *fiber*.

Proposition 4.2. *The flow of every acyclic network of N oscillators with no inputs can be represented by a hierarchy of skew-products with one-dimensional fibers.*

The proof is straightforward consequence of the partial ordering $\gtrsim$; see [27].

Next we generalize the notion of skew products to acyclic networks with stimuli. Such networks can also be represented by a directed graph of the type described above, except that some of the nodes correspond to stimuli and others to oscillators. If i is a stimulus and j an oscillator, then $i \to j$ if and only if oscillator j receives stimulus i. Clearly, since no arrow can terminate at a stimulus, a network driven by stimuli is acyclic if and only if the corresponding network without stimuli is acyclic.

Consider now a single oscillator driven by a single stimulus. Let Ω denote the set of all Brownian paths defined on $[0,\infty)$, and let $\sigma_t : \Omega \to \Omega$ be the time shift. Then the dynamics of the stochastic flow discussed in Sect. 4.2.2 can be represented as the skew-product on $\Omega \times S^1$ with

$$\Phi_t : (\omega, x) \mapsto (\sigma_t(\omega), F_{0,t;\omega}(x)) .$$

Similarly, a network of N oscillators driven by q independent stimuli can be represented as a skew-product with base Ω^q (equipped with the product measure) and fibers $\mathbb{T}^N$.

Proposition 4.3. *The dynamics of an acyclic network driven by q stimuli can be represented by a hierarchy of skew-products over Ω^q with one-dimensional fibers.*

The proof is again straightforward; see [27].

4.3.2 Lyapunov Exponents of Acyclic Networks

Consider a network of N oscillators driven by q independent stimuli. As before, let $\omega \in \Omega^q$ denote a q-tuple of Brownian paths, and let $F_{0,t;\omega}$ denote the corresponding stochastic flow on $\mathbb{T}^N$. Let $\lambda_\omega(x,v)$ denote the Lyapunov exponent defined in (4.6). The following is the main result of this section:

Theorem 4.2. *Consider a network of N oscillators driven by q independent stimuli, and let μ be a stationary measure for the stochastic flow. Assume*

(a) the network is acyclic, and
(b) μ has a density on $\mathbb{T}^N$.

Then $\lambda_\omega(x,v) \le 0$ for a.e. $\omega \in \Omega^q$ and μ-a.e. x.

One way to guarantee that condition (b) is satisfied is to set ϵ_i to a very small but strictly positive value if oscillator i is not originally thought of as receiving a stimulus, so that $\epsilon_i > 0$ for all i. Such tiny values of ϵ_i have minimal effect on the network dynamics. Condition (b) may also be satisfied in many cases where some $\epsilon_i = 0$ if suitable hypoellipticity conditions are satisfied, but we do not pursue this here [36].

Before proceeding to a proof, it is useful to recall the following facts about Lyapunov exponents. For a.e. ω and μ-a.e. x, there is an increasing sequence of subspaces $\{0\} = V_0 \subset V_1 \subset \cdots \subset V_r = \mathbb{R}^N$ and numbers $\lambda_1 < \cdots < \lambda_r$ such that $\lambda_\omega(x, v) = \lambda_i$ for every $v \in V_i \setminus V_{i-1}$. The subspaces depend on ω and x, but the exponents λ_j are constant a.e. if the flow is ergodic. We call a collection of vectors $\{v_1, \cdots, v_N\}$ a *Lyapunov basis* if exactly $\dim(V_i) - \dim(V_{i-1})$ of these vectors are in $V_i \setminus V_{i-1}$. If $\{v_j\}$ is a Lyapunov basis, then for any $u, v \in \{v_j\}$, $u \neq v$,

$$\lim_{t\to\infty} \frac{1}{t} \log |\sin \angle(DF_{0,t;\omega}(x)u, DF_{0,t;\omega}(x)v)| = 0 . \tag{4.9}$$

That is, angles between vectors in a Lyapunov basis do not decrease exponentially fast; see e.g., [50] for a more detailed exposition.

Proof. Since the network is acyclic, it factors into a hierarchy of skew-products. Supposing the oscillators are labeled so that $i < j$ means oscillator i is upstream from or unrelated to oscillator j, the kth of these is a stochastic flow $F^{(k)}_{0,t;\omega}$ on $\mathbb{T}^k$ describing the (driven) dynamics of the first k oscillators. Let $\mu^{(k)}$ denote the projection of μ onto $\mathbb{T}^k$. Then $\mu^{(k)}$ is an stationary measure for $F^{(k)}_{0,t;\omega}$, and it has a density since μ has a density. We will show inductively in k that the conclusion of Theorem 4.2 holds for $F^{(k)}_{0,t;\omega}$.

First, for $k = 1$, $\lambda_\omega(x, v) \leq 0$ for a.e. ω and $\mu^{(1)}$-a.e. x. This is a consequence of Jensen's inequality; see (4.7) in Sect. 4.2.2.

Now assume we have shown the conclusion of Theorem 4.2 up to $k - 1$, and view $F^{(k)}_{0,t;\omega}$ as a skew-product over $\Omega^q \times \mathbb{T}^{k-1}$ with S^1-fibers. Choose a vector v_k in the direction of the S^1-fiber. Note that due to the skew-product structure, this direction is invariant under the variational flow $DF^{(k)}_{0,t;\omega}$. Starting with v_k, we complete a Lyapunov basis $\{v_1, \cdots, v_k\}$ at all typical points. Due to the invariance of the direction of v_k, we may once more use Jensen to show that $\lambda_\omega(x, v_k) \leq 0$ for a.e. x and ω. We next consider v_i with $i < k$. First, define the projection $\pi : \mathbb{T}^k \to \mathbb{T}^{k-1}$ onto the first k coordinates, and note that

$$|DF^{(k)}_{0,t;\omega}(x)v_i| = \frac{|\pi(DF^{(k)}_{0,t;\omega}(x)v_i)|}{|\sin \angle(v_k, DF^{(k)}_{0,t;\omega}(x)v_i)|} .$$

Due to (4.9), we have $\lambda_\omega(x, v_i) = \lim_{t\to\infty} \frac{1}{t} \log |\pi(DF^{(k)}_{0,t;\omega}(x)v_i)|$. But the skew-product structure yields $\pi(DF^{(k)}_{0,t;\omega}(x)v_i) = DF^{(k-1)}_{0,t;\omega}(\pi x)(\pi v_i)$, so by our induction hypothesis, $\lambda_\omega(x, v_i) \leq 0$. $\square$

Remark 4.1. Some remarks concerning Theorem 4.2:

(a) Our conclusion of $\lambda_{\max} \leq 0$ falls short of reliability (which corresponds to $\lambda_{\max} < 0$). This is because our hypotheses allow for freely-rotating oscillators, i.e., oscillators that are not driven by either a stimulus or another oscillator, and clearly $\lambda_{\max} = 0$ in that case. When no freely-rotating oscillators are present, typically one would expect $\lambda_{\max} < 0$. We do not have a rigorous proof, but this intuition is supported by numerical simulations.
(b) An analogous result in the context of uncertainty propagation was obtained by Varigonda et. al.; see [44].

4.3.3 Modular Decompositions

Next, we describe how the ideas from the preceding section can be used to analyze the reliability of more general networks, by decomposing large networks into smaller subunits. Consider a graph with nodes $\{1, \cdots, N\}$, and let $\sim$ be an equivalence relation on $\{1, \cdots, N\}$. The quotient graph defined by $\sim$ has as its nodes the equivalence classes $[i]$ of $\sim$, and we write $[i] \to [j]$ if there exists $i' \in [i]$ and $j' \in [j]$ such that $i' \to j'$. The following is a straightforward generalization of Proposition 4.3:

Proposition 4.4. *In a network of oscillators driven by q independent stimuli, if an equivalence relation leads to an acyclic quotient graph, then the dynamics of the network can be represented by a hierarchy of skew-products over Ω^q, with the dimensions of the fibers equal to the sizes of the corresponding equivalence classes.*

Proposition 4.4 has a natural interpretation in terms of network structure: observe that an equivalence relation on the nodes of a network partitions the nodes into distinct *modules*. Introducing directed edges between modules as above, we obtain what we call a *quotient network*. Assume this quotient network is acyclic, and let $M_1, M_2, \cdots, M_p$ be the names of the modules, ordered so that M_i is upstream from or unrelated to M_j for all $i < j$. Let k_j be the number of nodes in module M_j. For $s_i = k_1 + k_2 + \cdots + k_i$, let $F_{0,t,\omega}^{(s_i)}$ denote, as before, the stochastic flow describing the dynamics within the union of the first i modules; we do not consider $F_{0,t,\omega}^{(s)}$ except when $s = s_i$ for some i. The dynamics of the entire network can then be built up layer by layer as follows: we begin with the stochastic flow $F_{0,t,\omega}^{(k_1)}$, then proceed to $F_{0,t,\omega}^{(k_1+k_2)}$, which we view as a skew-product over $F_{0,t,\omega}^{(k_1)}$. This is followed by $F_{0,t,\omega}^{(k_1+k_2+k_3)}$, which we view as a skew-product over $F_{0,t,\omega}^{(k_1+k_2)}$, and so on.

Let $\lambda_1^{(1)}, \cdots, \lambda_{k_1}^{(1)}$ denote the Lyapunov exponents of $F_{0,t,\omega}^{(k_1)}$. Clearly, these are the Lyapunov exponents of a network that consists solely of module M_1 and the stimuli that feed into it. If $\lambda_{\max}^{(1)} \equiv \max_j \lambda_j^{(1)} > 0$, we say *unreliability is produced within* M_1. We now wish to view M_1 as part of the larger network. To do so, for

$i > 1$ let $\lambda_1^{(i)}, \cdots, \lambda_{k_i}^{(i)}$ denote the *fiber Lyapunov exponents*[3] in the skew-product representation of $F_{0,t,\omega}^{(s_i)}$ over $F_{0,t,\omega}^{(s_i-1)}$, and let $\lambda_{\max}^{(i)} = \max_j \lambda_j^{(i)}$. Then $\lambda_{\max}^{(i)} > 0$ has the interpretation that *unreliability is produced within module M_i as it operates within the larger network* (but see the remark below).

The proof of the following result is virtually identical to that of Theorem 4.2:

Proposition 4.5. *Suppose for a driven network there is an equivalence relation leading to an acyclic quotient graph. Then, with respect to any ergodic stationary measure μ, the numbers $\lambda_j^{(i)}, 1 \le i \le p, 1 \le j \le k_i$, are precisely the Lyapunov exponents of the network.*

Proposition 4.5 says in particular that if, in each of the p skew-products in the hierarchy, the fiber Lyapunov exponents are ≤ 0, i.e., if no unreliability is produced within any of the modules, then $\lambda_{\max}$ for the entire network is ≤ 0. Conversely, if unreliability is produced within any one of the modules as it operates within this network, then $\lambda_{\max} > 0$ for the entire network.

Remark 4.2. Some comments on Proposition 4.5:

(a) The idea of "upstream" and "downstream" for acyclic networks extends to modules connected by acyclic graphs, so that it makes sense to speak of a module as being downstream from another module, or a node as being downstream from another node (meaning the modules in which the nodes reside are so related).
(b) Note that any network can be decomposed into modules connected by an acyclic graph, but the decomposition may be trivial, i.e., the entire network may be a single module.[4] If the decomposition is nontrivial and $\lambda_{\max} > 0$ for the network, Proposition 4.5 enables us to localize the *source* of the unreliability, i.e., to determine in which module unreliability is produced via their fiber Lyapunov exponents. In particular, modules that are themselves acyclic cannot produce unreliability.
(c) It is important to understand that while fiber Lyapunov exponents let us assess the reliability of a module M as it operates within a larger network, i.e., as it responds to inputs from upstream modules and external stimuli, this is not the same as the reliability of M when it operates *in isolation,* i.e., when driven by external stimuli alone. Nevertheless, for many concrete examples, there is reason to think that the two types of reliability may be related. See Part II of [27] for details.

[3]The fiber exponent can be defined exactly as in (4.6), but with the tangent vector v chosen to lie in the subspace tangent to each fiber; note these subspaces are invariant due to the skew product structure.

[4]It is straightforward to show that there is always a *unique* modular decomposition connected by an acyclic graph that is "maximal" in the sense that it cannot be refined any further without introducing cycles into the quotient graph.

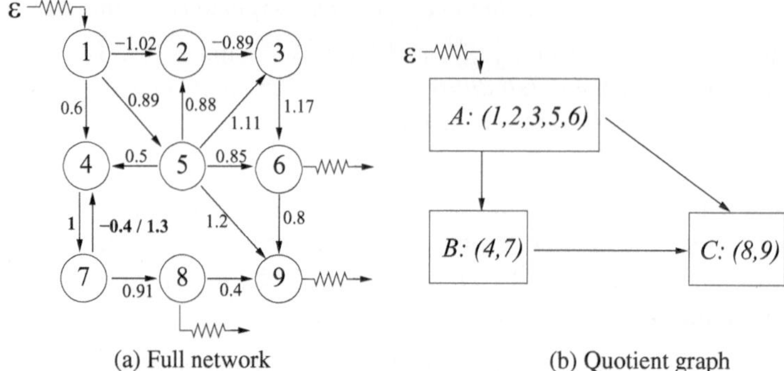

Fig. 4.4 Example of a larger network and its quotient graph. In (**a**) full network, we have labeled the edges with a sample of coupling constants;. The ω_i are drawn randomly from $[0.95, 1.05]$. Panel (**b**) quotient graph shows a modular decomposition. Figure adapted from [27]

(d) On a more practical level, the skew product structure implies that $DF_{0,t;\omega}$ is block-lower-triangular, and this fact together with Proposition 4.5 give us a more efficient way to numerically compute Lyapunov exponents of networks with acyclic quotients.

Example. To illustrate the ideas above, consider the network in Fig. 4.4a. Here, a single external input drives a network with nine nodes. The network can be decomposed into modules connected by an acyclic graph, as shown in Fig. 4.4b. Observe that Module A and Module C are both acyclic, and thus by Proposition 4.5 they cannot generate unreliability. From Proposition 4.5, it follows that whether the overall network is reliable hinges on the behavior of Module B. In [27], the reliability properties of 2-oscillator circuits like Module B are studied using ideas outlined in Sect. 4.4.1, and it is shown that one can indeed give qualitative predictions of the reliability of the network in Fig. 4.4 via modular decomposition.

4.4 Reliable and Unreliable Behavior in Recurrent Networks

Theorem 4.2 highlights the importance of feedback in reliability studies. This section examines some examples of recurrent networks.

4.4.1 Unreliability in a Two-oscillator Circuit

To better understand what can occur in a system with feedback, we have studied the simplest circuit with recurrent connections, namely a two-oscillator system driven by a single stimulus:

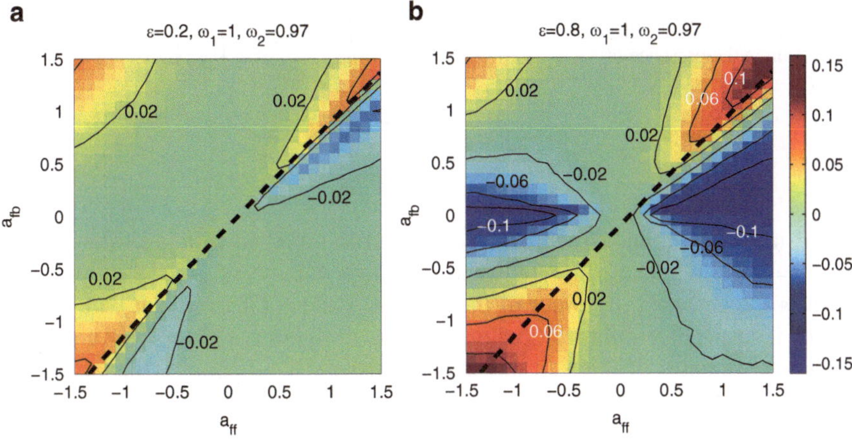

Fig. 4.5 Lyapunov exponent $\lambda_{\max}$ versus coupling strengths in the two-cell network. In all plots, we use $\omega_1 = 1$. The *dashed curve* shows the bifurcation curve $a^*_{\mathrm{fb}}(a_{\mathrm{ff}})$. (**a**) $\epsilon = 0.2$, (**b**) $\epsilon = 0.8$. Figure adapted from [27]

$\varepsilon\ I(t) \rightsquigarrow \theta_1 \;\; \overset{a_{\mathrm{ff}}}{\underset{a_{\mathrm{fb}}}{\rightleftarrows}} \;\; \theta_2$

Equation (4.2) here simplifies to

$$\begin{aligned} \dot{\theta}_1 &= \omega_1 + a_{\mathrm{fb}}\, z(\theta_1)\, g(\theta_2) + \varepsilon z(\theta_1)\, I(t)\,, \\ \dot{\theta}_2 &= \omega_2 + a_{\mathrm{ff}}\, z(\theta_2)\, g(\theta_1)\,, \end{aligned} \tag{4.10}$$

where we have written a_{ff} and a_{fb} (for "feed-forward" and "feedback") instead of a_{12} and a_{21}.

Figure 4.5 shows the maximal Lyapunov exponent $\lambda_{\max}$ as a function of a_{ff} and a_{fb}. In Fig. 4.5a, the stimulus amplitude is $\epsilon = 0.2$; in Fig. 4.5b, it is turned up to $\epsilon = 0.8$. In both figures, it can be seen that there are large regions of both reliable and unreliable behavior. There is quite a bit of structure in both plots. For example, in Fig. 4.5b, there is a region of strong reliability along the $a_{\mathrm{fb}} = 0$ axis. This is exactly as expected, since Theorem 4.2 guarantees that $\lambda_{\max} < 0$ when $a_{\mathrm{fb}} = 0$ (there are no freely-rotating oscillators here), and since $\lambda_{\max}$ should depend continuously on a_{ff} and a_{fb}, we would expect it to remain negative for some range of a_{fb}. This "valley" of negative $\lambda_{\max}$ is also present in Fig. 4.5a, but to a far lesser degree because ϵ is smaller, and the Jensen inequality argument given in Sect. 4.3.2 suggests that the greater ϵ is, the more negative $\lambda_{\max}$ should be.

A second, clearly visible structure occurs near the diagonal $\{a_{\mathrm{ff}} = a_{\mathrm{fb}}\}$ in Fig. 4.5a: one can clearly see $\lambda_{\max} > 0$ on one side and $\lambda_{\max} < 0$ on the

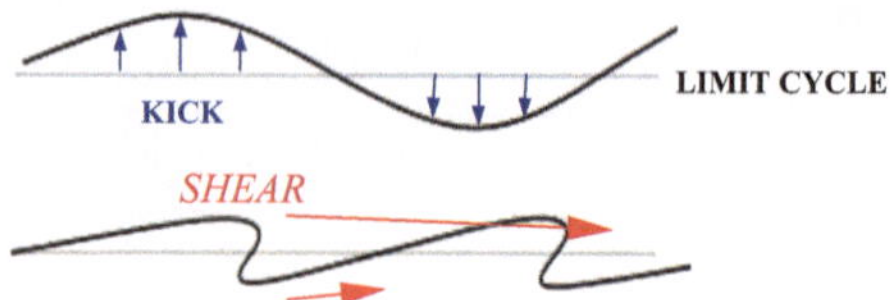

Fig. 4.6 The stretch-and-fold action of a kick followed by relaxation in the presence of shear

other; in Fig. 4.5b, this structure has expanded and merged with the valley around $\{a_{\text{fb}} = 0\}$. To make sense of what is going on there, it is necessary to first discuss the *unforced dynamics* of the two-cell system. Observe that with $\epsilon = 0$, (4.10) is just a deterministic flow on $\mathbb{T}^2$. It is straightforward to show that in the parameter regimes of interest, this 2D flow has no fixed points. One would thus expect essentially two types of behavior: either the flow has one or more limit cycles, or it is essentially quasiperiodic. In [27], an analysis of this ODE and an associated circle diffeomorphism (defined via a Poincaré section) shows that for $\epsilon = 0$ and $\omega_1 > \omega_2$, the a_{ff}-a_{fb} space is dominated by a large region, roughly equal to $\{a_{\text{fb}} > a_{\text{ff}}\}$, over which the 2D flow is quasiperiodic. As a_{fb} decreases, a bifurcation occurs in which the system acquires an attracting limit cycle. One can further prove that the corresponding critical value $a^*_{\text{fb}}(a_{\text{ff}})$ of a_{fb} occurs near a_{ff}, and that for $a_{\text{fb}} < a^*_{\text{fb}}(a_{\text{ff}})$, the behavior of the system is dominated by a large region with a single attracting limit cycle. That is to say, when $\epsilon = 0$ and $a_{\text{fb}} < a^*_{\text{fb}}(a_{\text{ff}})$, the two oscillators are phase-locked in a 1:1 resonance. The critical value $a^*_{\text{fb}}(a_{\text{ff}})$ can be numerically computed, and is shown as the dashed line in Fig. 4.5.

Since the noise amplitude in Fig. 4.5a is fairly small, the structure near the diagonal suggests that the onset of unreliability is connected with the onset of phase-locking in the unforced system. In [27], it was proposed that a dynamical mechanism called *shear-induced chaos* can explain this phenomenon, and a number of its predictions have been checked numerically there. Shear-induced chaos, a version of which was first studied numerically by Zaslavsky [52] and in a general, rigorous developed theory by Wang and Young [45–48], is a general mechanism for producing chaotic behavior when a dissipative system meeting certain dynamical conditions is subjected to external forcing. We provide a brief summary below; see [25, 26, 30] and references therein for more details.

Brief Summary of Shear-induced Chaos. First, recall that in order to generate positive Lyapunov exponents in a dynamical system, it is necessary to have a way of stretching and folding phase space. Shear-induced chaos is a general mechanism for accomplishing this using two main ingredients:

(i) an attracting limit cycle where the surrounding flow exhibits *shear*, and
(ii) a source of external perturbation that forces trajectories off the limit cycle.

By shear, we mean a differential in the velocity as one moves transversally to the limit cycle, as illustrated in the bottom panel of Fig. 4.6. The perturbation in (ii) can take a variety of forms (deterministic or random), the simplest being a sequence of brief "kicks" applied periodically at a fixed time interval; such periodic kicks can be modeled by applying a fixed (deterministic) mapping at a fixed interval.

Figure 4.6 illustrates the basic geometric ideas: imagine a set of initial conditions along the limit cycle, and that at a certain time a single kick is applied, moving most of the initial conditions off the cycle. Because the limit cycle is attracting, the curve of trajectories will fall back toward the cycle as they evolve. But because of shear, the curve will be stretched and folded as the trajectories fall back toward the cycle. If this process is iterated periodically, it is easy to see that it can lead to the formation of Smale horseshoes, which is well known to be a source of complex dynamical behavior in deterministic dynamical systems. However, horseshoes can coexist with attracting fixed points, so that the associated chaotic behavior may only be transient, i.e., the Lyapunov exponents may still be < 0. In [45–48], it is proved that for periodically-kicked dissipative oscillators, sustained chaotic behavior characterized by positive Lyapunov exponents, exponential decay of correlations, and the existence of SRB measures (see Sect. 4.2.2) are guaranteed whenever certain conditions are met.[5] In non-technical terms, the conditions are that the limit cycle possesses sufficient shear, that the damping is not too strong, and that the perturbations are sufficiently large and avoids certain "bad" phase space directions (associated with the "strong stable manifolds" of the limit cycle).

The theorems in [45–48] apply to periodically-kicked oscillators, and the proof techniques do not carry over to the stochastic setting. Nonetheless, the underlying ideas suggest that the shear-induced chaos, as a general dynamical mechanism for producing instabilities, is valid for other types of forcings as well, including stochastic forcing. A systematic numerical study [25] has provided evidence supporting this view, and a recent analysis of a specific SDE with shear has found positive Lyapunov exponents [8].

As explained in detail elsewhere (see, e.g., [25]), shear-induced chaos makes a number of testable qualitative predictions. First, the more shear is present in the vicinity of a limit cycle, the more effective the stretching is, so that all else being equal, increasing shear would lead to a more positive Lyapunov exponent. Similarly, if the limit cycle were strongly attracting, any perturbations would be quickly damped out, reducing the amount of phase space stretching and decreasing the exponent. Finally, as mentioned above, the direction of kicking relative to the geometry of the flow is also important. If these conditions are met, then a system would have the tendency to generate chaotic behavior; the exact nature of the external perturbations (provided they are sufficiently strong) will affect quantitative details, but not gross qualitative features.

Returning now to the two-oscillator system, it has been shown (see Part I of [27]) that as the limit cycle emerges from the bifurcation at $a_{\text{fb}} \sim a^*_{\text{fb}}(a_{\text{ff}})$, there is a great deal of shear in the vicinity of the limit cycle. Moreover, the forcing term in (4.10) is such that it can take advantage of the shear to produce stretching and folding.

[5]These results have been extended to certain nonlinear parabolic PDEs [32] and periodically-kicked homoclinic loops [37].

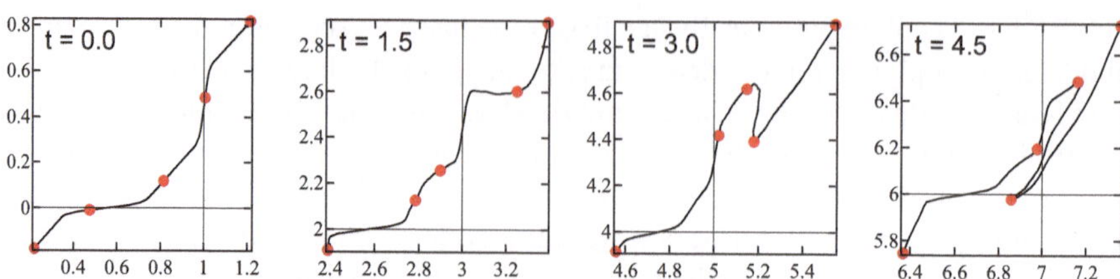

Fig. 4.7 Folding action caused by white noise forcing and shear near the limit cycle (with $a_{fb} > a_{fb}^*$). At $t = 0$, the *curve* shown is the lift of the limit cycle γ to $\mathbb{R}^2$. The remaining panels show lifts of the images $F_{0,t;\omega}(\gamma)$ at increasing times. The parameters are $\omega_1 = 1$, $\omega_2 = 1.05$, $a_{ff} = 1$, $a_{fb} = 1.2$, and $\epsilon = 0.8$. Note that it is not difficult to find such a fold in simulations: very roughly, 1 out of 4 realizations of forcing gives such a sequence for $t \in [0, 5]$. Figure adapted from [27]

Figure 4.7 illustrates how the folding and stretching occurs. These numerical results show that shear-induced stretching and folding do occur in the system (4.10).

While in this context, there are no theorems linking shear-induced folding with positive exponents, the various features seen around the diagonal in Fig. 4.5a can be readily explained using the ideas of shear-induced chaos. First, consider the phase-locked side of the a_{fb}^*-curve, i.e., $a_{fb} < a_{fb}^*$. Observe that as a_{fb} decreases, λ_{max} becomes more negative for some range of a_{fb}. This is consistent with increasing damping as the limit cycle (initially weak right after the bifurcation) becomes more strongly attracting. As we move farther away from the a_{fb}^*-curve still, λ_{max} increases and remains for a large region close to 0. Intuitively, this is due to the fact that for these parameters the limit cycle is very robust. The damping is so strong that the forcing cannot (usually) deform the limit cycle appreciably before it returns near its original position. That is to say, the perturbations are negligible, and the value of λ_{max} is close to the value for the unforced flow (which for a limit cycle is always $\lambda_{max} = 0$). On the other side of the a_{fb}^*-curve, where the system is essentially quasiperiodic, regions of unreliability are clearly visible. These regions in fact begin slightly on the phase-locked side of the curve, where a weakly attractive limit cycle is present. The fact that λ_{max} is more positive before the limit cycle is born than after can be attributed to the weaker-to-nonexistent damping before its birth. Thus, the general progression in Fig. 4.5a of λ_{max} from roughly 0 to definitively negative to positive as we cross the a_{fb}^*-curve from below consistent with the mechanism of shear-induced chaos.

In Fig. 4.5b, where the stimulus amplitude is increased to $\epsilon = 0.8$, the picture one obtains clearly continues some of the trends seen in Fig. 4.5a: there are still regions of $\lambda_{max} > 0$ near the diagonal, and a valley of $\lambda_{max} < 0$ around $\{a_{fb} = 0\}$. But while the valley around $\{a_{fb} = 0\}$ can be explained on the basis of Theorem 4.2 (which is valid regardless of the magnitude of ϵ), the behavior around the diagonal now likely involves more global effects as the system takes larger excursions from the limit cycle due to the increased forcing amplitude.

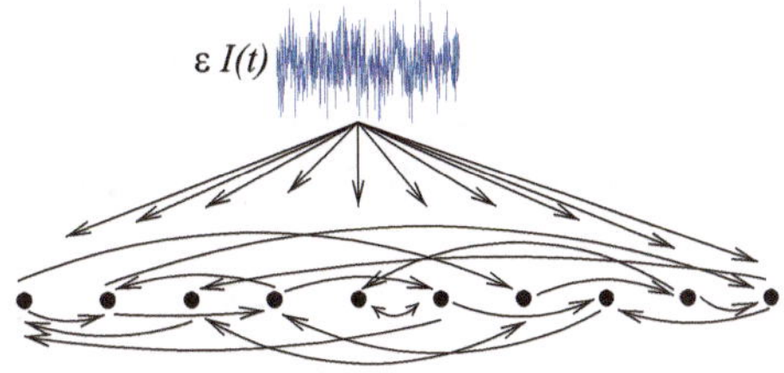

Fig. 4.8 Schematic of a single-layer network

4.4.2 Single-Layer Networks

Lest the reader think that all recurrent networks are unreliable, for our last example we examine a large recurrent network that is strongly, robustly reliable. The network is shown schematically in Fig. 4.8. It is a sparsely-coupled recurrent network in which each neuron receives exactly κ inputs from other neurons (usually κ is 10 or 20 % of the network size N). The intrinsic frequencies ω_i are drawn randomly from $[1-\rho, 1+\rho]$, and the nonzero coupling constants a_{ji} are drawn randomly from $[a(1-\rho), a(1+\rho)]$; the heterogeneity parameter ρ is taken to be 0.1. All oscillators receive the same external drive of amplitude ϵ. This architecture is motivated by layered models often found in neuroscience.

The Lyapunov exponent $\lambda_{\max}$ of such a single-layer network is plotted against $A := \kappa \cdot a$ in Fig. 4.9a. Two values of ϵ are used. As can be seen, as ϵ increases, $\lambda_{\max}$ decreases. This is not unexpected: just as for the single oscillator in Sect. 4.2.2, we expect the magnitude of $\lambda_{\max}$ to increase with increasing ϵ (more on this below). Next, observe that $\lambda_{\max}$ is most negative when $A = 0$. This is also expected: with no coupling, the "network" is just a collection of uncoupled oscillators, each of which has $\lambda_{\max} < 0$ by Jensen's inequality; by continuity, this persists for a range of A. Finally, as $|A|$ increases, $\lambda_{\max}$ increases, suggesting that network interactions generally have a destabilizing effect in this network. However, even when A is quite large,[6] $\lambda_{\max}$ remains < 0 for $\epsilon = 2.5$.

In [28], the following qualitative explanation was proposed: consider first the system with $a = \rho = 0$. In this case, we have a collection of uncoupled, identical theta neurons. Since $\lambda_{\max} < 0$ for single theta neurons (Sect. 4.2.2), the ensemble will become entrained to the common input, and thus synchronize with each other. If we now allow slightly nonzero couplings and heterogeneous frequencies, i.e., $a \approx 0$ and $\rho \approx 0$, then by continuity we expect $\lambda_{\max}$ to still be < 0, and that the oscillators will remain nearly synchronized much of the time. Recall now that in (4.2), the phase response curve $z(\theta) = O(\theta^2)$ for $\theta \sim 0$. That is, around the time a neuron spikes, it is very insensitive to its inputs. Thus the near-synchrony of the neurons will lead to an attenuation in the *effective* strength of the coupling. This provides

[6]A rough estimate shows that when $A = 2$, each kick should be sufficient to drive the oscillator roughly 1/3 of the way around its cycle.

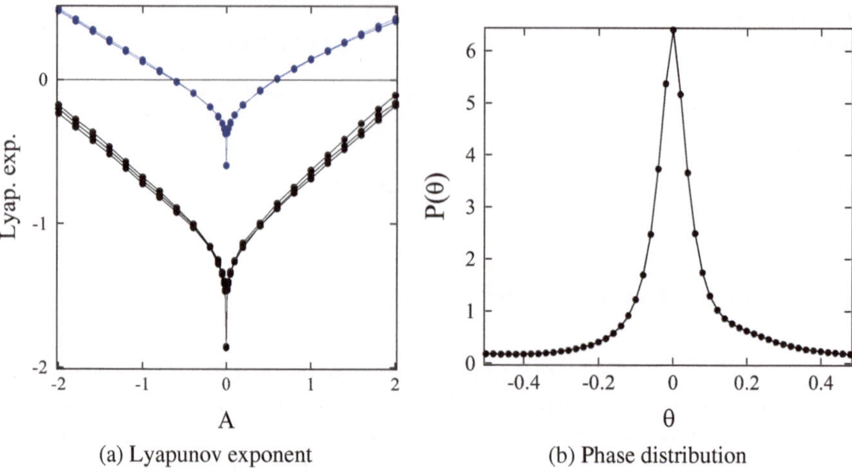

(a) Lyapunov exponent (b) Phase distribution

Fig. 4.9 Behavior of single-layer network. In (**a**), we plot the maximum Lyapunov exponent $\lambda_{\max}$ versus $A = \kappa \cdot a$; the parameters are $N = 100$, $\kappa = 10$, and $\epsilon = 1.5$ (*upper curve*) and 2.5 (*bottom*). Panel (**b**) shows the phase distribution of neurons at the arrival of synaptic impulses for a single-layer network with $N = 100$. Figure adapted from [28]

an explanation for why A can be made so large in Fig. 4.9 without causing $\lambda_{\max}$ to become positive.

This explanation also leads to a number of testable predictions. First, if we examine the phases of the oscillators when a spike arrives, we should observe a highly-clustered distribution. This has been checked numerically; an example is shown in Fig. 4.9b. Second, anything that makes it harder for the neurons to synchronize should lead to an *increase* in $\lambda_{\max}$. For example, if we were to increase the amount of heterogeneity ρ in the system, $\lambda_{\max}$ should also increase. This is corroborated by the following results:

Heterogeneity ρ	0	0.01	0.1	0.3
$\lambda_{\max}$	−1.9	−1.7	−0.70	−0.18

As the degree of heterogeneity in the network increases, the oscillators become harder to entrain, and accordingly $\lambda_{\max}$ increases as well.

Concluding Discussion

The work surveyed here has shown that the reliability of neuronal networks can be fruitfully formulated and studied within the framework of random dynamical systems. Particularly useful is the maximum Lyapunov exponent of a system, as a summary statistic for detecting reliability in numerical simulations. The results reviewed here show that:

(a) *RDS theory, in particular the maximum Lyapunov exponent* $\lambda_{\max}$ *and results linking* $\lambda_{\max}$ *to the structure of sample measures, provide a useful framework for studying reliability.*
(b) *Acyclic networks of pulse-coupled theta neurons are never unreliable.* This result highlights the importance of feedback in producing unreliability. The underlying ideas also generalize to the setting of modules connected by acyclic graphs, providing a way to analyze larger networks via modular decompositions.
(c) *Recurrent networks can be reliable or unreliable.* In particular, a system as small as a two-oscillator circuit can become unreliable; moreover, there is evidence that unreliability in the two-cell circuit can be explained using the ideas of shear-induced chaos. At the same time, large recurrent networks can be robustly reliable as a result of (i) entrainment to a common input, and (ii) the strong refractory effect of the phase response of Type I theta neurons.

There are a number of additional issues relevant to neuroscience that have not been discussed here. We highlight three that are perhaps the most relevant from a biological point of view:

Noise. Neurons and synapses are well known to behave in a noisy fashion, both in vivo and in vitro. In the context of our model, one can represent the effects of noise by adding stochastic forcing terms that vary from trial to trial. While such a model no longer fits exactly in the framework of standard RDS theory (though the theory of RDS with inputs from Chap. 2 may be relevant), it can be studied numerically via direct measures of reliability such as cross-trial variance. In [28], the theoretical ideas surveyed here (Lyapunov exponents, random attractors) are used to carry out an analysis of the effects of noise on reliability, and to explain the different effects of correlated versus independent noise.

Pooled-response Reliability. As mentioned in Sect. 4.2.2, even when $\lambda_{\max} > 0$ there can still be a great deal of structure in the sample distribution μ_ω, suggesting that by suitable projections or pooling of neuronal outputs, one can obtain responses that have some degree of reliability; this has been studied numerically in [28], and has received some attention in the experimental literature [9].

Beyond Lyapunov Exponents. As previously mentioned, Lyapunov exponents do not always capture what one wants to know about neuronal response, and other aspects of sample measures and random attractors may be more relevant in studies of network reliability. Exactly which aspects matter depends, of course, on the application at hand. Some steps in this direction have been taken in [22].

Acknowledgements The work described in this review were supported in part by the Burroughs-Wellcome Fund (Eric Shea-Brown) and the NSF (Lai-Sang Young and Kevin K Lin).

References

1. L. Arnold, *Random Dynamical Systems* (Springer, New York, 2003)
2. W. Bair, E. Zohary, W.T. Newsome, Correlated firing in macaque visual area MT: time scales and relationship to behavior. J. Neurosci. **21**(5), 1676–1697 (2001)
3. P.H. Baxendale, Stability and equilibrium properties of stochastic flows of diffeomorphisms, in *Progress in Probability*, vol. 27 (Birkhauser, Basel, 1992)
4. M. Berry, D. Warland, M. Meister, The structure and precision of retinal spike trains. Proc. Natl. Acad. Sci. **94**, 5411–5416 (1997)
5. E. Brown, P. Holmes, J. Moehlis, Globally coupled oscillator networks, in *Problems and Perspectives in Nonlinear Science: A celebratory volume in honor of Lawrence Sirovich*, ed. by E. Kaplan, J.E. Marsden, K.R. Sreenivasan (Springer, New York, 2003), pp. 183–215
6. H.L. Bryant, J.P. Segundo, Spike initiation by transmembrane current: a white-noise analysis. J. Physiol. **260**, 279–314 (1976)
7. R. de Reuter van Steveninck, R. Lewen, S. Strong, R. Koberle, W. Bialek, Reproducibility and variability in neuronal spike trains. Science **275**, 1805–1808 (1997)
8. R.E.L. Deville, N. Sri Namachchivaya, Z. Rapti, Stability of a stochastic two-dimensional non-Hamiltonian system. SIAM J. Appl. Math. **71**, 1458–1475 (2011)
9. A.S. Ecker, P. Berens, G.A. Keliris, M. Bethge, N.K. Logothetis, A.S. Tolias. Decorrelated neuronal firing in cortical microcircuits. Science **327**, 584–587 (2010)
10. J.-P. Eckmann, D. Ruelle, Ergodic theory of chaos and strange attractors. Rev. Mod. Phys. **57**, 617–656 (1985)
11. G.B. Ermentrout, Type I membranes, phase resetting curves and synchrony. Neural Comput. **8**, 979–1001 (1996)
12. G.B. Ermentrout, D. Terman, *Foundations of Mathematical Neuroscience* (Springer, Berlin, 2010)
13. A.A. Faisal, L.P.J. Selen, D.M. Wolpert, Noise in the nervous system. Nat. Rev. Neurosci. **9**, 292–303 (2008)
14. J.-M. Fellous, P.H.E. Tiesinga, P.J. Thomas, T.J. Sejnowski, Discovering spike patterns in neuronal responses. J. Neurosci. **24**, 2989–3001 (2004)
15. D. Goldobin, A. Pikovsky, Antireliability of noise-driven neurons. Phys. Rev. E **73**, 061906-1–061906-4 (2006)
16. J. Hunter, J. Milton, P. Thomas, J. Cowan, Resonance effect for neural spike time reliability. J.Ñeurophysiol. **80**, 1427–1438 (1998)
17. P. Kara, P. Reinagel, R.C. Reid, Low response variability in simultaneously recorded retinal, thalamic, and cortical neurons. Neuron **27**, 636–646 (2000)
18. Yu. Kifer, *Ergodic Theory of Random Transformations* (Birkhauser, Basel, 1986)
19. P.E. Kloeden, E. Platen, *Numerical Solution of Stochastic Differential Equations* (Springer, New York, 2011)
20. E. Kosmidis, K. Pakdaman, Analysis of reliability in the Fitzhugh–Nagumo neuron model. J.C̃omput. Neurosci. **14**, 5–22 (2003)
21. H. Kunita, *Stochastic Flows and Stochastic Differential Equations*. Cambridge Studies in Advanced Mathematics, vol. 24 (Cambridge University Press, Cambridge, 1990)
22. G. Lajoie, K.K. Lin, E. Shea-Brown, Chaos and reliability in balanced spiking networks with temporal drive. Phys. Rev. E **87** (2013)
23. Y. Le Jan. Équilibre statistique pour les produits de difféomorphismes aléatoires indépendants. Ann. Inst. H. Poincaré Probab. Stat. **23**(1), 111–120 (1987)
24. F. Ledrappier, L.-S. Young, Entropy formula for random transformations. Probab. Theory Relat. Fields **80**, 217–240 (1988)
25. K.K. Lin, L.-S. Young, Shear-induced chaos. Nonlinearity **21**, 899–922 (2008)
26. K.K. Lin, L.-S. Young, Dynamics of periodically-kicked oscillators. J. Fixed Point Theory Appl. **7**, 291–312 (2010)

27. K.K. Lin, E. Shea-Brown, L.-S. Young, Reliability of coupled oscillators. J. Nonlin. Sci. **19**, 497–545 (2009)
28. K.K. Lin, E. Shea-Brown, L.-S. Young, Spike-time reliability of layered neural oscillator networks. J. Comput. Neurosci. **27**, 135–160 (2009)
29. K.K. Lin, E. Shea-Brown, L.-S. Young, Reliability of layered neural oscillator networks. Commun. Math. Sci. **7**, 239–247 (2009)
30. K.K. Lin, K.C.A. Wedgwood, S. Coombes, L.-S. Young, Limitations of perturbative techniques in the analysis of rhythms and oscillations. J. Math. Biol. **66**, 139–161 (2013)
31. T. Lu, L. Liang, X. Wang, Temporal and rate representations of time-varying signals in the auditory cortex of awake primates. Nat. Neurosci. **4**, 1131–1138 (2001)
32. K. Lu, Q. Wang, L.-S. Young, *Strange Attractors for Periodically Forced Parabolic Equations*. Memoirs of the AMS, American Mathematical Society, (Providence, Rhode Island, 2013)
33. Z. Mainen, T. Sejnowski, Reliability of spike timing in neocortical neurons. Science **268**, 1503–1506 (1995)
34. G. Murphy, F. Rieke, Network variability limits stimulus-evoked spike timing precision in retinal ganglion cells. Neuron **52**, 511–524 (2007)
35. H. Nakao, K. Arai, K. Nagai, Y. Tsubo, Y. Kuramoto. Synchrony of limit-cycle oscillators induced by random external impulses. Phys. Rev. E **72**, 026220-1–026220-13 (2005)
36. D. Nualart, *The Malliavin Calculus and Related Topics* (Springer, Berlin, 2006)
37. W. Ott, Q. Wang, Dissipative homoclinic loops of two-dimensional maps and strange attractors with one direction of instability. Commun. Pure Appl. Math. **64**, 1439–1496 (2011)
38. K. Pakdaman, D. Mestivier, External noise synchronizes forced oscillators. Phys. Rev. E **64**, 030901–030904 (2001)
39. J. Ritt, Evaluation of entrainment of a nonlinear neural oscillator to white noise. Phys. Rev. E **68**, 041915–041921 (2003)
40. J. Teramae, T. Fukai, Reliability of temporal coding on pulse-coupled networks of oscillators. arXiv:0708.0862v1 [nlin.AO] (2007)
41. J. Teramae, D. Tanaka, Robustness of the noise-induced phase synchronization in a general class of limit cycle oscillators. Phys. Rev. Lett. **93**, 204103–204106 (2004)
42. A. Uchida, R. McAllister, R. Roy, Consistency of nonlinear system response to complex drive signals. Phys. Rev. Lett. **93**, 244102 (2004)
43. B.P. Uberuaga, M. Anghel, A.F. Voter, Synchronization of trajectories in canonical molecular-dynamics simulations: observation, explanation, and exploitation. J. Chem. Phys. **120**, 6363–6374 (2004)
44. S. Varigonda, T. Kalmar-Nagy, B. LaBarre, I. Mezić, Graph decomposition methods for uncertainty propagation in complex, nonlinear interconnected dynamical systems, in *43rd IEEE Conference on Decision and Control*, Paradise Island, Bahamas (2004)
45. Q. Wang, L.-S. Young, Strange attractors with one direction of instability. Commun. Math. Phys. **218**, 1–97 (2001)
46. Q. Wang, L.-S. Young, From invariant curves to strange attractors. Commun. Math. Phys. **225**, 275–304 (2002)
47. Q. Wang, L.-S. Young, Strange attractors in periodically-kicked limit cycles and Hopf bifurcations. Commun. Math. Phys. **240**, 509–529 (2003)
48. Q. Wang, L.-S. Young, Toward a theory of rank one attractors. Ann. Math. **167**, 349–480 (2008)
49. A. Winfree, *The Geometry of Biological Time* (Springer, New York, 2001)
50. L.-S. Young, Ergodic theory of differentiable dynamical systems, in *Real and Complex Dynamics*, NATO ASI Series (Kluwer, Dordrecht, 1995), pp. 293–336
51. L.-S. Young, What are SRB measures, and which dynamical systems have them? J. Stat. Phys. **108**(5), 733–754 (2002)
52. G. Zaslavsky, The simplest case of a strange attractor. Phys. Lett. **69A**(3), 145–147 (1978)
53. C. Zhou, J. Kurths, Noise-induced synchronization and coherence resonance of a Hodgkin-Huxley model of thermally sensitive neurons. *Chaos* **13**, 401–409 (2003)

Chapter 5
Coupled Nonautonomous Oscillators

Philip T. Clemson, Spase Petkoski, Tomislav Stankovski, and Aneta Stefanovska

Abstract First, we introduce nonautonomous oscillator—a self-sustained oscillator subject to external perturbation and then expand our formalism to two and many coupled oscillators. Then, we elaborate the Kuramoto model of ensembles of coupled oscillators and generalise it for time-varying couplings. Using the recently introduced Ott-Antonsen ansatz we show that such ensembles of oscillators can be solved analytically. This opens up a whole new area where one can model a virtual physiological human by networks of networks of nonautonomous oscillators. We then briefly discuss current methods to treat the coupled nonautonomous oscillators in an inverse problem and argue that they are usually considered as stochastic processes rather than deterministic. We now point to novel methods suitable for reconstructing nonautonomous dynamics and the recently expanded Bayesian method in particular. We illustrate our new results by presenting data from a real living system by studying time-dependent coupling functions between the cardiac and respiratory rhythms and their change with age. We show that the well known reduction of the variability of cardiac instantaneous frequency is mainly on account of reduced influence of the respiration to the heart and moreover the reduced variability of this influence. In other words, we have shown that the cardiac function becomes more autonomous with age, pointing out that nonautonomicity and the ability to maintain stability far from thermodynamic equilibrium are essential for life.

The authors contributed equally to this work and have therefore decided for an alphabetical order.

P.T. Clemson · S. Petkoski · T. Stankovski · A. Stefanovska (✉)
Physics Department, Lancaster University, Lancaster, UK
e-mail: aneta@lancaster.ac.uk

P.E. Kloeden and C. Pötzsche (eds.), *Nonautonomous Dynamical Systems in the Life Sciences*, Lecture Notes in Mathematics 2102, DOI 10.1007/978-3-319-03080-7_5,

Keywords Nonautonomous coupled oscillators • Networks of oscillators • Coupling function • Dynamical Bayesian inference • Kuramoto model • Time series analysis • Cardio-respiratory interactions • Ageing

5.1 Introduction

Searching for the basic particle of a living system one arrives at two important molecules: DNA, which specifies the structural property of a cell, and ATP, which serves as the primary energy currency of the cell. While the role of DNA has been excessively studied over the last decades, the role of ATP is still largely unknown.

To understand the role of ATP a dynamical approach is needed. Moreover, since ATP enables a cell to exchange energy and matter with its surroundings, the theory of thermodynamically open systems must be applied. It is in this endeavour that one unavoidably identifies the need to mathematically describe life as being composed of linked nonautonomous systems. Furthermore, recent experiments have shown that the activity of mitochondria, the main producers of ATP in human cells, is oscillatory [46], thus pointing to the oscillatory nature of the underlying dynamics.

The link between the mitochondrial function and disfunction on one hand and the cardiovascular system on the other, has been frequently reported in recent years. For example, Dai et al. [15] review the evidence supporting the role of mitochondrial oxidative stress, mitochondrial damage and biogenesis as well as the crosstalk between mitochondria and cellular signaling in cardiac and vascular ageing. Dromparis and Michelakis [17] highlight the profound impact of mitochondria on vascular function in both health and disease based on their role in integrating metabolic, oxygen, or external signals with inputs from other cellular organelles, as well as local and systemic signals. Given that the mitochondrial activity has been shown to be oscillatory [46], the need for an appropriate theory of nonautonomous oscillatory systems in further studies of mitochondrial dynamics and its connection with the cardiovascular dynamics is obvious.

Oscillatory activity has long been identified on the higher level of organisation in living systems (e.g. [27]). For example, it has been shown that the cardiovascular system, which can be perceived as a system that transports substances needed for the generation of ATP, is characterised by many interacting oscillatory processes [81, 82]. Similarly, the neuronal activity controlled by the brain, which can be perceived as a control system of the information transfer between the ensembles of cells and ensembles of systems of cells in the body, is characterised by many interacting waves [12].

The accumulated evidence of oscillatory dynamics at various levels of complexity in living systems is now imposing an urgent need for a theory of nonautonomous coupled oscillators and ensembles of coupled nonautonomous oscillators. In this chapter, we present a new framework for nonautonomous oscillatory systems. In Sect. 5.2 we define the basic properties of a nonautonomous self-sustained oscillator and then discuss the properties that arise in case of two coupled and interacting nonautonomous oscillators. In the case of weakly interacting oscillators, application of the phase approximation leads to the Kuramoto model (KM) [45] which has been extensively used to describe globally coupled phase oscillators. In Sect. 5.3 we show how the KM can be expanded to include time-varying parameters [62]. We also review current efforts to explain certain forms of nonautonomicity. In Sect. 5.4 we review methods used in the inverse approach to nonautonomous dynamics. Namely, with the rapid developments of sensors and computational facilities we are now able to collect—most often non-invasively—time series of almost any dynamical processes of interest. In the last decades, various methods have been proposed for extracting information about the workings of the underlying dynamics [1,7,19]. This is usually achieved not through experimental perturbations, but by observing the spontaneous dynamics of a system over a finite time period. We now point to novel methods suitable for the reconstruction of nonautonomous dynamics, in particular the recently expanded Bayesian method [18,78,79]. In the final Sect. 5.5, we present data from a real living system. We study the effect of ageing on cardio-respiratory interactions and extract time-dependent coupling functions. We show that the well known reduction of the variability of instantaneous cardiac frequency with age is mainly on account of the reduced influence of the respiration on the heart and, moreover, the reduced variability of this influence. In other words, we have shown that the cardiac function is becoming more autonomous with age. We therefore emphasise that nonautonomicity and the ability to sustain a stable functioning far from thermodynamic equilibrium are vital for living systems.

5.2 Coupled Nonautonomous Self-sustained Oscillators with Time-varying Couplings and Frequencies

5.2.1 Introduction

Physicists usually try to study isolated systems, free from external influences, that can be described precisely by well-defined equations. In practice, of course, this ideal is seldom completely realised and it is normally necessary to take account of a variety of external perturbations. When these perturbations are parametric, i.e. tending to alter the parameters or even the functional relationships of the modeling equations, a wide range of often counter-intuitive effects can arise. These include the occurrence of noise-induced phase transitions [30], or spontaneous shifts in synchronization ratio in cardio-respiratory interactions [85]. Consequently, particular care is needed in analysing the underlying physics. Such phenomena are especially important in relation to oscillatory systems, whose frequency or amplitude may be modified by external fields. One approach to the problem involves focusing on the idealised model system but, at the same time, accepting that it is *nonautonomous*, i.e. that one or more of its parameters may be subject to external modulation. Without some knowledge of the form of modulation, little more can be said other than admitting to the corresponding inherent uncertainty in the analysis. It often happens, however, that the external field responsible for the nonautonomicity may itself be deterministic, e.g. periodic. At the other extreme, it might be either chaotic or stochastic. In each of these cases, it is possible to perform a potentially useful analysis.

5.2.2 Nonautonomous Systems: A Physics Perspective

Nonautonomous (Greek: *auto*-"self" + *nomos*-"law") systems (NA) are those whose law of behaviour is influenced by external forces. From a dynamical point of view, a set of differential equations is nonautonomous if they include an explicit time-dependence (TD). The external influence can be formulated in different ways, for instance, it could be a periodic force, a quasi-periodic function or a noisy process. It could also affect the systems in a various ways i.e. it might be additive, could enter in the definition of a parameter, or might modulate the functional relationships that define the interactions between systems. By focussing our attention on only one or a few components of a high dimensional autonomous dynamical system, we are actually dealing with nonautonomous differential equations because of the time-variability embedded within their interactions with the rest of the system.

Often in the literature, and especially in inverse problems, the nonautonomous dynamics have been associated or referred to as non-stationary. This poses a significant confusion, misunderstanding and even spurious results. Therefore, we first outline the differences between non-stationary statistics and nonautonomous dynamics. Stationarity is a statistical property of the output signal and as such is characterized by the application of tools from statistical mechanics [86]. The definition of stationarity is closely related to the time of observation, especially in inverse problems where what seems to be non-stationary on short time scales can be stationary on longer ones. The solution of an autonomous dynamical systems $\dot{\mathbf{x}}(t) = f(\mathbf{x}(t))$ depends only on the time difference $(t - t_0)$ between the current state $\mathbf{x}(t)$ and the initial condition $\mathbf{x}(t_0)$. It therefore follows that the statistical behaviour of a bounded-space solution, if far enough from the initial condition, must be time-independent. In contrast, when a process is bounded and non-stationary, then it is clearly impossible to represent the driving dynamics with autonomous equations. For this reason, nonautonomous dynamics $\dot{\mathbf{x}}(t) = f(\mathbf{x}(t), t)$ must constitute the core mechanism underlying a non-stationary output signal. On the other hand, for an appropriate time-dependence of the external dynamical field, it is possible that nonautonomous dynamics may be perfectly stationary in the statistical sense. Hence nonautonomous dynamics can act as a functional "generator" for both stationary and non-stationary dynamics.

Nonautonomous dynamical systems have attracted considerable attention from mathematicians, with much effort being expended on the development of a solid formalism [43, 66]. Chapter 1 succinctly outlines the main concepts from mathematical theory of deterministic nonautonomous dynamical systems, describing in detail nonautonomous differential equations, stability and bifurcation theory, and the nature of nonautonomous attractors. The treatment of pullback attractors with a fixed target set and progressively earlier starting time $t_0 \to -\infty$ gives additional insight for the analysis of nonautonomous attractors, which is greatly important for many physical systems. The proposed theory has been found useful in number of applications, including switching and control systems [41] and complete (dissipative) synchronization [40,42]. This recently established mathematical theory promises many applications in more complex nonautonomous systems.

In the physics community, there seems to have been a degree of reluctance to address the problem as it really is and, in general, the issue has been sidestepped by reducing the nonautonomous equation to an autonomous one by the addition of an extra variable to play the role of time-dependence in $\mathbf{f}(\mathbf{x}(t), t)$. This approach is not mathematically justified because the new dimension is not bounded in time (as $t \to \infty$), and because attractors cannot be defined easily. Certain transformations can be employed to bound the extra dimension, but this approach does not work in the general case. However, the procedure of reduction to the autonomous form has been safely employed in some situations—especially in studies closely related with experiments, where the dynamical behaviour is observed for finite length of time. One particular example of this kind is the geometric singular perturbation theory applied to slow-fast dynamics—an approach discussed succinctly in relation to canards in Chap. 3.

There are two cases for the treatment of nonautonomous dynamics that recur in the physics literature: (1) where the dynamical field is a periodic function of t (i.e. $\mathbf{x} = \mathbf{f}(\mathbf{x}, \sin(t))$, often referred as an "oscillating external perturbation"); and (2) when the dynamical field is stochastic (the noise being the time-dependent part). Figure 5.1a and b illustrate a simple example of such nonautonomous systems. The first case is obviously one where an extra variable is often substituted, and the latter case involves the application of the mathematical instruments of stochastic dynamics. These can be seen as the two limiting-cases of an external perturbation from a system with either one degree of freedom, or with an infinite number of degrees of freedom. In between these two extremes there is a continuum of cases where the time dependence is neither precisely periodic, nor purely stochastic. An example of an intermediate case of this kind would be a dynamical system $\dot{\mathbf{x}} = \mathbf{f}(\mathbf{x}(t), g(t))$, where $g(t)$ is the n-th component of a chaotic (low dimensional) dynamical system.

5.2.3 Single Nonautonomous Self-sustained Oscillator

The nonautonomous systems constitute a vast and very general class of systems. Motivated largely by biological systems, here we will concentrate on nonautonomous self-sustained oscillatory systems. To date, limited work has been done in this area. Anishchenko et al. [6] mainly focus on the case in which limit cycles are induced by external nonautonomous fields. In what follows, however, we direct our

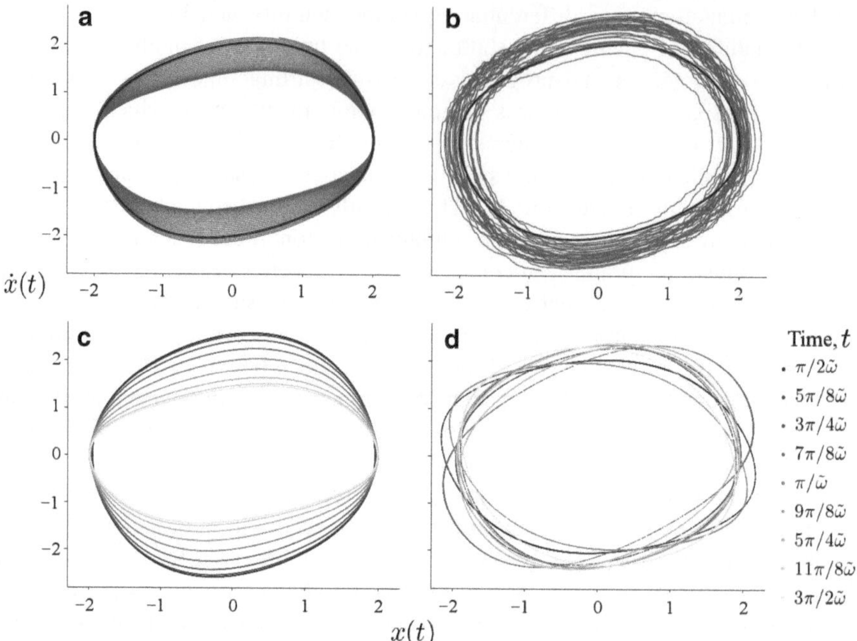

Fig. 5.1 Phase portrait of nonautonomous van der Pol oscillator with time-varying frequency, for: (**a**), (**c**) $f(t) = \tilde{A}\sin(\tilde{\omega}t)$, $\tilde{A} = 0.3$, $\tilde{\omega} = 0.01$; (**b**), (**d**) f(t) is uncorrelated Gaussian noise, $\tilde{A} = 0.6$. The *top plots* show the time evolution of a single trajectory of the system in *grey*, while the *black line* is the phase portrait in the autonomous ($\tilde{A} = 0$) case. The *bottom plots* show the positions of 10,000 trajectories with various initial conditions and at snapshots in time, given in the legend, which correspond to points in the phase of sin($\tilde{\omega}$ t). The system is given as: $\ddot{x} - \mu(1 - x)\dot{x} + [\omega + f(t)]^2 x = 0$, where $\omega = 1$ and $\mu = 0.2$

attention to self-sustained oscillators which exhibit stable limit cycles in the absence of the nonautonomous contribution. This means that an oscillator can still be treated as being self-sustained at all times, even though its characteristics (frequency, shape of limit cycle, etc,...) are time-varying.

Let us consider an oscillator $d\mathbf{x}/dt = \mathbf{f}(\mathbf{x}(t))$ with a stable periodic solution $\mathbf{x}(t) = \mathbf{x}(t + T)$ in an absence of external influence, characterized by a period T. The field $\mathbf{f}(\mathbf{x}(t), t)$ can be set to be an explicit function of the time. This will be the case, for instance, if one or more of the parameters that characterize $\mathbf{f}$ are bounded (periodic or non-periodic) functions of time. The periodic solution $\mathbf{x}(t)$ is, in general, lost and the definition of the period T becomes somewhat "blurred". An example of such nonautonomous oscillator is presented on Fig. 5.1a. In the absence of a periodic solution $\mathbf{x}(t) = \mathbf{x}(t + T)$, the definition of period could be replaced by the concept of "instantaneous period" (and correspondingly "instantaneous frequency"): at any instant of time τ the instantaneous period $T(\tau)$ of the dynamics is the period of the limit cycle solution of $\mathbf{f}(\mathbf{x}(t), \tau)$, where τ is fixed.

Defining unambiguously the phase of a nonautonomous oscillator can be a difficult and nontrivial task. In an autonomous system, the phase over the limit cycle is defined as that quantity which increases by 2π during each cycle of the dynamics. A nonautonomous version of the phase function can be found by exploiting weak external perturbations and the instantaneous period definition. Following the concept of phase reduction proposed by Kuramoto [45] one can then derive the notion of phase. This procedure, however, does not hold in the general case. In particular, if the external nonautonomous source is strong, the separation of amplitude and phase dynamics becomes very difficult. In terms of analysis, there has been some notable progress in the development of techniques that can estimate the instantaneous phases from time series. The most used methods are based on the Hilbert transform [65] and the synchrosqueezed wavelet transform [16] which can decompose the time-varying phases from relatively complex signals.

5.2.4 Coupled Nonautonomous Oscillators

When two or more systems coexist in the same environment, they often interact and tend to influence each others' dynamical fields. The couplings can be manifested as linear or nonlinear parametric connectivity or as functional relationships. Due to the couplings the systems can go through qualitative states of collective behaviour, including synchronization, oscillation death (Bar-Eli effect), clustering, etc. Even outside these states, the interactions can cause some important dynamical properties to vary—for example inducing time-variability of the frequency of an oscillator. A more difficult problem is faced where two or more interacting oscillatory systems are subject to external deterministic influences, a scenario that often arises in practice, e.g. in physiology including cellular dynamics, blood circulation, and brain dynamics. In such cases, the interacting systems (e.g. cardio-respiratory) are influenced by other oscillatory processes as well as by noise. These nonautonomous influences can perturb the dynamical properties of the interacting system and can cause transitions between qualitative states to appear.

In what follows we are going to concentrate on one of the most popular qualitative states of interaction—synchronization. It is defined as the mutual adjustment of rhythms due to weak interactions between oscillatory systems [65]. When the oscillators are weakly nonlinear and the couplings are weak as well, the synchronization phenomenon can be described qualitatively and sufficiently well by the corresponding phase dynamics. This is often referred to as phase synchronization [65,68]. To set up a general description of synchronization between nonautonomous systems, two nonautonomous oscillators are set to interact through coupling function $\mathbf{g}_1, \mathbf{g}_2$ parameterized by the coupling parameters ϵ_1, ϵ_2

$$\begin{aligned}\dot{\mathbf{x}}_1 &= \mathbf{f}_1(\mathbf{x}_1, t) + \epsilon_1(t)\,\mathbf{g}_1(\mathbf{x}_1, \mathbf{x}_2, t) \\ \dot{\mathbf{x}}_2 &= \mathbf{f}_2(\mathbf{x}_2, t) + \epsilon_2(t)\,\mathbf{g}_2(\mathbf{x}_1, \mathbf{x}_2, t)\,.\end{aligned}$$

When the frequency mismatch is relatively small, one can observe for which parameter values the system is synchronized and does not exhibit phase-slips [65], i.e. when $|\psi(\phi_1, \phi_2, t)| <$ constant, where $\phi_1(\mathbf{x}_1(t), t)$ and $\phi_2(\mathbf{x}_2(t), t)$ are the instantaneous phases of the two oscillators respectively, and the phase difference is defined as[1] $\psi(\phi_1, \phi_2, t) \equiv \phi_2(\mathbf{x}_2(t), t) - \phi_1(\mathbf{x}_1(t), t)$.

The synchronization condition $|\psi(\phi_1, \phi_2, t)| <$ constant will be satisfied if there exists a stable solution for the dynamics $d\psi(\phi_1, \phi_2, t)/dt$. Because the velocity field is a function of time, the existence of a stable equilibrium $\psi_{eq}(t)$ satisfying $d\psi(\phi_1, \phi_2, t)/dt = 0$ does not mean that the relative phase remains constant. Not even the existence of a time-dependent stable root can guarantee an absence of phase-slips. However, if $\psi_{eq}(t)$ changes in time slowly enough for the solution $\psi(t)$ to remain continuously within its attracting basin, then the phase difference will vary with time (as imposed by the nonautonomous source) while the system remains within the state of synchronization.

The Poincaré oscillators are chosen as an example of a nonautonomous limit-cycle system whose dynamical field can be made explicitly time-dependent. This isochronous oscillator in polar coordinates (r, ϕ) rotates at a constant-frequency, attracted with exponential velocity towards the radius, $\dot{r} = r(1-r); \dot{\phi} = \omega$. In terms of Euclidean coordinates, a model of two weakly interacting Poincaré oscillators takes the form

$$\begin{aligned} \dot{x}_i &= -q_i x_i - \omega_i(t)\, y_i + \varepsilon_i(t)\, g_i(x_i, x_j, t) + \xi_i(t), \\ \dot{y}_i &= -q_i y_i + \omega_i(t)\, x_i + \varepsilon_i(t)\, g_i(y_i, y_j, t) + \xi_i(t), \\ q_i &= \left(\sqrt{x_i^2 + y_i^2} - 1\right) \quad i, j = 1, 2, \end{aligned} \tag{5.1}$$

where ω_i are angular frequencies, ε_i are the coupling amplitudes and $g_i(x_i, x_j, t)$, $g_i(y_i, y_j, t)$ are the coupling functions. We considered the case where the frequency parameter of the first oscillator consists of a leading constant part and a small nonautonomous term e.g. $\omega_1(t) = \omega_1 + \tilde{A}_1 \sin(\tilde{\omega}_1 t)$, where $\tilde{A}_1$ and $\tilde{\omega}_1$ are small compared to ω_1. Note that, in the absence of the nonautonomous terms ($\tilde{A}_1 = 0$), the oscillators generate self-sustained oscillations [5, 6]. The coupling functions were linear and autonomous $g_i(x_i, x_j, t) = x_i - x_j$ and $g_i(y_i, y_j, t) = y_i - y_j$. The phases were evaluated as $\phi_i = \arctan \frac{y_i}{x_i}$, with arctan defined as a four-quadrant operation.

For certain parameters the systems oscillate in synchrony. Due to the nonautonomous influence imposed on the frequency parameter, periodic modulations are introduced both on the amplitude and the phase dynamics. If the effect from the periodic influence is increased (through $\tilde{A}$) the oscillators can lose synchrony. For some intervals within the period of the nonautonomous modulation (the light gray

[1]The following statement holds also for higher frequency ratios in the form $\psi = n\phi_2 - m\phi_1$ where n and m are integer numbers.

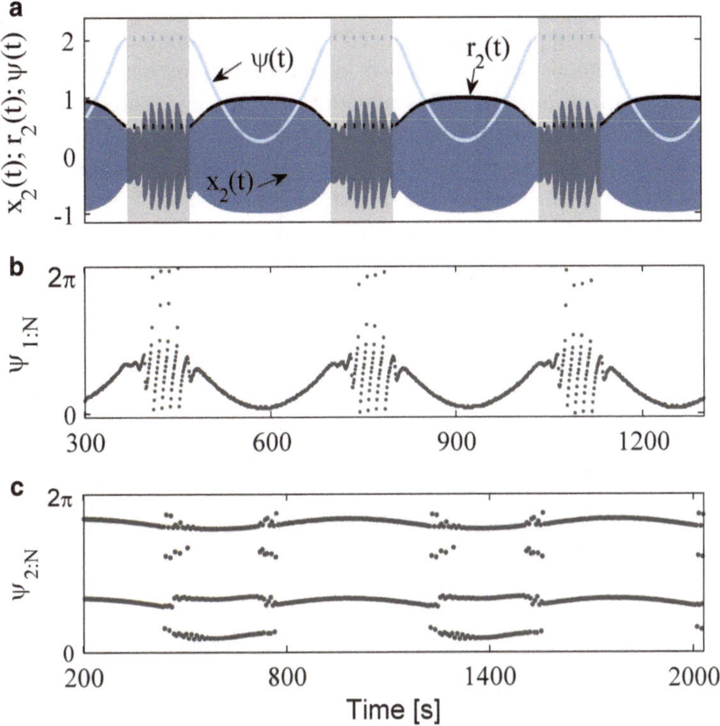

Fig. 5.2 Intermittent synchronization transitions for unidirectionally coupled ($1 \rightarrow 2$) Poincaré oscillators (5.1). The frequency of the first oscillator is nonautonomous: $\omega_1(t) = \omega_1 + \widetilde{A}_1 \sin(\widetilde{\omega}_1 t)$. (**a**) $r_2(t)$, $\psi(t)$ and $x_2(t)$ from numerical simulation. The *light gray regions* indicate the non-synchronous state. The *dashed lines* of $\psi(t)$, $r_2(t)$ within this state indicate existence of phase-slips. (**b**) 1:N synchrogram for the case under (**a**). (**c**) 2:N synchrogram showing synchronization transitions from 2 : 2 to 2 : 3 ratio

regions in Fig. 5.2a) the conditions for synchronization do not hold: $(r_{\text{eq}}(t), \psi_{\text{eq}}(t))$ is unstable or does not exist, a continuously-running phase appears and the two oscillators lose synchrony. More precisely, they go in and out of synchrony as time passes, i.e. there is intermittent synchronization.

The existence of synchronization and the corresponding transitions are investigated by application of a method for the detection of phase synchronization—the synchrogram [65], Fig. 5.2b and c. They are constructed by plotting the normalized relative phase of one oscillator within m cycles of the other oscillator, according to $\Psi_m(t_k) = \frac{1}{2\pi}\phi(t_k) \bmod 2\pi m$, where t_k is the time of the k-th marked event of the first oscillator, and $\phi(t_k)$ is the instantaneous phase of the second oscillator at time t_k. The synchrogram provides a qualitative measure where (for autonomous systems) the appearance of horizontal lines is normally taken to correspond to the synchronous state. The method clearly detects synchronization consistently with our analysis. The synchrograms show, however, that synchronization is now characterized by a smooth curve rather than a horizontal line, owing to the continuously changing phase shift induced by the nonautonomous modulation.

The nonautonomous source can also induce transitions between different frequency synchronization ratios. This situation is often encountered in high order interactions of open oscillatory systems—an obvious example being the cardiorespiratory system. A numerical example of this kind is presented on Fig. 5.2c. The synchrogram shows consecutive transitions from 2:2 (or 1:1) to 2:3 frequency locking, with short non-synchronized epochs in between. The external influence causes the system to not only lose and gain synchrony, but also induces transitions between different synchronization states.

So far we have concentrated on one nonautonomous frequency parameter and how it affects the interactions only. An equally important parameter that defines the interactions is the coupling amplitude. If influenced by nonautonomous sources, the coupling variability can also perturb the coupled systems, consequently leading to variability and transitions of the qualitative states. In interactions of many systems, like in networks of oscillators, the time-variability of the coupling strength can change the structural connectivity within the network. Another important property that characterizes the interactions among oscillators is the coupling function. It defines the functional law of the interactions and the law through which the interactions undergo transitions to synchronization.

It was recently shown [78, 79] that, as opposed to closed autonomous oscillators, the coupling function in open oscillatory systems can vary in time, both in intensity and form. Indeed, time-varying coupling functions have already been identified in living systems. For example, the functional relationships that characterise cardiorespiratory interactions are in fact time-varying. In Sect. 5.5 we elaborate on this in some detail and show how these relationships change with age. The time-variability of the form of the coupling function is important because it alone can be the cause for synchronization transitions. The external forces can influence several parametric and functional properties of the interacting systems at the same time, leading to relatively complex dynamics that is largely difficult to decipher. There is therefore a clear need for improved analysis techniques, formalism and understanding of such systems. The application of current methods of analysis will be presented in Sect. 5.4.

5.2.5 Summary

In this section we outlined a general description of nonautonomous oscillatory systems. First, we discussed the difference between non-stationary and nonautonomous systems and how they are treated. In the general class of nonautonomous systems we focused on self-sustained nonautonomous oscillators. We point out that the phase of such oscillators cannot be defined uniquely. However, in the presence of slow external forces we show that the problem is tractable. Furthermore, we reviewed the interactions and the state of synchronization, and how they are affected by nonautonomous perturbations. By analysis of two coupled Poincaré oscillators subject to periodic nonautonomous sources, we illustrated the onset of phase synchronization. In this case, the phase difference that defines the state of synchronization becomes time-varying, leading to qualitative transitions and intermittent synchronization. Transitions between synchronized and non-synchronized states, as well as between different synchronization ratios, were also demonstrated.

5.3 Ensembles of Nonautonomous Oscillators

After introducing the concept of nonautonomous oscillators in Sect. 5.2, here the focus will be turned to the effects of nonautonomicity in a population of coupled oscillators. The dynamics of such systems will be analyzed in case of parametric perturbations. Hence, the external influence in these systems can be directed either to the natural frequencies of the single oscillators or to the coupling strengths. Nevertheless, due to the assumption of the thermodynamical limit, the analysis of ensembles of coupled oscillators is not just a trivial extension of the case with two coupled oscillators. Thereby, the focus in the following analysis is put on the changes of the mean-field dynamics, as a result of the frequency, strength and the distribution of the external field.

Systems consisting of large numbers of interacting oscillating subsystems are pervasive in science and nature, and have been the essential modelling tools in physics, biology, chemistry and social science [89]. In the case of weakly interacting units, application of the phase approximation leads to the Kuramoto model for globally coupled phase oscillators [45]. It represents a mainstream approach today in tackling wide diversity of significant problems, and the variety of these issues, spanning from Josephson-junctions arrays [94] to travelling waves [29], has led to many extensions and generalizations of the basic model [4, 65, 88]. However, although biological examples are known to have provided the original motivation lying behind this model, neither the original model [45], nor most of its extensions [4], have incorporated a fundamental property of living systems—their inherent time-variability.

In this chapter first the original KM will be described and then its recent generalization that allows for time-varying parameters [62] will be introduced, together with the other studies of the original model that explain certain forms of nonautonomicity.

5.3.1 *The Kuramoto Model*

Kuramoto showed that the long-term dynamics for any system of weakly coupled, nearly identical limit-cycle oscillators, are given by the following universal ordinary differential equations (ODEs)

$$\dot{\theta}_i = \omega_i + \frac{K}{N}\sum_{j=1}^{N}\sin(\theta_j - \theta_i), \quad i = 1,\ldots,N. \tag{5.2}$$

Here $K \geq 0$ is the coupling strength, while ω_i are the natural frequencies of the uncoupled oscillators. They are randomly distributed according to some unimodal probability density function $g(\omega)$ with half-width, half-maximum γ. This model corresponds to the simplest possible case of equally weighted, all-to-all, purely sinusoidal coupling and it can be applied to any system whose states can each be captured by a single scalar phase θ_i.

A centroid of the phases as given in the complex plane complex defines an complex order parameter

$$z = re^{i\psi} = \frac{1}{N}\sum_{j=1}^{N} \mathrm{e}^{i\theta_j}. \tag{5.3}$$

It is then introduced into the governing equation (5.2), so that it becomes

$$\dot{\theta}_i = \omega_i - K\,r\sin(\theta_i - \psi), \tag{5.4}$$

where r and ψ are mean field and phase respectively. For a coupling strength larger than some critical value, some of the oscillators become locked to each other resulting in a bump in the distribution of phases and thus non-zero value for r.

In the thermodynamic limit $N \to \infty$ the state of the system (5.4) is described by a continuous PDF $\rho(\theta, \omega, t)$ which gives the proportion of oscillators with phase θ at time t, for fixed ω [52].The number of oscillators is conserved and since ω is fixed, the following continuity equation emerges for every ω

$$\frac{\partial\rho}{\partial t} = -\frac{\partial}{\partial\theta}\{[\omega + \frac{K}{2i}(z\mathrm{e}^{-i\theta} - z^*\mathrm{e}^{i\theta})]\rho\}, \tag{5.5}$$

where the velocity along θ is substituted from the governing equations (5.4). The definition (5.3) is also included in (5.3), rewritten using $\frac{1}{N}\sum_j \sin(\theta_j - \theta_i) = \mathrm{Im}\{z\mathrm{e}^{-i\theta_i}\}$, thus becoming

$$z = \int_0^{2\pi}\int_{-\infty}^{\infty} \rho(\omega, \theta, t) g(\omega)\mathrm{e}^{i\theta} d\theta d\omega. \tag{5.6}$$

The last two equations self-consistently give the stationary mean field behaviour of the autonomous model. The stability analysis of this behaviour in the general case is thoroughly discussed in Ref. [31]. However, analytical description of the dynamics of oscillator ensembles, remains an important and interesting problem in bulk of the situations, while the closed solution for the mean field exists only in case of Lorentzian natural frequencies' distribution for the simplest KM, (5.4).

5.3.2 The Nonautonomicity in the Kuramoto Model

Non-constant collective rhythms in the inverse problem are very often a result of external influence. Nevertheless, they can simply follow from asymmetrically-coupled ensembles [54, 72] or from populations with multimodally distributed natural frequencies [3, 10]. Multimodal distribution of the parameters is common cause for the complex collective behaviour of these and should not be confused with NA [63].

Many studies have been performed on coupled oscillators influenced by external dynamics. Noise is the first form of external influence introduced by Sakaguchi [70] and it is thoroughly studied since then. Its effect into increasing of the heterogeneity is similar to increasing the width of the frequency distribution and the bifurcation analysis for this case was performed in [87]. Likewise, driving by an external periodic force [74] is a long-explored model. Each of the oscillators in this case become additionally driven by an external frequency Ω. This leads to the mean behaviour directly characterized by the interplay between the external pacemaker and the mean field of all other oscillators. Hence, the system corresponds to the case of oscillator driven by an external force.

A generalization of the KM that allowed certain time-varying frequencies and couplings have been also numerically explored in [14] or applied in certain models of brain dynamics [71]. However, in none of them were the dynamics described analytically, nor a qualitative description was given for slow or fast varying cases.

Frequency adaptation as discussed in [92] also assumes non-constant natural frequencies, but without external influence. It is similar to the models with inertia [2] and its dynamics, apart from the stable incoherence, are characterized by either synchronization or bistable regime of both synchronized and incoherent states. In addition, the model with drifting frequencies [69] assumes frequency dynamics formulated as an *Ornstein-Uhlenbeck process*, but it also leads to time-independent mean fields, resembling the simple KM under influence of colored noise.

Alternately-switching connectivity [47, 77] or periodic couplings [48], are some of examples that explore phase oscillators with varying coupling strengths. Yet, most of the discussions in these are concerned with the networks and graph theory properties of the system, the analysis in mostly numerical and only Heaviside step functions are considered for the interaction between oscillators.

Nevertheless, none of these models for group dynamics can exhibit the deterministic and stable TD dynamics of many real physical, chemical, biological, or social systems that can never be completely isolated from their surroundings. These systems do not reach equilibrium but, instead, exhibit complex dynamical behavior that stems from some external system.

5.3.3 The Kuramoto Model with Time-dependent Parameters

Thermodynamical assumption of the KM, makes the introduction of NA into system's parameters a special case of interest. Unlike the case of two or finite number of interacting oscillators, external influence in the KM affects the whole population and continuously alters its group behaviour. The non-equilibrium dynamics that arise from the NA influenced parameters was addressed in a recent generalization of the KM with TD parameters [62]. It introduced an external, explicitly TD, bounded function $x(t)$ that modulates the frequencies or couplings of the original KM. In the most general case, the strengths of the interactions I_i are distributed according to a PDF $h(I)$ and depending on which parameter is influenced two generalized models emerge

$$A: \quad \dot{\theta}_i = \omega_i + I_i x(t) + K\, r(t) \sin(\psi - \theta_i), \tag{5.7}$$

$$B: \quad \dot{\theta}_i = \omega_i + [K + I_i x(t)]\, r(t) \sin(\psi - \theta_i). \tag{5.8}$$

For description of each oscillator of the NA KM, beside the natural frequency ω_i and the coupling strength K_i, one should know the strength I_i and the form of the external forcing, $x(t)$. Additionally, for each oscillator of the above models, at any given time there exists a correspondence between the fixed and TD parameters, such that $\tilde{I}_i(t) = I_i x(t)$. Thereafter, in the limit $N \to \infty$ the population can be described either by a continuous PDF $\rho(\theta, \omega, I, t)$ which assumes fixed parameters, or by its counterpart $\tilde{\rho}(\theta, \omega, \tilde{I}, t)$ with TD parameters. However, since the latter would further complicate the continuity equation for fixed volume by including gradients along the TD variables also, the distribution for the fixed ω and I is chosen. Hence the continuity equation for every fixed ω and I is given by

$$A: \quad \frac{\partial \rho}{\partial t} = -\frac{\partial}{\partial \theta}\{[\omega + I x(t) + \frac{K}{2i}(ze^{-i\theta} - z^* e^{i\theta})]\rho\}, \tag{5.9}$$

$$B: \quad \frac{\partial \rho}{\partial t} = -\frac{\partial}{\partial \theta}\{[\omega + \frac{K + I x(t)}{2i}(ze^{-i\theta} - z^* e^{i\theta})]\rho\}, \tag{5.10}$$

where the velocity along θ is substituted from the governing equations ((5.7), (5.8)).

Since $\rho(\theta, \omega, I, t)$ is real and 2π periodic in θ, it allows a Fourier expansion. The same would also hold for $\tilde{\rho}(\theta, \omega, \tilde{I}, t)$. Next, we apply the Ott and Antonsen ansatz [58] in its coefficients, such that $f_n(\omega, I, t) = [\alpha(\omega, I, t)]^n$. Thus,

$$\rho(\theta, \omega, I, t) = \frac{1}{2\pi}\{1 + \{\sum_{n=1}^{\infty} [\alpha(\omega, I, t)]^n e^{in\theta} + \text{c.c.}\}\}, \tag{5.11}$$

where c.c. is the complex conjugate. Substituting (5.11) into the continuity equations ((5.9), (5.10)), it follows that this special form of ρ is their particular solution as long as $\alpha(\omega, I, t)$ evolves with

$$A: \quad \frac{\partial \alpha}{\partial t} + i[\omega + Ix(t)]\alpha + \frac{K}{2}(z\alpha^2 - z^*) = 0, \tag{5.12}$$

$$B: \quad \frac{\partial \alpha}{\partial t} + i\omega\alpha + \frac{K + Ix(t)}{2}(z\alpha^2 - z^*) = 0, \tag{5.13}$$

for models A and B respectively. The same ansatz implemented in (5.6), reduces the order parameter to

$$z^* = \int_{-\infty}^{+\infty} \int_{-\infty}^{+\infty} \alpha(\omega, I, t) g(\omega) h(I) d\omega dI. \tag{5.14}$$

Equations ((5.12), (5.13)) give the evolution for the parameter α which is related to the complex mean field through the integral equation (5.14) and they all hold for any distributions $g(\omega)$ and $h(I)$, and for any forcing $x(t)$. Despite this, the integrals in (5.14) can be analytically solved for certain polynomial or multimodal-δ distributions $g(\omega)$ and $h(I)$, leading to direct evolution of the mean field.

5.3.3.1 Low-dimensional Dynamics

In order to be obtained evolution of the mean field, the integral (5.14) should be solved. Therefore the natural frequencies follow a Lorentizan distribution, and $\alpha(\omega, I, t)$ is continued to the complex ω-plane so $g(\omega)$ can be written as $g(\omega) = \frac{1}{2\pi i}[\frac{1}{\omega-(\hat{\omega}-i\gamma)} - \frac{1}{\omega-(\hat{\omega}+i\gamma)}]$ with poles $\omega_{p1,2} = (\hat{\omega} \pm i\gamma)$.

For the model A (5.7) with forcing strengths proportional to frequencies, $\tilde{\omega}(t) = \omega[1 + \varepsilon x(t)]$ with a constant ε. This means that $I = \varepsilon\omega$ and $h(I) = g(\varepsilon\omega)$. Hence, the integration in (5.14) is now only over ω, and by using the residue theorem it yields $z^* = \alpha(\hat{\omega} \mp i\gamma, t)$. This is substituted in (5.12) returning

$$\dot{r} = -r[\gamma|1 + \varepsilon x(t)| + \frac{K}{2}(r^2 - 1)], \qquad \dot{\psi} = \hat{\omega}[1 + \varepsilon x(t)]. \tag{5.15}$$

The simplest case of model A, (5.7), is when the external forcing is identical for each oscillator, $h(I) = \delta(I - \varepsilon)$. This leads to trivial dynamics, since the original model is invariant to equal shift of the natural frequencies. On contrary, the model B with identical forcing to each oscillator yields TD mean field parameters given by

$$\dot{r} = -r[\gamma + \frac{K}{2}[1 + \varepsilon x(t)](r^2 - 1)], \qquad \dot{\psi} = \hat{\omega}. \tag{5.16}$$

The similar approach is used for obtaining the low-dimensional dynamics of other cases of Models A and B which include forcing strengths with polynomial Lorentzian-like distributions. Thus, Model A with an independent Lorentzian distribution of forcing strengths evolves as

$$\dot{r} = -r[\gamma + \gamma_I|x(t)| + \frac{K}{2}(r^2 - 1)], \; \dot{\psi} = \hat{\omega} + \hat{I}x(t), \tag{5.17}$$

where $\hat{I}$ and γ_I are the mean and half-width of $h(I)$ respectively. However, for a Lorentzian distributed forcing strengths of Model B, contour integration cannot be applied to (5.14). Namely, the integration contour should be such that if $\alpha(\omega, I, t)$ is analytic and $|\alpha| \leq 1$ everywhere inside the contour at $t = 0$, this would also hold for all $t > 0$. For this to happen, one of the requirements from [53] is $|\alpha| \leq 0$, for $|\alpha| = 1$, but this cannot be proven to hold [62].

The integral (5.14) has straightforward solution for multimodal δ-distributed external strengths. Hence, for bimodal function $h(I) = \frac{1}{2}[\delta(I - \hat{I} - \gamma_I) + \delta(I - \hat{I} + \gamma_I)]$, the complex parameter z becomes

$$z^* = \frac{1}{2}[\alpha_1(\hat{\omega} - i\gamma, \hat{I} - \gamma_I, t) + \alpha_2(\hat{\omega} - i\gamma, \hat{I} + \gamma_I, t)]. \tag{5.18}$$

The dynamics on the other hand is consistently described by the evolutions of $\alpha_{1,2}$ obtained from (5.12) as

$$\frac{\partial \alpha_{1,2}}{\partial t} = -\{i[\hat{\omega} + (\hat{I} \mp \gamma_I)x(t)] - \gamma\}\alpha_{1,2} + \frac{K}{4}[\alpha_1 + \alpha_2 - \alpha_{1,2}^2(\alpha_1 + \alpha_2)^*], \tag{5.19}$$

and from (5.13) as

$$\frac{\partial \alpha_{1,2}}{\partial t} = -(i\hat{\omega} - \gamma)\alpha_{1,2} + \frac{1}{4}K[1 + (\hat{I} \mp \gamma_I)x(t)][\alpha_1 + \alpha_2 - \alpha_{1,2}^2(\alpha_1 + \alpha_2)^*]. \tag{5.20}$$

Choi et al. [13] carried out a bifurcation analysis near the limit $rK \ll 1$ for this case of model A with cosine forcing.

The plots in Fig. 5.3 show the observed NA mean field for different cases of both models. The plots Fig. 5.3a, b are for cosine forcing and the plot Fig. 5.3c shows chaotic forcing. A theorem in [59] states that ((5.12), (5.13)) asymptotically capture all macroscopic behavior of the system as $t \to \infty$. Similarly, the incoherent and partly synchronized states both belong to the manifold defined by ((5.12), (5.13)) [58], and the initial incoherent state is set with uniformly distributed phases at time $t = 0$. Thus, the ansatz ((5.12), (5.13)) and the evolutions ((5.15)–(5.20)) should continuously describe the NA system, as confirmed by Fig. 5.3.

5.3.3.2 Slow/Fast Reduction

Since the evolutions ((5.15)–(5.20)) are nonlinear and include explicit dependence on time, the classical bifurcation analysis cannot be applied for these cases, as was

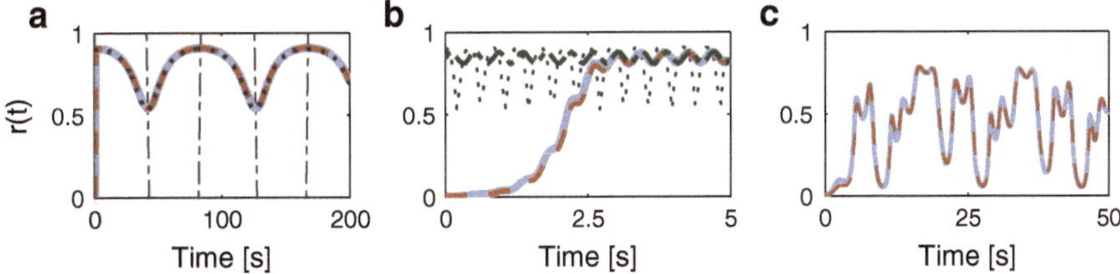

Fig. 5.3 The time-varying mean field for model B resembles the externally applied cosine (**a**),(**b**), or chaotic forcing for model B (**c**). Numerical simulations of the full system (*light blue*) are in agreement with the low-dimensional dynamics (*dashed red*). Adiabatic (*dotted brown*) and non-adiabatic evolutions (*dashed green*) confirm the limits of the reduced dynamics (see text for details). (**a**),(**b**) Constant forcing, $K = 7, \varepsilon = 0.6$, $\Omega = 0.075$ and $\Omega = 15$ respectively; (**c**) bimodal δ distributed forcing, $K = 5$, $\gamma = 1$, $\gamma_I = 1.2$ and $\hat{I} = 0.5$

also discussed in Chap. 1. In addition, these are Bernoulli ODEs which can not be explicitly solved, yielding to integro-differential equations. Still, the dynamics could be reduced for the slow and fast limits of the external forcing, relative to the inherent time-scale of the dynamics of the autonomous system. This is a similar approach to the simplification of the dynamics introduced in Chap. 3. The systems analyzed there evolve on multiple time scales, a fast and a slow, thus allowing for the proposed reduction.

Here, we address the impact of the external forcing to the nonautonomous system. The plots in Fig. 5.3a, b show that oscillations of the mean field follow the frequency of the forcing, but the challenge is to describe the magnitude of these oscillations and whether it adiabatically follows the strength of the forcing. The low-frequency filtering and different responses depending on the frequency of the external forcing are also obvious. These two characteristics of population models are well known and are a direct consequence of their intrinsic transient dynamics. Accordingly, the NA dynamics can be reduced depending on the period of the external field $T = 2\pi/\Omega$, relative to the system's transition time [62].

The exponential damping rate of the original system is defined by $\tau = 1/|K/2-\gamma|$ [58]. For a system far from incoherence, $K = 2\gamma + O(2\gamma)$, $\tau \approx 1/O(\gamma)$ holds. This means that the transition time depends only on the width of the distribution of natural frequencies, γ. Thereafter for this case, the system's response is adiabatic for slow external fields, $\Omega \ll \gamma$, and non-adiabatic for fast, $\Omega \gg \gamma$. To make the analysis independent on γ, it is removed by scaling the time and the couplings in the autonomous system, $t = t/\gamma$, $K = K/2\gamma$ and $\tau = 1/|K-1|$ (the scaled variables keep the same letters in the further analysis).

For model B, (5.8), with $x(t) = \varepsilon \cos \Omega t$, after the initial transition and in the absence of bifurcations, the amplitude of the mean field consists of a constant term r_0 and a TD term $\Delta r(t)$. For the non-adiabatic response, simulations, grey lines in Fig. 5.3b, show that $\Delta r(t) \sim 1/\Omega$ and $r_0 \gg \Delta r(t)$. Thereafter r_0 can be expressed as averaged over one period $T = 2\pi/\Omega$ of the oscillations of $\Delta r(t)$. Proceeding with averaging of both sides of (5.15) for one period, the term $\Delta r(t) \cos \Omega t$ in the integral vanishes only if $\Delta r(t) \sim \sin \Omega t$, which is self-consistently proved as the obtained form of $\Delta r(t)$ for non-adiabatic response, (5.22), follows this assumption. Thus $r_0 = \sqrt{1 - 1/K}$. Further, $r(t) \approx r_0$ and $\frac{dr}{dt} = \frac{d\Delta r}{dt}$ is applied to (5.15) and then it is integrated. From there $\Delta r(t) = -r_0 \frac{\varepsilon}{\Omega} \sin \Omega t$, and the magnitude of the NA response is

$$\Delta_{\text{fast}} = 2\frac{\varepsilon}{\Omega}\sqrt{1 - \frac{1}{K}}. \tag{5.21}$$

Hence the long-term non-adiabatic evolution follows

$$r_{\text{fast}}(t) = \left(1 + \frac{\varepsilon}{\Omega} \sin \Omega t\right) \sqrt{1 - \frac{1}{K}}. \tag{5.22}$$

The adiabatic behavior emerges through the introduction of a slow time-scale $t' = \Omega t$, such that the system is constant on the fast time-scale t, and changes only in t'. Hence the l.h.s. of (5.16) is zero, whence

$$r_{\text{slow}}(t) = \sqrt{1 - \frac{1}{K(1 + \varepsilon \cos \Omega t)}}, \tag{5.23}$$

while, for the magnitude of the NA part, one obtains

$$\Delta_{\text{slow}} = \sqrt{1 - \frac{1}{K(1+\varepsilon)}} - \sqrt{1 - \frac{1}{K(1-\varepsilon)}}. \tag{5.24}$$

The adiabatic responses can also be obtained from the self-consistency of (5.6) and (5.9) for stationary states of the mean field. Namely, assuming very slow dynamics of the external forcing, the system can be treated as quasistationary. This is similar to assuming stationarity on a fast time scale. Thus one obtains $r = \sqrt{1 - 2\gamma(t)/K(t)}$, corresponding to the result (5.23).

The reduced dynamics, Fig. 5.3a–c, are in line with the above analysis, confirming the interplay between external and internal time scales of the NA system. The magnitudes of the slow/fast responses to cosine forcing are given in Fig. 5.4 for model A, (5.7). It confirms the obtained dependence of Δ on the frequency and amplitude of the external field. The low-frequency filtering is also obvious, and the transient behavior for slow and fast forcing can be seen.

However, for coupling close to critical, the system's transition time increases and $\tau \to \infty$ when $K \approx K_c$. As a result, the slow reduction fails when r is close to 0, unlike the case $K = K_c + \mathrm{O}(K_c)$ given in Fig. 5.3a, b for r far from 0.

The analysis of the reduced dynamics is shown only for simple periodic forcing. Still, this does not decrease the generality of the reduction, since any external field can be represented by its Fourier components. As a result, this method could be of great importance in modeling systems with multiple time-scales of oscillation and interaction.

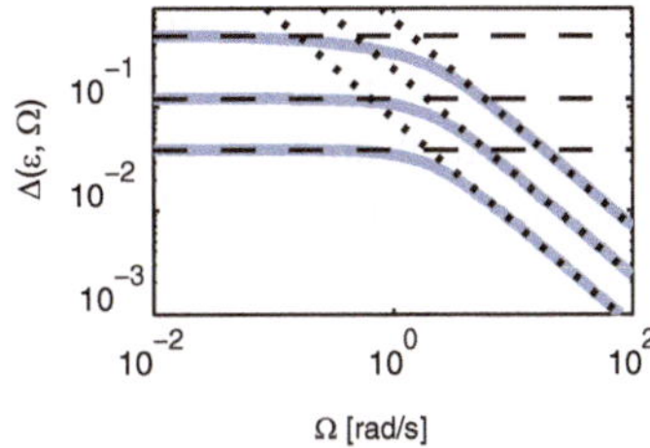

Fig. 5.4 Magnitude of the response, Δ, of the NA model B to the cosine forcing. External forcing strengths follow the distribution of frequencies, $K = 3$ and $\Omega \in [10^{-2}, 10^{2}]$. Non-adiabatic (*dotted black*), (5.21), and adiabatic, (*dashed black*), (5.24), evolution for $\epsilon \in [0:1;\ 0:3;\ 0:9]$, compared with the real dynamics (*light blue*), (5.16)

5.3.4 *Summary*

In summary, different models of ensembles of interacting phase oscillators influenced by external systems are presented. The main focus is put on the nonautonomous, stationary, time-dependent dynamics of interacting oscillators subject to continuous, deterministic perturbation. Thus, this represents a generalization of the case of two coupled oscillators, as discussed in Sect. 5.2, for a large number of interacting units. It consists of the dynamics of an external system superimposed on the original collective rhythm and have been missing from earlier models and extensions described in Sect. 5.3.2, possibly leading to an incorrect interpretation of some real dynamical systems. The impact of the forcing to the original system is also explained and the effect of its dynamics, amplitude and distribution is evaluated. Hence, the generalization of the KM that encompasses NA systems [62] offers possibility for direct tackling the NA in the interacting oscillators. In particular, it allows reconstruction of the stable, time-varying mean field. As a result, a large range of systems explained by the Kuramoto model—spanning from a single cell up to the level of brain dynamics—can be described more realistically.

5.4 Nonautonomous Systems as Inverse Problems

5.4.1 *Introduction*

The theory of nonautonomous systems is able to explain a wide range of phenomena, especially in living systems. However, there is very little literature available on how to analyse the observables of these systems for the inverse problems encountered in the real world.

Time series analysis has been used for decades as a non-invasive tool for extracting information about the workings of unknown dynamical systems [1,7,19].

This is achieved not through experimental perturbations, but by observing the spontaneous dynamics of a system over a finite time period. In the case of nonautonomous systems, this form of analysis is even more important given that in real world systems there is rarely control over the initial time t_o.

5.4.2 Time-Delay Embedding

For deterministic systems, phase space is the most common domain of analysis. In this representation, systems in or close to equilibrium are confined to either a limit cycle or fixed point, while non-equilibrium systems such as those exhibiting chaos occupy an area of phase space known as a strange attractor. The stability of these systems can also be quantified in phase space by the use of Lyapunov exponents, which determine whether two nearby trajectories will converge, diverge or remain at the same separation over time.

The theory of transforming time series data to phase space was developed by Floris Takens [91] and Ricardo Mañé [50]. The procedure involves the construction of an embedding vector for each point in time

$$\mathbf{x}(t_i) = [x(t_i), x(t_i + l\Delta t), \ldots, x(t_i + (d-1)l\Delta t)], \tag{5.25}$$

where d is the *embedding dimension* and l is an integer, both of which must be chosen prior to embedding. The dimensions of the reconstructed attractor are therefore composed of time-delayed versions of the data in $x(t)$.

For the choice of l, the embedding theorem specifies no conditions. Technically any delay time $l\Delta t$ (so long as it is not exactly equal to the period of an oscillatory mode) should give a "correct" reconstruction of the attractor, preserving all of its local properties. However, for the purpose of improved statistics, the best time delays are neither extremely short or extremely large [11,39,76].

For the estimation of d, the embedding theorem specifies the following condition

$$d \geq 2D + 1, \tag{5.26}$$

where D is the smallest theoretical dimension of phase space for which the trajectories of the system will not overlap [1]. An appropriate value of d can be estimated empirically from a time series by using the false nearest neighbours method [37,67].

While embedding works well for the case of autonomous systems, the theory does not consider nonautonomous systems and the possibility of time-dependent attractors. As discussed in Chap. 3, the time-dependent components in these systems are incorporated into extra dimensions in phase space, essentially resulting in a more complex, time-independent attractor. For example, the nonautonomous van der Pol oscillator shown in Fig. 5.1a, c becomes equivalent to the following four-dimensional system of equations

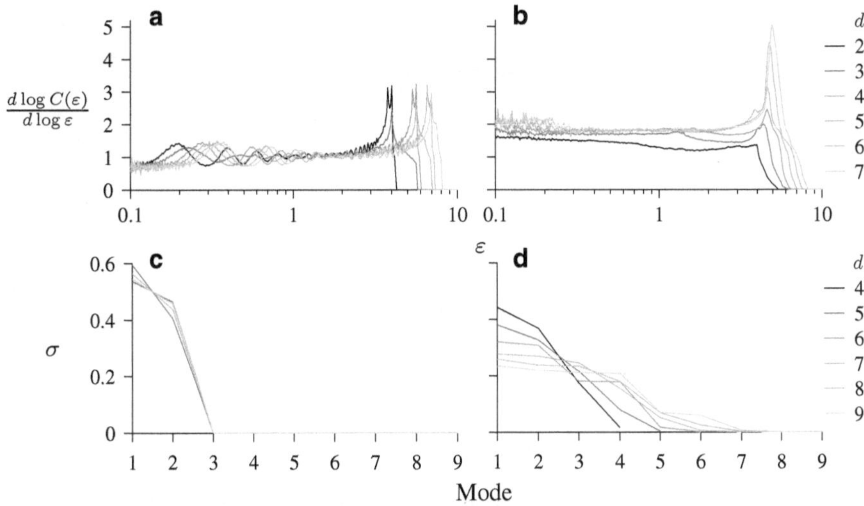

Fig. 5.5 Analysis of time series from the system in Fig. 5.1a, c using time-delay embedding: (**a**),(**c**) The autonomous system; (**b**),(**d**) The nonautonomous system. The *top plots* show the derivative of the correlation integrals [24, 25] with respect to the length scale ε for the embedding dimensions $d = 2, 3, \ldots, 7$. The *bottom plots* show the normalised exponents from Karhuen-Loève decomposition [36, 49] for $d = 4, 5, \ldots, 9$. The results for the autonomous system are given in (**a**) and (**c**) for $l\Delta t = 1.5s$, while the results for the nonautonomous system are given in (**b**) and (**d**) for $l\Delta t = 2.1s$. In each case the time delay was estimated from the first minimum of the mutual information [21]

$$\begin{aligned}
\dot{x} &= y \\
\dot{y} &= (1 - x^2)y + [\omega + \tilde{A}u]^2 x \\
\dot{u} &= v \\
\dot{v} &= -\tilde{\omega}^2.
\end{aligned} \tag{5.27}$$

This problem is demonstrated in Fig. 5.5. In Fig. 5.5a and b the plateau in the derivative of the correlation integral indicates the dimension of the attractor [83, 84]. For the nonautonomous system, the lines for different embedding dimensions converge to a value $\sim$ 2, despite the fact that the time-dependent attractor, as seen in Fig. 5.1c, is really a one-dimensional limit cycle. In Fig. 5.1c, the distribution of exponents from Karhunen-Loève decomposition quickly converges as the embedding dimension is increased for the autonomous system. However, in Fig. 5.1d the decomposition is highly dependent on the dimension for the nonautonomous system, which again suggests the incorporation of much higher-dimensional dynamics. When embedded in phase space, the deterministic motion of the attractor in Fig. 5.5c is therefore treated in a similar way to the genuinely stochastic motion of the attractor in Fig. 5.5d.

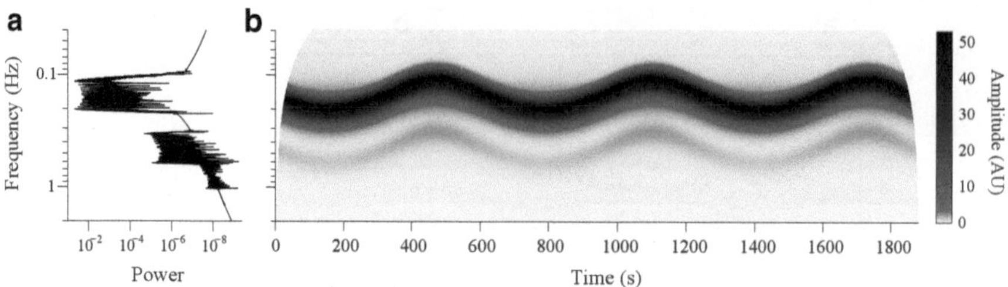

Fig. 5.6 Time-frequency analysis of a time series from the nonautonomous system in Fig. 5.1a, c. The Fourier transform power spectrum is shown in (**a**), while the amplitude of the Morlet wavelet transform is shown in (**b**) using arbitrary units (AU)

5.4.3 *Time-Frequency Analysis*

At present there exists no theory to extract a time-dependent nonautonomous attractor from a single time series. For inverse problems, the analysis of nonautonomous systems in phase space is therefore impractical.

Similarly to phase space analysis, transformations to the frequency domain remove time-dependent information and therefore suffer from related problems. The conversion of data from the time domain to the frequency domain is achieved via the well known Fourier transform,

$$F(\omega) = \frac{1}{L} \int_{-\frac{L}{2}}^{\frac{L}{2}} f(t) e^{-\frac{2\pi i \omega t}{L}} dt, \tag{5.28}$$

where $f(t)$ is the time series, ω is the angular frequency of the components in the frequency domain, and L is the length of the time series.

Figure 5.6a shows the Fourier transform of the nonautonomous van der Pol system. The wide distribution of peaks provides no information about the underling system, which comprises of a simple time-dependent limit cycle oscillator. However, the gaps between these peaks suggests some deterministic structure, rather than a single stochastic oscillation. Combining this information with the fact that only one oscillation is observed in the time domain indicates that the other components must be harmonics, which identifies the presence of nonlinearity in the system.

The combination of information from both the time and frequency domains is a powerful tool for inverse problems involving dynamical systems. However, it also has inherent limitations due to the uncertainty principle associated with the time and frequency resolution. The optimal transformation to the time-frequency domain therefore requires an adaptive basis, which allows components at all times and frequencies to be extracted at this maximal resolution. This optimum is provided by the wavelet transform,

$$W(s,t) = \int_{-L/2}^{L/2} \Psi(s, u-t) f(u) du, \tag{5.29}$$

where $\Psi(s, u-t)$ is the wavelet defined at scale s and time t. The closest wavelet to the basis used in the Fourier transform is the Morlet wavelet,

$$\Psi(s,u) = s^{-1/2} e^{-\frac{u^2}{2s^2}} e^{-\frac{2\pi i \omega_c u}{s}}, \tag{5.30}$$

where ω_c is the *central frequency*, which defines the time/frequency resolution ratio. With this basis the frequency is given by $1/s$.

The amplitude of the wavelet transform shown in Fig. 5.6b reveals a less complex distribution of peaks than the Fourier transform. This is because the wavelet transform is able to track the nonstationary frequency distribution of the time series. However, due to the fact that wavelets are a linear basis, the additional harmonic component is still observed.

Harmonics that are observed in the wavelet transform do not pose a problem because time-dependent *phase* information is preserved. Methods have now been developed which exploit this information to separate the underlying nonlinear components from the harmonics [73]. In addition, by analysing these phases it is possible to detect the nature of the interactions between the real nonlinear components.

5.4.4 Interacting Systems

For inverse problems involving coupled systems, the interactions can be time-dependent if the nature of the coupling is nonautonomous. In particular, the problem of time-varying couplings between two van der Pol oscillators is considered,

$$\begin{aligned} \dot{x}_{1,2} &= y_{1,2}, \\ \dot{y}_{1,2} &= \mu \left(1 - x_{1,2}^2\right) y_{1,2} + \omega_{1,2}^2 x_{1,2} + \gamma_{1,2}(t) \left(y_{1,2} - y_{2,1}\right)^2 + \xi \eta_{1,2}(t). \end{aligned} \tag{5.31}$$

Here $\eta_{1,2}(t)$ are uncorrelated Gaussian noises with $\xi = 0.05$. The parameters $\gamma_{1,2}$(t) determine the strength of the time-dependent quadratic couplings between the two systems.

Previously, the most common way of observing this time-dependence was by tracking epochs of synchronization [85] but this is in fact a consequence of interaction rather than a cause. Other methods have now been developed which are able to decipher the more subtle effects of these interactions in order to directly reveal the couplings $\gamma_{1,2}$(t).

The first of these methods follows directly from time-frequency analysis. The bispectrum was introduced by Hasselmann et al. [28] and arises from high-order statistics [57]. In the frequency domain, the bispectrum can be thought of as an estimate of the third-order statistic and describes the nonlinear (or more specifically, quadratic) properties of a time series [33, 56]. By using the Fourier bispectrum, the nonlinear couplings between oscillations at different frequencies can be detected, although there is no easy way to track the changes in these couplings due to the limits of time-frequency resolution. The wavelet transform again provides the optimal solution to this limit and following from (5.29), the wavelet bispectrum is given by Jamšek et al. [34, 35],

$$B_W(s_1, s_2) = \int_L W(s_1, t)W(s_2, t)W^*(s_3, t)dt, \tag{5.32}$$

where $s_3 = 1/(\frac{1}{s_1} + \frac{1}{s_2})$. The instantaneous biamplitude can now be defined as $A(s_1, s_2, t) = |W(s_1, t)W(s_2, t)W^*(s_3, t)|$ to give a time-dependent bispectrum. In addition, the instantaneous biphase $\phi_W(s_1, s_2, t) = \theta(s_1, t) + \theta(s_2, t) - \theta(s_3, t)$ can also be defined, where $\theta(s, n)$ are the phases of the wavelet components. Whenever a coupling between the scales s_1 and s_2 is active, the biphase remains constant [35].

Wavelets are not the only available tool for the analysis of interactions between coupled systems. Couplings can also be detected using methods based around Granger causality [35, 60, 61, 93]. In this case, a coupling is said to exist if one system gives information about the state of the other system after some time lag. Using the observed time series from the two systems $x_{1,2}$, the appropriate measure for this principle is given by the conditional mutual information (CMI),

$$I(x_1(t); x_2(t+\tau))|x_2(t)) = H(x_1(t)) + H(x_2(t)) - H(x_1(t), x_2(t+\tau)|x_2(t)), \tag{5.33}$$

where $H(x_{1,2})$ is the Shannon entropy and $H(x_1(t), x_2(t)|x_1(t+\tau))$ is the conditional entropy [61]. Here the CMI gives a measure of the information flow in the direction $x_1 \rightarrow x_2$ for the time lag τ. Alternating the time series indices $1, 2$ gives the CMI for the opposite direction.

The most recent approach applies the Bayesian theorem [8], in which information about the couplings can be propagated in time. The method uses a set of periodic basis functions which are inferred using the phases of the two oscillators. The dependence of one oscillator on the other can be detected by inferring the most likely parameters for the basis given the data from the phases and generating the coupling functions. For the full details of the method see [18, 78, 79]. However, the key to Bayesian inference is that the inferred parameters from a previous time window are assumed as prior information in the calculation of the parameters in the next window. Consequently, the inference of the couplings is formulated in time and is ideal for the application to nonautonomous coupled systems.

Figure 5.7 shows these methods applied to the coupled van der Pol oscillators. The couplings were made very weak so that the additive noise prevented synchronization without having a strong effect on the limit cycle oscillations. Both the wavelet biamplitude Fig. 5.7b and biphase Fig. 5.7c are able to detect the coupling

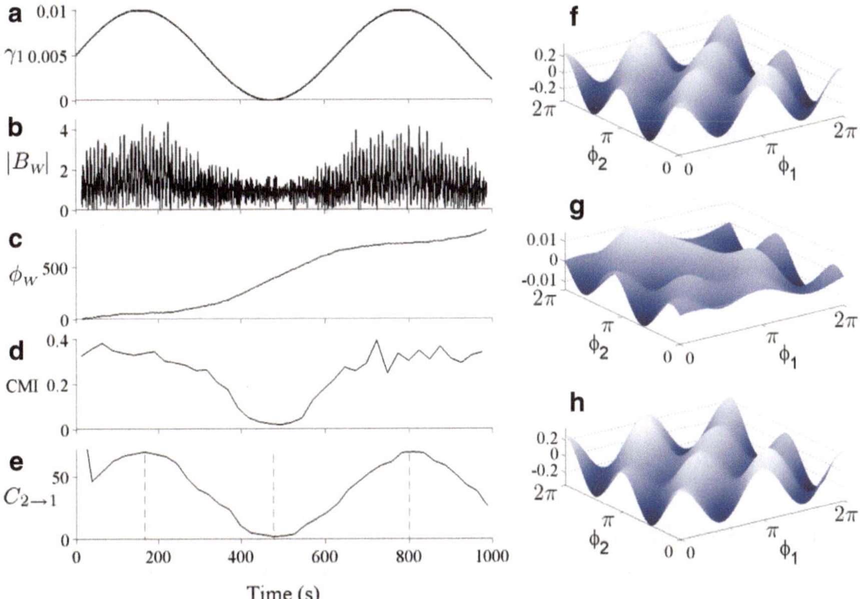

Fig. 5.7 Time series analysis of the van der Pol oscillators defined by (5.31) with a very weak one-directional, time-dependent coupling $0 \leq \gamma_1 \leq 0.01$, $\gamma_2 = 0$. (**a**) The strength of the coupling against time. (**b**) The wavelet biamplitude and (**c**) biphase for the frequency pair ω_1, $(\omega_1 + \omega 2)/2$. (**d**) The CMI of the extracted phases as calculated with a 25 s moving window using a time delay $\tau = 0.1$. (**e**) The inferred coupling parameter from the Bayesian method using a 25 s window, with the coupling functions (**f**), (**g**) and (**h**) for the corresponding direction of coupling at the times 175, 475 and 800 s respectively. The parameters used were $\mu = 0.2$, $\omega_1 = 2$ and $\omega_2 = 2.7$. The phases were extracted by applying the protophase to phase transformation [44]

for the chosen frequency pair, although there are many other combinations for the oscillations frequencies and their harmonics and each gives a slightly different result. The CMI in Fig. 5.7d is also able to trace the general shape of the coupling strength in time, although again slightly different results can be obtained depending on the time delay used. In Fig. 5.7e, the Bayesian method is able to track the coupling with a high degree of accuracy after the initial window where the prior information is unknown. Additionally, the coupling functions Fig. 5.7f and g were derived from completely different time windows but are an almost identical match, which demonstrates the robust dependence on the coupling parameters.

5.4.5 Summary

Previous methods used in inverse problems of deterministic dynamical systems have relied on time-independent representations. Analyses performed in phase space and the frequency domain provide measures of the dimensionality and complexity of a system. However, when the same analyses are applied to nonautonomous systems the results can be misleading, causing the system to appear equivalent to a more complex or even stochastic system.

In order to understand the true nature of a nonautonomous system from a single time series it is important to use methods that are able to track time-dependence. This allows the separation of the purely time-dependent parts of the system so that a simple nonautonomous limit cycle oscillator is not mistaken for a chaotic or stochastic system.

5.5 Living Systems as Coupled Nonautonomous Oscillators

5.5.1 Introduction

Spontaneous oscillations are abundant in nature, from the cellular level [22] to oscillations of whole populations [51]. No matter at which level of complexity we observe the dynamics of a living system the net balance of concentrations associated with its functioning will be a dynamical process rather than a static one, and will be associated with a dynamical equilibrium. This is a consequence of a continuous exchange of energy and matter of each living unit with its environment, or its direct perturbation from the environment.

The recognition that living systems are characterised by many interacting rhythms can be traced back to at least as early as to the experiments by Hales [27]. He observed that the heart rate, seen as pulsations of blood flow and blood pressure, is modulated by the respiratory rhythm and in this way introduced what has been studied much later as coupled oscillatory processes. The interaction between the heart beat and respiratory rhythm is known as respiratory sinus arrythmia and its precise physiological mechanisms are still subject of investigation. Below we will briefly discuss some of the current progress.

The brain waves are another types of rhythms associated with functioning of a human organisms, which were discovered as soon as the electrical activity of the brain was non-invasively recorded [9]. Almost 100 years later their neurophysiological mechanisms are still poorly understood, owing to the immense number of connections between neurons in the brain and the resulting complex spatio-temporal dynamics. Although intracellular oscillations were reported as early as in 1975 [22], we still cannot link the basic cellular oscillations to those observed in the ensembles of neurones in the brain.

One of the obstacles in understanding the functioning of living systems has been a lack of appropriate physical framework as well as a mathematical description of nonautonomous systems far from thermodynamical equilibrium. However, these areas have faced a huge development in the last decades which is now coming to fruition. There are several milestones in this development, both in theory and numerical approaches. One line of research resulted in the introduction of the theory of cooperative phenomena by Haken [26], the theory of entrainment and phase resetting by Winfree [95] and the phase dynamics approach by Kuramoto [45]. The other line of research resulted in developments in the theory of stochastic dynamics [55]. At the same time various numerical approaches were proposed to deal with the growing amount of time series that are now easy to record. Generally, due to their immense complexity living systems are most comfortably treated as stochastic and their statistical properties are elaborated in various ways [80].

Taking the nonautonomicity as one of the fundamental properties of living systems, we adopt another approach. Namely, paying particular attention to the effect of interactions by inferring it as a time evolution, we can reduce an immensely complex dynamics to a mainly deterministic dynamics of interacting units. In this way we can treat a human as a network of interacting ensembles of oscillators. Armed with the methods presented in the Sects. 5.2–5.4 we will take cardio-respiratory interactions to illustrate our point.

5.5.2 Dynamical Inference of Cardio-respiratory Coupling Function

We now take real data recorded from healthy subjects and focus on cardio-respiratory coupling functions and their change with age. The major results of this study, where 197 subjects of all age spanning from 16 to 90 years were included, have been published earlier [32, 75]. The electrical activity of the heart (ECG) was recorded with electrodes placed over bony prominence: two over the shoulders and one over the lover left rib. The respiration was recorded using an elastic belt with an attached Biopac TSD201 Respiratory Effort Transducer (Biopac Systems Inc., CA, USA) positioned around the chest. The signals were recorded continuously and simultaneously for 30 min with subjects lying relaxed and supine in a quiet environment at normal room temperature. In a sense, we have reproduced Hales' experiment, asking two questions—

(i) What is the coupling function that maintains the respiratory sinus arrythmia? and
(ii) What happens to the coupling function with ageing?

Both instantaneous frequencies, cardiac (IHR) and respiratory (IRF), were extracted using synchrosqueezed wavelet transform (SWT, for details see [32]). It is obvious from Fig. 5.8a that both are not constant, but are varying in time. The variations can be equally considered as either resulting from stochastic or deterministic modulation/perturbation. To date, both approaches have been applied and two major conclusions can be drawn—

(i) The spectral peaks of both the cardiac and respiratory instantaneous frequencies contain several time-varying oscillations. This means that several oscillatory

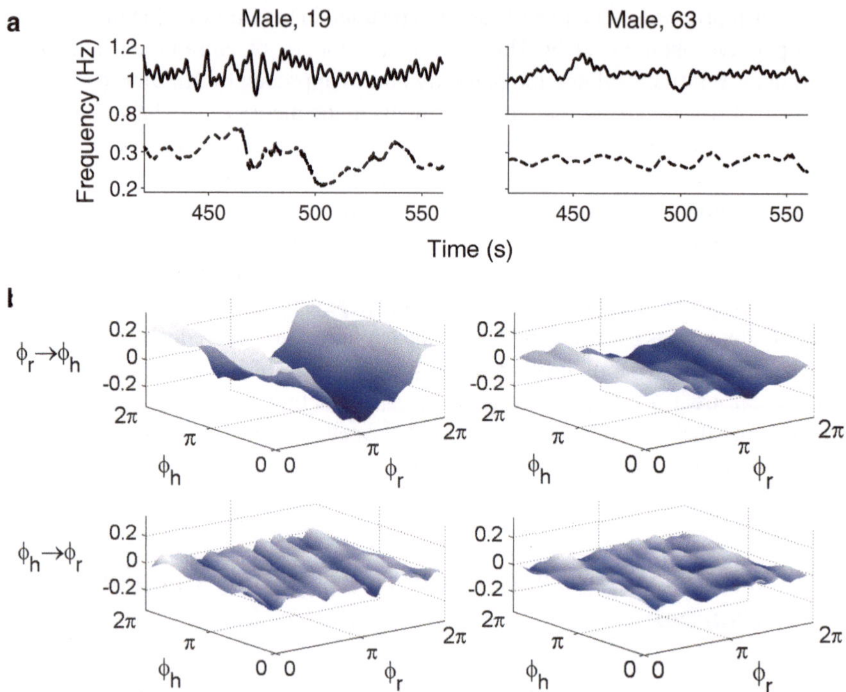

Fig. 5.8 (**a**) Instantaneous cardiac and respiratory frequencies for a young and an aged subject and (**b**) the corresponding coupling functions

processes are perturbing/ modulating the beating of the heart [81, 82]. A similar situation holds for the instantaneous respiratory frequency [38]. The power of these perturbations/modulations is reduced with age [32, 75]. Hence, one can treat both, the cardiac and the respiratory process as resulting from thousands of highly 1:1 synchronized oscillators that are continuously perturbed by other processes. Because all cells in the ensembles of the "heart" and "respiration" are synchronized most of the time—with a pacemaker in case of the heart and with a similar mechanism in case of the lungs—we will consider their macroscopic behaviour only, reducing each to a single oscillator. They also perturb each other and below we will extract the extent of this perturbation.

(ii) The instantaneous frequencies are highly complex and stochastic processes. To date various methods to extract complexity have been applied, especially to the instantaneous heart frequency which is also known as heart rate variability (HRV). In addition, it has been shown that the complexity highly significantly reduces with age [23, 75].

5.5.2.1 A Model of Cardio-respiratory Interactions

We now model the cardio-respiratory system as a pair of nonautonomous, noisy coupled oscillators with phases $\phi_{h,r}$, where h implies heart and r implies respiration

$$\begin{aligned}\dot{\phi}_h(t) &= 2\pi\omega_h(t) + q_h(\phi_h, \phi_r, t) + \xi_h(t)\\ \dot{\phi}_r(t) &= 2\pi\omega_r(t) + q_r(\phi_h, \phi_r, t) + \xi_r(t).\end{aligned} \tag{5.34}$$

The time-derivatives of the phases $\dot{\phi}_{h,r}/2\pi$ correspond to IHF and IRF, $\omega_{h,r}$ denote natural frequencies, and $\xi_{h,r}(t)$ is assumed to be white Gaussian noise: $\langle\xi_i(t)\xi_j(\tau)\rangle = \delta(t-\tau)E_{ij}$. Note that the coupling functions $q_{h,r}$ are time-dependent.

The model (5.34) explicitly includes only the phases of the cardiac and respiratory activities. However, it is known that the IHF and IRF are modulated by other processes as well, e.g. modulation of IHF at low frequencies around 0.1 Hz [82] and here we incorporate these external influences by considering nonautonomicity and time-dependent parameters.

5.5.2.2 Extraction of Nonlinear Interactions Using Dynamical Inference

In brief, the Bayesian method, extended to account for nonautonomicity [18,78,79] as discussed in Sect. 5.4, is applied to the phases extracted using SWT. Thus, we model the right hand sides of (5.34) as a sum of Fourier basis functions multiplied by some coefficients, $\dot{\phi}_{h,r} = C_{h,r;1}f_1(\phi_h, \phi_r) + C_{h,r;2}f_2(\phi_h, \phi_r) + \ldots + \xi_{h,r}(t)$, and find most probable values of these coefficients $C_{h,r;i}$ in each time window. Fourier series up to second order, $\sin(m\phi_h - n\phi_r)$, $\cos(m\phi_h - n\phi_r)$ with $n = -2, -1, 0, 1, 2$ and $m = -2, -1, 0, 1, 2$, were used.

The inference was performed within non-overlapping windows of time length 50 s, chosen to incorporate at least ten cycles of the slower oscillatory process and thus provide enough information for accurate inference. While further details of analyses and the results obtained for all subjects can be found in [32], in Fig. 5.8, here we present results characteristic of a young and an older subject.

To gain an insight into the nature of cardio-respiratory interactions, we have reconstructed the time-varying coupling functions $q_{h,r}(\phi_h, \phi_r, t)$ in (5.34). Figure 5.8b shows the coupling functions $q_{h,r}$ typical of a young and an older subject. It is obvious that the heart coupling function, q_h, changes markedly with age. We now show that especially it is its time-variability that changes, whereas the respiratory coupling function, q_r, seems to be irregular and unaffected by age.

5.5.3 Summary

What have we learned by considering cardiac and respiratory oscillatory properties as nonautonomous? We have confirmed that the variability of heart rate and hence its complexity reduce with age. But in addition, we have shown that this reduction is mainly on account of reduced influence of the respiration to the heart and moreover the reduced variability of this influence. In other words, we have shown that the cardiac function is becoming more autonomous with age, pointing out that nonautonomicity and the ability to sustain stable functioning far from thermodynamic equilibrium are essential for life.

5.6 Outlook

In this chapter we first provided a general description of nonautonomous oscillatory systems, discussing their properties and the ways in which they are usually treated. We proceeded to discuss the difference between non-stationary and nonautonomous dynamics. Namely, the notation and relationship between the two is often misinterpreted, especially when certain phenomena are investigated from an experimental point of view. In Sect. 5.2 we point out that non-stationarity is a statistical measure that one obtains from a signal, while the nonautonomous dynamics can act as a functional generator of non-stationary signals.

Nonautonomous systems are a very broad class of systems and in this chapter we narrowed down our interest to self-sustained nonautonomous oscillators. Furthermore, we studied the interactions and particularly the synchronization state of such oscillators under nonautonomous perturbations. We first used a model consisting of two coupled Poincaré limit-cycle oscillators with periodically varying frequency parameters. By the use of numerical simulations and methods for phase synchronization detection, we presented how the nonautonomous sources can affect the interacting dynamics. Moreover, we show that the phase difference can become a time-varying process leading to qualitative transitions in the synchronization ratios, or intermittent synchronization with transitions between synchronized and non-synchronized states. Transitions between synchronization ratios and between synchronized and non-synchronized states are frequently observed in analyses of biological oscillators, e.g. cardio-respiratory interactions [32, 38, 82], which we briefly discuss in Sect. 5.5.

In the case of a population of nonautonomous oscillators (Sect. 5.3), the external forcing is superimposed on the dynamics of the original system—for any parameter of forcing. We have shown numerically and analytically [62] that this influence can be quantified for periodic forcing depending on the frequency of the external forcing compared to the system's transient time, i.e. the system's homogeneity. We show that this new approach could be of great importance in modelling systems with multiple time scales of oscillations and interactions, such as the human cardiovascular system [82], interactions between the cardiovascular system and the brain [81], or interactions between inhibitory neurons in the cortex [20].

The generalization of the KM that encompasses NA systems is directly applicable to any thermodynamically open system. For example, the observed time variations of brain dynamics can be easily explained as a consequence of TD frequencies or couplings of the single neurons, where the source of the external variation could be due to anaesthesia [71], event related [64], or due to some influence from another part of the brain, or the cardiovascular system [81]. In the brain dynamics, these findings could be used to explain how slow-varying signals from the cardio-vascular system [82] could modulate membrane potentials of the neuronal populations, leading to modulated spiking activity. Analogously, the same slow signals would have a greater influence on the group dynamics of the neurons, while the faster signals from the brain would mostly influence more homogeneous and more synchronized neurons.

An important issue for the further research is the bifurcation analysis for the TD KM described by its low-dimensional dynamics. This is not a trivial problem because of the explicit dependence of the model parameters (frequencies, couplings) on time. Unfortunately, in such a case the classical approach used for autonomous systems cannot be applied. However the system's dynamics are deterministic and could experience different macroscopic states at different points in the time, depending on the system's parameters. Hence, one possible way out—of a big importance for the theory of NA systems—is determining the times or other space points where these transitions occur.

We also discussed recent developments in numerical methods for the analysis of data recorded from nonautonomous systems (Sect. 5.4). With the rapid increase in computational facilities today, the range of methods used for the time series analysis of nonautonomous systems continues to grow. However, unlike other types of deterministic systems there is still no theorem to embed the time series of a nonautonomous system in phase space. Such a theorem would have to track the time-dependent attractor of the system, which in the absence of an analytic solution can currently only be found by observing many trajectories with different initial conditions.

We have also illustrated that the coupling function between the cardiac and respiratory activity becomes weaker with ageing. At the same time new aspects of mitochondrial function and their role in diseases of both systemic and pulmonary vessels have been continuously revealed [17]. One of the challenges is now to relate the mitochondrial oscillations to those in the cardiovascular system and see if and how they change with ageing. Undoubtedly, the theory of nonautonomous oscillators in this endeavour will be crucial.

Following from the theory of self-sustained nonautonomous oscillators a new class named chronotaxic systems was also recently defined [90]. It provides formalism for deterministic systems that can maintain stable frequencies. The chronotaxic attractor in these systems is point pullback attractor related to the ones discussed in Chap. 1. It has already been shown that the heart can be readily described as a chronotaxic system. Furthermore, we expect many systems to be identified as chronotaxic, despite the fact that some were previously characterised as stochastic. Additional work, however, is needed to generalise the theory of chronotaxic systems including how they should be tackled in an inverse approach.

Acknowledgements This work was supported by the Engineering and Physical Sciences Research Council (UK) [Grant No. EP/100999X1]. Our grateful thanks are due to A. Duggento, D. Iatsenko, P.V.E. McClintock and Y. Suprunenko for many useful discussions.

References

1. H.D.I. Abarbanel, R. Brown, J.J. Sidorowich, L.S. Tsimring, The analysis of observed chaotic data in physical systems. Rev. Mod. Phys. 65(4), 1331–1392 (1993)
2. J.A. Acebrón, R. Spigler, Adaptive frequency model for phase-frequency synchronization in large populations of globally coupled nonlinear oscillators. Phys. Rev. Lett. **81**(11), 2229–2232 (1998)
3. J.A. Acebrón, L.L. Bonilla, S. De Leo, R. Spigler, Breaking the symmetry in bimodal frequency distributions of globally coupled oscillators. Phys. Rev. E **57**(5), 5287–5290 (1998)
4. J.A. Acebrón, L.L. Bonilla, C.J. Pérez Vicente, F. Ritort, R. Spigler, The Kuramoto model: a simple paradigm for synchronization phenomena. Rev. Mod. Phys. **77**, 137–185 (2005)
5. A.A. Andronov, A.A. Vitt, S.E. Khaikin, *The Theory of Oscillators* (Dover, New York, 2009)
6. V. Anishchenko, T. Vadivasova, G. Strelkova, Stochastic self-sustained oscillations of non-autonomous systems. Eur. Phys. J. Spec. Top. **187**, 109–125 (2010)
7. A. Bahraminasab, F. Ghasemi, A. Stefanovska, P.V.E. McClintock, H. Kantz, Direction of coupling from phases of interacting oscillators: a permutation information approach. Phys. Rev. Lett. **100**(8), 084101 (2008)
8. T. Bayes, An essay towards solving a problem in the doctrine of chances. Philos. Trans. **53**, 370–418 (1763)
9. H. Berger, Ueber das Elektroenkephalogramm des Menschen. Arch. Psychiatr. Nervenkr. **87**, 527–570 (1929)
10. L.L. Bonilla, J.C. Neu, R. Spigler, Nonlinear stability of incoherence and collective synchronization in a population of coupled oscillators. J. Stat. Phys. **67**, 313–330 (1992)
11. R. Brown, P. Bryant, H.D.I. Abarbanel, Computing the Lyapunov spectrum of a dynamical system from an observed time series. Phys. Rev. A **43**(6), 2787–2806 (1991)
12. G. Buzsáki, A. Draguhn, Neuronal oscillations in cortical networks. Science **304**, 1926–1929 (2004)
13. M.Y. Choi, Y.W. Kim, D.C. Hong, Periodic synchronization in a driven system of coupled oscillators. Phys. Rev. E **49**(5), 3825–3832 (1994)
14. D. Cumin, C.P. Unsworth, Generalising the kuramoto model for the study of neuronal synchronisation in the brain. Physica D 226, 181–196 (2007)
15. D.-F. Dai, P.S. Rabinovitch, Z. Ungvari, Mitochondria and cardiovascular aging. Circ. Res. **110**, 1109–1124, (2012)
16. I. Daubechies, J. Lu, H.-T. Wu, Synchrosqueezed wavelet transforms: an empirical mode decomposition-like tool. Appl. Comput. Harmon. Anal. **30**(2), 243–261 (2011)
17. P. Dromparis, E.D. Michelakis, Mitochondria in vascular health and disease. Annu. Rev. Physiol. **75**, 95–126 (2013)
18. A. Duggento, T. Stankovski, P.V.E. McClintock, A. Stefanovska, Dynamical Bayesian inference of time-evolving interactions: from a pair of coupled oscillators to networks of oscillators. Phys. Rev. E **86**, 061126 (2012)
19. J.P. Eckmann, D. Ruelle, Ergodic theory of chaos and strange attractors. Rev. Mod. Phys. **57**(3), 617–656 (1983)

20. G.B. Ermentrout, M. Wechselberger, Canards, clusters, and synchronization in a weakly coupled interneuron model. SIAM J. Appl. Dyn. Syst. **8**, 253–278 (2009)
21. A.M. Fraser, H.L. Swinney, Independent coordinates for strange attractors from mutual information. Phys. Rev. A **33**(2), 1134–1140 (1986)
22. G. Gerisch, U. Wick, Intracellular oscillations and release of cyclic-AMP from dictiostelium cells. Biochem. Biophys. Res. Commun. **65**(1), 364–370 (1975)
23. A.L. Goldberger, L.A.N. Amaral, J.M. Hausdorff, P.C. Ivanov, C.K. Peng, H.E. Stanley. Fractal dynamics in physiology: alterations with disease and aging. Proc. Natl. Acad. Sci. USA **99**(Suppl. 1), 2466–2472 (2002)
24. P. Grassberger, I. Procaccia, Characterization of strange attractors. Phys. Rev. Lett. **50**(5), 346–349 (1983)
25. P. Grassberger, I. Procaccia, Measuring the strangeness of strange attractors. Physica D **9**, 189–208 (1983)
26. H. Haken, Cooperative phenomena in systems far from thermal equilibrium and in nonphysical systems. Rev. Mod. Phys. **47**, 67–121 (1975)
27. S. Hales, *Statistical Essays II, Hæmastatisticks* (Innings Manby, London, 1733)
28. K. Hasselmann, W. Munk, G. MacDonald, Bispectra of ocean waves, in *Time Series Analysis* (Wiley, New York, 1963), pp. 125–139
29. H. Hong, S.H. Strogatz, Kuramoto model of coupled oscillators with positive and negative coupling parameters: an Example of conformist and contrarian oscillators. Phys. Rev. Lett. **106**(5), 054102 (2011)
30. W. Horsthemke, R. Lefever, *Noise Induced Transitions* (Springer, Berlin, 1984)
31. D. Iatsenko, S. Petkoski, P.V.E. McClintock, A. Stefanovska, Stationary and traveling wave states of the Kuramoto model with an arbitrary distribution of frequencies and coupling strengths. Phys. Rev. Lett. **110**(6), 064101 (2013)
32. D. Iatsenko, A. Bernjak, T. Stankovski, Y. Shiogai, P.J. Owen-Lynch, P.B.M. Clarkson, P.V.E. McClintock, A. Stefanovska, Evolution of cardio-respiratory interactions with age. Philos. Trans. R. Soc. A **371**(1997), 20110622 (2013)
33. J. Jamšek, A. Stefanovska, P.V.E. McClintock, I. A. Khovanov, Time-phase bispectral analysis. Phys. Rev. E **68**(1), 016201 (2003)
34. J. Jamšek, A. Stefanovska, P.V.E. McClintock, Wavelet bispectral analysis for the study of interactions among oscillators whose basic frequencies are significantly time variable. Phys. Rev. E **76**, 046221 (2007)
35. J. Jamšek, M. Paluš, A. Stefanovska, Detecting couplings between interacting oscillators with time-varying basic frequencies: instantaneous wavelet bispectrum and information theoretic approach. Phys. Rev. E **81**(3), 036207 (2010)
36. K. Karhunen, Zur spektraltheorie stochastischer prozesse. Ann. Acad. Sci. Fenn. A1, Math. Phys. **37** (1946)
37. M.B. Kennel, R. Brown, H.D.I Abarbanel. Determining embedding dimension for phase-space reconstruction using a geometrical construction. Phys. Rev. A **45**(6), 3403–3411 (1992)
38. D.A. Kenwright, A. Bahraminasab, A. Stefanovska, P.V.E. McClintock, The effect of low-frequency oscillations on cardio-respiratory synchronization. Eur. Phys. J. B. **65**(3), 425–433 (2008)
39. H.S. Kim, R. Eykholt, J.D. Salas, Nonlinear dynamics, delay times and embedding windows. Physica D **127**(1–2), 48–60 (1999)
40. P.E. Kloeden, Synchronization of nonautonomous dynamical systems. Electron. J. Differ. Equ. **1**, 1–10 (2003)
41. P.E. Kloeden, Nonautonomous attractors of switching systems. Dyn. Syst. **21**(2), 209–230 (2006)
42. P.E. Kloeden, R. Pavani, Dissipative synchronization of nonautonomous and random systems. GAMM-Mitt. **32**(1), 80–92 (2009)
43. P.E. Kloeden, M. Rasmussen, *Nonautonomous Dynamical Systems*. AMS Mathematical Surveys and Monographs (American Mathematical Society, New York, 2011)

44. B. Kralemann, L. Cimponeriu, M. Rosenblum, A. Pikovsky, R. Mrowka, Phase dynamics of coupled oscillators reconstructed from data. Phys. Rev. E **77**(6, Part 2), 066205 (2008)
45. Y. Kuramoto, *Chemical Oscillations, Waves, and Turbulence* (Springer, Berlin, 1984)
46. F.T. Kurz, M.A. Aon, B. O'Rourke, A.A. Armoundas, Spatio-temporal oscillations of individual mitochondria in cardiac myocytes reveal modulation of synchronized mitochondrial clusters. Proc. Natl. Acad. Sci. USA **107**, 14315–14320 (2010)
47. S.P. Kuznetsov, A. Pikovsky, M. Rosenblum, Collective phase chaos in the dynamics of interacting oscillator ensembles. Chaos **20**, 043134 (2010)
48. S.H. Lee, S. Lee, S.-W. Son, P. Holme, Phase-shift inversion in oscillator systems with periodically switching couplings. Phys. Rev. E **85**, 027202 (2006)
49. M. Loève, *Fonctions aleatoires de second ordre*. C.R. Acad. Sci. Paris **222**(1946)
50. R. Mañé, On the dimension of the compact invariant sets of certain non-linear maps, in *Dynamical Systems and Turbulence*, ed. by D.A. Rand, L.S. Young. Lecture Notes in Mathematics, vol. 898 (Springer, New York, 1981)
51. R.M. May, Biological populations with nonoverlapping generations — stable points, stable cycles, and chaos. Science **186**(4164), 645–647 (1974)
52. R. Mirollo, S.H. Strogatz, The spectrum of the partially locked state for the Kuramoto model. J. Nonlinear Sci. **17**(4), 309–347 (2007)
53. E. Montbrio, D. Pazo, Shear diversity prevents collective synchronization. Phys. Rev. Lett. **106**(25), 254101 (2011)
54. E. Montbrio, J. Kurths, B. Blasius, Synchronization of two interacting populations of oscillators. Phys. Rev. E *70*(5), 056125 (2004)
55. F. Moss, P.V.E. McClintock (ed.), *Noise in Nonlinear Dynamical Systems*, vols. 1–3 (Cambridge University Press, Cambridge, 1989)
56. C.L. Nikias, M.R. Raghuveer, Bispectrum estimation: a digital signal processing framework. IEEE Proc. **75**(7), 869–891 (1987)
57. C.L. Nikias, A.P. Petropulu, *Higher-Order Spectra Anlysis: A Nonlinear Signal Processing Framework* (Prentice-Hall, Englewood Cliffs, 1993)
58. E. Ott, T.M. Antonsen, Low dimensional behavior of large systems of globally coupled oscillators. Chaos **18**(3), 037113 (2008)
59. E. Ott, T.M. Antonsen, Long time evolution of phase oscillator systems. Chaos **19**(2), 023117 (2009)
60. M. Paluš, From nonlinearity to causality: statistical testing and inference of physical mechanisms underlying complex dynamics. Contemp. Phys. **48**(6), 307–348 (2007)
61. M. Paluš, A. Stefanovska, Direction of coupling from phases of interacting oscillators: an information-theoretic approach. Phys. Rev. E **67**, 055201(R) (2003)
62. S. Petkoski, A. Stefanovska, Kuramoto model with time-varying parameters. Phys. Rev. E **86**, 046212 (2012)
63. S. Petkoski, D. Iatsenko, L. Basnarkov, A. Stefanovska, Mean-field and mean-ensemble frequencies of a system of coupled oscillators. Phys. Rev. E **87**(3), 032908 (2013)
64. G. Pfurtschelle, F.H. Lopes da Silva, Event-related eeg/meg synchronization and desynchronization: basic principles. SIAM J. Appl. Dyn. Syst. **110**, 1842–1857 (1999)
65. A. Pikovsky, M. Rosenblum, J. Kurths, *Synchronization — A Universal Concept in Nonlinear Sciences* (Cambridge University Press, Cambridge, 2001)
66. M. Rasmussen, *Attractivity and Bifurcation for Nonautonomous Dynamical Systems* (Springer, Berlin, 2007)
67. C. Rhodes, M. Morari, False-nearest-neighbors algorithm and noise-corrupted time series. Phys. Rev. E **55**(5), 6162–6170 (1997)
68. M.G. Rosenblum, A.S. Pikovsky, J. Kurths, Phase synchronization of chaotic oscillators. Phys. Rev. Lett. **76**(11), 1804–1807 (1996)
69. J. Rougemont, F. Naef, Collective synchronization in populations of globally coupled phase oscillators with drifting frequencies. Phys. Rev. E **73**, 011104 (2006)
70. H. Sakaguchi, Cooperative phenomena in coupled oscillator sytems under external fields. Prog. Theor. Phys. **79**(1), 39–46 (1988)

71. J.H. Sheeba, A. Stefanovska, P.V.E. McClintock, Neuronal synchrony during anesthesia: a thalamocortical model. Biophys. J. **95**(6), 2722–2727 (2008)
72. J.H. Sheeba, V.K. Chandrasekar, A. Stefanovska, P.V.E. McClintock, Asymmetry-induced effects in coupled phase-oscillator ensembles: routes to synchronization. Phys. Rev. E **79**, 046210 (2009)
73. L.W. Sheppard, A. Stefanovska, P.V.E. McClintock, Detecting the harmonics of oscillations with time-variable frequencies. Phys. Rev. E **83**, 016206 (2011)
74. S. Shinomoto, Y. Kuramoto, Phase transitions in active rotator systems. Prog. Theor. Phys. **75**(5), 1105–1110 (1986)
75. Y. Shiogai, A. Stefanovska, P.V.E. McClintock, Nonlinear dynamics of cardiovascular ageing. Phys. Rep. **488**, 51–110 (2010)
76. M. Small, *Applied Nonlinear Time Series Analysis: Applications in Physics, Physiology and Finance* (World Scientific, Singapore, 2005)
77. P. So, A. Bernard, B.C. Cotton, E. Barreto, Synchronization in interacting populations of heterogeneous oscillators with time-varying coupling. Chaos **18**, 037114 (2008)
78. T. Stankovski, A. Duggento, P.V.E. McClintock, A. Stefanovska, Inference of time-evolving coupled dynamical systems in the presence of noise. Phys. Rev. Lett. 109, 024101 (2012)
79. T. Stankovski, *Tackling the Inverse Problem for Non-Autonomous Systems: Application to the Life Sciences* Springer Theses (Springer, Cham, 2013)
80. H.E. Stanley, L.A.N. Amaral, A.L. Goldberger, S. Havlin, P.C. Ivanov, C.K. Peng, Statistical physics and physiology: monofractal and multifractal approaches. Physica D **270**(1–2), 309–324 (1999)
81. A. Stefanovska, Coupled oscillators: complex but not complicated cardiovascular and brain interactions. IEEE Eng. Med. Bio. Mag. **26**(6), 25–29 (2007)
82. A. Stefanovska, M. Bračič, Physics of the human cardiovascular system. Contemp. Phys. **40**(1), 31–55 (1999)
83. A. Stefanovska, P. Krošelj, Correlation integral and frequency analysis of cardiovascular functions. Open Syst. Inf. Dyn. **4**, 457–478 (1997)
84. A. Stefanovska, S. Strle, P. Krošelj, On the overestimation of the correlation dimension. Phys. Lett. A **235**(1), 24–30 (1997)
85. A. Stefanovska, H. Haken, P.V.E. McClintock, M. Hožič, F. Bajrović, S. Ribarič, Reversible transitions between synchronization states of the cardiorespiratory system. Phys. Rev. Lett. **85**(22), 4831–4834 (2000)
86. R.L. Stratonovich, *Topics in the Theory of Random Noise: General Theory of Random Processes, Nonlinear Transformations of Signals and Noise*. Mathematics and Its Applications (Gordon and Breach, New York, 1963)
87. S.H. Strogatz, R.E. Mirollo, Stability of incoherence in a population of coupled oscillators. J. Stat. Phys. **63**(3–4), 613–635 (1991)
88. S.H. Strogatz, From Kuramoto to Crawford: exploring the onset of synchronization in populations of coupled oscillators. Physica D **143**, 1–20 (2000)
89. S.H. Strogatz, *Sync: The Emerging Science of Spontaneous Order* (Hyperion, New York, 2003)
90. Y.F. Suprunenko, P.T. Clemson, A. Stefanovska, Chronotaxic systems: a new class of self-sustained nonautonomous oscillators. Phys. Rev. Lett. **111**(2), 024101 (2013)
91. F. Takens, Detecting strange attractors in turbulence, in *Dynamical Systems and Turbulence*, ed. by D.A. Rand, L.S. Young. Lecture Notes in Mathematics, vol. 898 (Springer, New York, 1981)
92. D. Taylor, E. Ott, J.G. Restrepo, Spontaneous synchronization of coupled oscillator systems with frequency adaptation. Phys. Rev. E **81**(4), 046214 (2010)
93. M. Vejmelka, M. Paluš, Inferring the directionality of coupling with conditional mutual information. Phys. Rev. E. **77**(2), 026214 (2008)
94. K. Wiesenfeld, P. Colet, S.H. Strogatz, Synchronization transitions in a disordered josephson series array. Phys. Rev. Lett. **76**(3), 404–407 (1996)
95. A.T. Winfree, *The Geometry of Biological Time* (Springer, New York, 1980)

Chapter 6
Multisite Mechanisms for Ultrasensitivity in Signal Transduction

Germán A. Enciso

Abstract One of the key aspects in the study of cellular communication is understanding how cells receive a continuous input and transform it into a discrete, all-or-none output. Such so-called ultrasensitive dose responses can also be used in a variety of other contexts, from the efficient transport of oxygen in the blood to the regulation of the cell cycle and gene expression. This chapter provides a self contained mathematical review of the most important molecular models of ultrasensitivity in the literature, with an emphasis on mechanisms involving multisite modifications. The models described include two deeply influential systems based on allosteric behavior, the MWC and the KNF models. Also included is a description of more recent work by the author and colleagues of novel mechanisms using alternative hypotheses to create ultrasensitive behavior.

Keywords Systems biology • Ultrasensitivity • Allostery • Cooperativity • Signal transduction

6.1 Introduction: Ultrasensitive Dose Responses

Chemical reaction networks (CRN) lie at the heart of many biochemical processes inside the cell. They have been extensively modeled to understand the behavior of specific systems, and they have also been systematically studied at the theoretical level [5, 14, 23]. Although often implicitly assumed to converge globally towards a unique equilibrium in chemical engineering and other applications, CRNs can have exceedingly complex dynamical behavior. Moreover, many biological systems have arguably evolved towards precisely such relatively rare complex examples, driven

G.A. Enciso (✉)
Mathematics Department, University of California, Irvine, Irvine, CA, USA
e-mail: enciso@uci.edu

P.E. Kloeden and C. Pötzsche (eds.), *Nonautonomous Dynamical Systems in the Life Sciences*, Lecture Notes in Mathematics 2102, DOI 10.1007/978-3-319-03080-7_6,

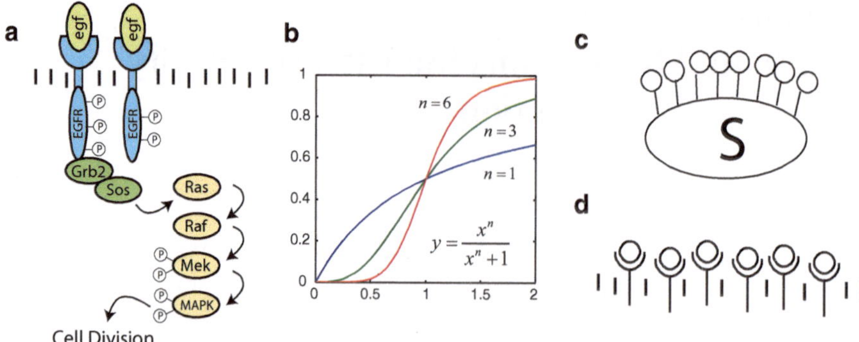

Fig. 6.1 (**a**) A sample signal transduction pathway involving the signaling protein egf and the output protein MAPK, which can trigger cell division. (**b**) The function $y = x^n/(x^n + 1)$ is the canonical example of an ultrasensitive function. The coefficient n quantifies the ultrasensitive behavior. (**c**) A protein can be activated through phosphorylation at multiple specific locations. (**d**) A receptor complex is activated through the collective binding of ligand to its multiple receptors

by a need to exhibit behaviors such as oscillations (e.g. circadian rhythms, cell cycle) and multistability (e.g. cell differentiation).

Here we focus on chemical reaction networks in the context of signal transduction, i.e. in the study of cell communication and the processing of information. Such systems have a parameter that is usually regarded as an input, and which corresponds e.g. to a signal molecule binding on the cell or to a component of a larger network. They also have an output, i.e. a molecule in the system that is thought to produce a downstream response and which represents the overall activity level of the network. For instance, a protein known as *epidermal growth factor*, or egf for short, is used as a messenger molecule to induce cells near the skin to divide after a wound (Fig. 6.1a). If a sufficient number of egf molecules bind to the membrane of a cell, a series of internal reactions takes place resulting in the activation of the output protein MAPK. This protein goes on to activate many transcription factors that can cause the cell to divide [2].

There are good reasons to study nonautonomous networks with inputs and outputs in biology, rather than autonomous networks. First, sometimes the full model would take place at a scale much larger than desired. If a hormone is used as a cell ligand input, the cell behavior might ultimately feed back into the tissues that produce the hormone. But modeling the full system would involve including other parts of the body, which is much larger in scope than an intra-cellular model. Another reason is that understanding subsets of a larger network, e.g. the way that Cdc28 affects Wee1 in the cell cycle, is often a fairly difficult and open problem in itself [31, 32] and a step towards an understanding of the complete network.

In the context of nonautonomous systems which is the main topic of this book, the goal would be to understand how the system responds over time given a time-varying input concentration. Although such time-varying inputs are ubiquitous in nature, the majority of experimental data in signal transduction measures the response of the system to a constant input concentration. A preliminary goal that is the focus of this chapter is to understand the so-called *dose response* of the system, that is, the steady state value of the output as a function of the input (assuming such a steady state is uniquely defined). A special case of high interest to many experimental biologists is that of *ultrasensitive*, or all-or-none behavior (Fig. 6.1b). Imagine that a cell is intended to divide in response to a sufficiently large egf stimulus, and to do nothing for a low egf stimulus. This essentially transforms a *continuous* egf input signal into a *binary*, all-or-none output, which should be reflected in the concentration of the output protein of this system, the protein MAPK. The all-or-none conversion of a continuous input into a binary output is quite common, not only in cell communication but also for the internal components of biochemical pathways such as the cell cycle, and simple networks that achieve it would likely be favored by evolution.

The most well-known mechanisms for ultrasensitivity involve *multisite systems*, in which one of the proteins has many identical modification sites. For instance, a very common protein modification known as phosphorylation involves the covalent attachment of a phosphate group to a specific location in the protein. Many proteins have not only one but multiple specific locations that can be phosphorylated at any given time (Fig. 6.1c). Multisite phosphorylation can cause dramatic changes to the shape and properties of the protein, to the point that it can effectively activate an otherwise inactive protein (or vice versa) [60]. In fact, the egf signal transduction cascade in Fig. 6.1a contains several proteins that are activated through multisite phosphorylation as indicated. Proteins can also be covalently modified in other ways, for instance they can be acetylated or methylated, also in very specific locations. These can act as the modification sites in other mathematical models [64].

An entirely different category of multisite protein modification is multisite ligand binding (Fig. 6.1d). For example, a group of membrane receptors can cluster together and trigger a downstream signal only when sufficiently many of them are bound to some signaling molecule [9, 19, 74]. A protein may bind to multiple nearby sites on a DNA molecule, in order to promote or prevent the expression of a gene [11, 28]. The direction of rotation of the flagellar motor in *E. coli* is controlled by the binding of a protein to one of 34 sites on a ring around the motor [20]. More often than not it is still unknown exactly how the different sites interact with each other and why there are many sites and not just fewer sites or even one.

This chapter will review the main mechanisms known to create ultrasensitive dose responses in biochemistry, as well as some newer mechanisms (some developed recently by the author) that have not been directly tested experimentally. The focus will be on multisite ultrasensitivity, but other well known mechanisms will also be discussed in a separate section. See also the lively review on cooperativity by Ferrell [25], and the recent more general review on signal transduction, including ultrasensitive responses, by Bluthgen et al. [7].

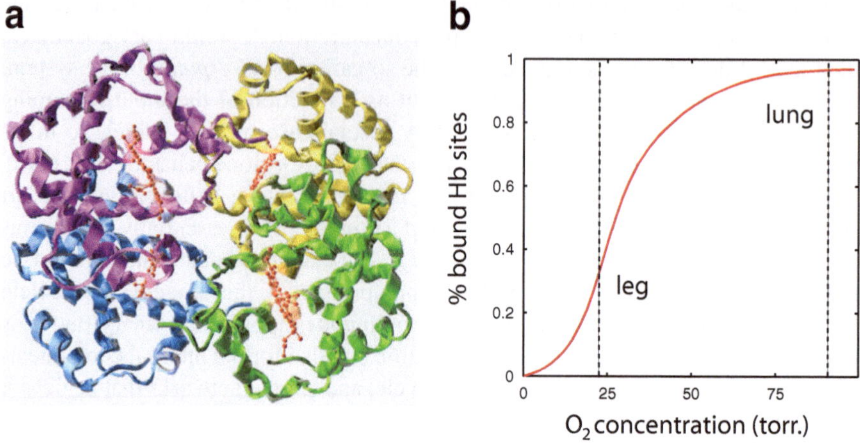

Fig. 6.2 (**a**) The structure of the hemoglobin protein as it binds to four O_2 molecules. From *Biology* by Brooker, Widmaier, Graham, and Stiling, copyright McGraw-Hill. (**b**) Hemoglobin has a high affinity to O_2 under high O_2 concentrations (such as in the lung) and a low affinity in low concentrations (such as in distant tissues), which allows for an efficient O_2 transport in the blood

6.2 Hemoglobin and Hill Functions

Early work on ultrasensitive behavior in biochemistry appears to have focused on hemoglobin, the molecule that transports much of the oxygen in the bloodstream (Fig. 6.2a). Physiologists were puzzled about the behavior of this protein: when oxygen concentration is low, it has a low binding affinity to oxygen. But when oxygen concentration increases, the affinity to oxygen grows fairly dramatically. This makes physiological sense: when the blood is in the lung, where oxygen abounds, hemoglobin captures as much of it as it can. When it is in the far reaches of the body (e.g. in a leg), the oxygen concentration is low and hemoglobin unloads its cargo (Fig. 6.2b). This leads to a much more efficient transport of oxygen than merely binding and unbinding at random times.

The open question was how hemoglobin can work in this efficient way. In 1910, a 23-year old scientist called A.V. Hill proposed a simple hypothetical reaction that could explain this [Hill, 36]. Each hemoglobin molecule would have n O_2-binding sites rather than one, and they would bind or unbind at the same time:

$$H + nO_2 \to C, \quad C \to H + nO_2. \tag{6.1}$$

Using mass action reaction kinetics [22, 23], the differential equation for the complex C is

$$\frac{dC}{dt} = \alpha H O_2^n - \beta C,$$

where α and β are the binding and unbinding reaction rates respectively. One mass conservation law that holds for this system is the preservation of the total amount of hemoglobin, whether bound or unbound to oxygen: $H + C = H_{tot}$. At steady state, one can set $\alpha H O_2^n = \beta C$ and replace H by $H_{tot} - C$ in order to solve for C as a function of the oxygen concentration:

$$C = H_{tot} \frac{O_2^n}{\frac{\beta}{\alpha} + O_2^n}. \tag{6.2}$$

As O_2 increases, the amount of oxygen bound to hemoglobin increases in an ultrasensitive way. The function $x^n/(K^n + x^n)$ became known as a Hill function, and it is one of the most important functions in mathematical biology. The exponent n in this function is known as the *Hill coefficient*, and it is a measure of the ultrasensitivity of the function (Fig. 6.1b). Incidentally, A.V. Hill went on to receive a Nobel prize in 1922 for his work on muscle physiology, and he is considered one of the founders of biophysics.

One should actually not take (6.1) too seriously, because a reaction involving such a large number of molecules is highly unlikely to take place. At most, this reaction can be thought of as a shorthand version of a reaction involving multiple steps. Depending on how the different steps are specifically spelled out, the high ultrasensitive behavior may or may not be preserved. Nevertheless this is an illustrative example of how one can derive a Hill function from first principles, involving the collective action of multiple individual sites. Hill functions are used in many contexts in mathematical biology, see for instance their application to PK-PD modeling in Chap. 7 of the present book by Koch and Schropp, as well as in Chap. 8 by Herrmann and Asai.

6.3 Cooperativity and the Adair Model

A ubiquitous concept in the study of multisite ultrasensitivity is that of *cooperativity*. A multisite protein is said to be cooperative if the modification of one of its sites (phosphorylation, ligand binding, etc) increases the rate of modification of its neighboring sites. The general idea is that cooperativity leads to ultrasensitive behavior, which is illustrated quantitatively in this section.

Suppose that a protein with n sites can be in states S_0 through S_n, where S_i represents the concentration of protein with exactly i modified sites. S_i turns into S_{i+1} at a linear rate equal to $a_i E S_i$, where E is the input concentration (enzyme, ligand, etc) (Fig. 6.3a). Assume that S_{i+1} turns back into S at a rate of $b_{i+1} S_i$. The differential equation for this system is

$$S_i' = a_i E S_{i-1} - b_i S_i - a_{i+1} E S_i + b_{i+1} S_{i+1}, \tag{6.3}$$

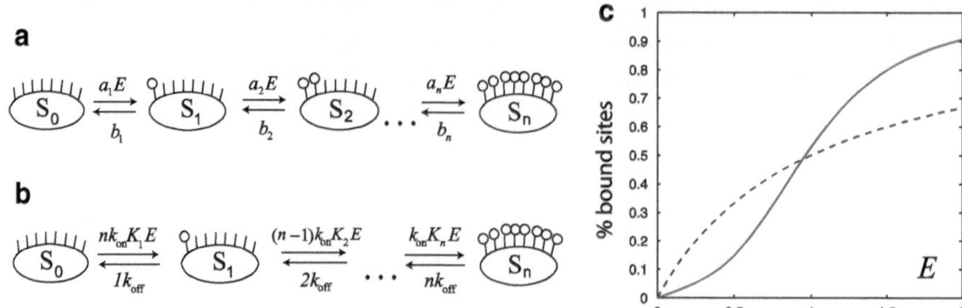

Fig. 6.3 (**a**) General sequential modification model. The corresponding ODE can be derived from the linear flow rates. (**b**) Adair model. In the special case $K_i = 1$, this model replicates a nonsequential ligand binding model with n independent sites. (**c**) Simulation of the Adair model for $n = 4$. If all $K_i = 1$, the fraction of bound sites is equal to $cE/(cE + 1)$, plotted (*dashed*) for $c := k_{on}/k_{off} = 1$. If $K_1 = K_2 = K_3 = 1$, $K_4 = 1000$, and $c = 0.2$, a more ultrasensitive behavior is obtained (*solid*)

for $i = 1 \dots n - 1$. For $i = 0$ and $i = n$ simply omit the first and last two terms, respectively. At steady state one can prove that $S_i = \frac{a_i}{b_i} E S_{i-1}$ for $i = 1 \dots n$ (prove first for $i = 1$, then through induction on i). Therefore

$$S_i = \frac{a_1 \dots a_i}{b_1 \dots b_i} E^i S_0, \quad i = 1 \dots n.$$

Defining the new parameters

$$A_0 := 1, \quad A_i := \frac{a_1 \dots a_i}{b_1 \dots b_i}, \quad i = 1 \dots n,$$

the total substrate concentration can be calculated as

$$S_{tot} = S_0 + \dots S_n = S_0 \sum_{i=0}^{n} A_i E^i.$$

Solving for S_0 it follows that

$$S_i = S_{tot} \frac{A_i E^i}{A_0 + A_1 E + A_2 E^2 + \dots + A_n E^n}, \quad i = 0, \dots, n. \tag{6.4}$$

A simple assumption made often in the literature is that the protein S is only active when it is fully modified. In that case, the dose response function is

$$f(E) = S_{tot} \frac{A_n E^n}{A_0 + A_1 E + A_2 E^2 + \dots + A_n E^n}.$$

By exploring different values for the parameters, one can observe that the ultrasensitive behavior of this dose response increases when the last few net modification constants a_i/b_i are larger than the first. This is because the middle terms in the denominator have a smaller influence and the function becomes similar to a Hill function with Hill coefficient n. For instance, set $n = 3, b_1 = b_2 = b_3 = 1$, and $a_1 = \varepsilon, a_2 = 1, a_3 = 1/\varepsilon$ for $\varepsilon < 1$. Then $A_1 = \varepsilon, A_2 = \varepsilon, A_3 = 1$, and

$$f(E) = \frac{E^3}{1 + \varepsilon E + \varepsilon E^2 + E^3}.$$

The so-called Adair model attempts to replicate the nonsequential behavior of multisite systems within the above framework (Fig. 6.3b). There is usually no order in which to modify or de-modify the sites, and any of the 2^n possible configurations can be found at any time. Suppose that each of the sites binds at a rate of $k_{on}E$ and unbinds at a rate of k_{off}. Then the rate at which S_0 flows into S_1 can be described as $nk_{on}E$, since there are n possible locations at which the modification can happen. Protein S_i can be de-modified at i different locations, so one can set $b_i := ik_{off}$. Similarly, set $a_1 := nk_{on}, a_2 := (n-1)k_{on}, \ldots, a_n := k_{on}$. If $c := k_{on}/k_{off}$, then $A_i = \binom{n}{i}c^i$. Also, at steady state

$$S_i = \binom{n}{i} c^i E^i S_0. \tag{6.5}$$

Using the binomial formula one also obtains

$$S_n = \frac{c^n E^n}{(1 + cE)^n}.$$

In the case of transport molecules such as hemoglobin, one might want to define another output such as the fraction of bound ligand, $Y(E)$. A calculation shows that at steady state

$$Y(E) = \frac{S_1 + 2S_2 + \ldots + nS_n}{nS_{tot}} = \frac{cE}{cE + 1}.$$

This is consistent with the idea of n sites independently binding and unbinding to the ligand, and it further confirms the intuition behind this particular choice of parameters.

Notice that in that model the sites are reacting with the ligand independently of each other. In the more general case of the Adair model, the sites are allowed to interact by setting $a_1 := nk_{on}K_1, \ldots, a_n := k_{on}K_n$, and $b_i = ik_{off}$. The abstract parameter $K_i \geq 1$ is intended to represent the cooperative effect of binding i sites beyond what would be expected from independent binding. Once again, it can be observed computationally that when the last few K_i are larger than the rest the ultrasensitive behavior increases. See Fig. 6.3c for a quantitative comparison of the system for $n = 4$ and $K_4 = 1$ as well as $K_4 = 1000$.

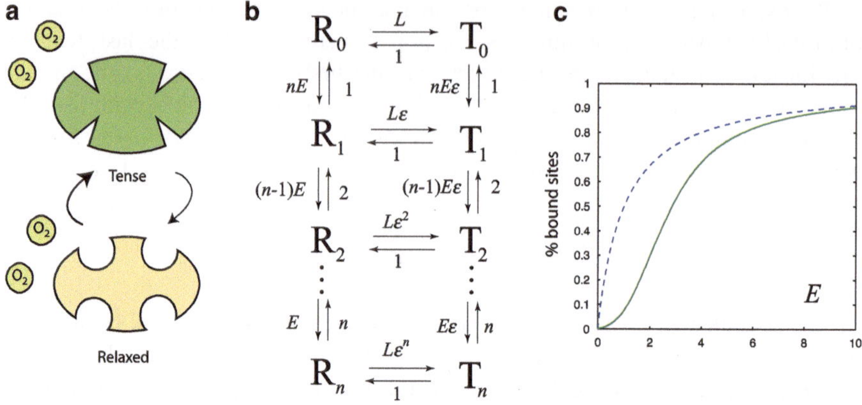

Fig. 6.4 (**a**) According to the MWC model, all four hemoglobin sites can change simultaneously from a low affinity (tense) state to a high affinity (relaxed) state. Oxygen binding traps the molecule in the relaxed state, allowing other oxygen molecules to bind at a higher rate. (**b**) Wire diagram of the MWC model. Here $L \gg 1$ and $\varepsilon \ll 1$. (**c**) Simulation of the model for $n = 4$ and $L = 100$. For $\varepsilon = 1$ the dynamics corresponds to that of the Adair model and the dose response is $E/(E+1)$ (*dashed*). For $\varepsilon = 0.01$ the ultrasensitivity increases (*solid*)

6.4 Allostery and the MWC Model

The contemporary theory of multisite ultrasensitivity was founded with an influential 1965 paper by Monod, Wyman and Changeux (MWC) [52]. To this day this paper receives around 200 citations every year according to Google scholar, and it has profoundly affected the way that biologists think of multisite interactions.

The main assumption in the paper is that the hemoglobin protein as a whole can spontaneously jump between a "tense" state of low O_2 affinity and a "relaxed" state of high O_2 affinity (Fig. 6.4a). The state is a global property of the protein, i.e. there cannot be both tense and relaxed sites simultaneously in the same protein. This invokes the concept of *allostery*, or the idea that there are strong internal interactions among the sites of a protein, such that changes in one site affect other sites as well.

The model reactions are described in Fig. 6.4b. T_i represents the concentration of the tense hemoglobin protein with exactly i bound oxygen ligands, and similarly R_i represents the relaxed protein with i bound ligands. The oxygen concentration is represented with the variable E. The relaxed molecule R_i has $n - i$ binding sites left and therefore its binding rate is $(n - i)E$. Its rate of unbinding is i, since each of its i sites are equally likely to unbind. This is reminiscent of the Adair model in the previous section, except that $K_i = 1$ in the MWC model. We are also using $k_{on} = k_{off} = 1$ for notational convenience.

A similar argument applies for the tense protein T_i, except that the binding rate is multiplied by a small number ε, which is intended to model the fact that the tense protein T_i cannot bind oxygen as well as R_i. R_0 spontaneously turns into T_0 at rate L, and back at rate 1. It is believed that for hemoglobin $L \approx 100$, i.e. in the absence of oxygen most of the protein is in the tense state. The remaining rates of exchange between R_i and T_i are determined by the detailed balance principle [22, 23].

At steady state, the detailed balance principle allows us to assume that each of the individual reactions is in equilibrium with its reverse reaction. So for instance,

$$(n-i)ER_i = (i+1)R_{i+1}, \quad (n-i)E\varepsilon T_i = (i+1)T_{i+1},$$

for $i = 1 \dots n$. It follows that

$$R_i = \binom{n}{i} E^i R_0, \quad T_i = \binom{n}{i} E^i \varepsilon^i T_0.$$

By the binomial formula $R_0 + \dots + R_n = R_0(E+1)^n$, and $T_0 + \dots + T_n = T_0(E\varepsilon+1)^n$. With some additional work one can also prove $R_1 + 2R_2 + \dots + nR_n = R_0 nE(E+1)^{n-1}$ by factoring out nE from each of the terms of the left hand side. Similarly for tense variables T_i. The total fraction of occupied sites (i.e. the output of this model) can then be computed as a function of E,

$$Y(E) = \frac{\sum_i iR_i + \sum_i iT_i}{\sum_i nR_i + \sum_i nT_i} = \frac{E(E+1)^{n-1} + LE\varepsilon(E\varepsilon+1)^{n-1}}{(E+1)^n + L(E\varepsilon+1)^n},$$

where $L = T_0/R_0$. This function is fairly ultrasensitive for small ε or for large n. In fact one can consider the limit case when the tense protein does not bind oxygen at all, i.e. $\varepsilon = 0$, in which case

$$Y(E) = \frac{E(E+1)^{n-1}}{(E+1)^n + L}.$$

This is similar to a Hill function with Hill coefficient n [41]. See Fig. 6.4c for a comparison of the dose response using $\varepsilon = 1$ or $\varepsilon = 0.01$.

The Monod-Wyman-Changeux model has proved hugely popular in pharmacology and molecular biology ever since its publication, because of its wide applicability in many enzymatic and other biophysical systems. In the mind of many biologists, multisite ultrasensitivity goes hand in hand with allostery and the MWC model, along with the concept of cooperativity.

6.5 The KNF Cooperative Model

The model by Koshland et al. [44], also known simply as KNF and the second of the two classical models in the field, is a spiritual heir to the Adair model described in Sect. 6.3. It also attempts to describe how cooperative interactions between sites can affect the dose response of a multisite biochemical system. But while the Adair

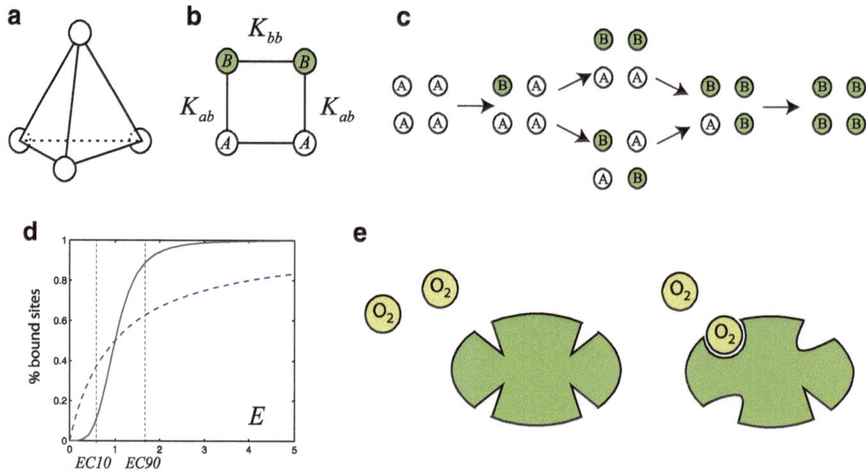

Fig. 6.5 (**a**) In the KNF model, sites are represented as nodes on a graph, and edges represent direct interactions between sites. (**b**) Given a configuration of unbound (A) and bound (B) sites, each edge is labeled as AA, BB, or AB. The AB and BB edges lower the overall energy of the molecule by K_{ab} and K_{bb} respectively, increasing its concentration at steady state relative to the fully unbound molecule. (**c**) The different conformations of a square molecule. (**d**) Simulation of the KNF model for $n = 4$, $K_{ab} = 1$. If $K_{bb} = 1$, the system reduces to the Adair model and the dose response is $cE/(1 + cE)$, displayed here for $c = 1$ (*dashed*). For $K_{bb} = 1000$ the dose response becomes ultrasensitive, here shown for $c = 0.001$ (*solid*). For this second graph the inputs EC10 and EC90 are shown, which yield a 10 and 90 % response respectively and can be used to quantify ultrasensitive behavior. (**e**) According to the induced fit hypothesis, the shape of the binding site changes upon binding, increasing its affinity. This in turn changes the shape of neighboring sites as well

model introduces cooperativity in the form of abstract constants $K_i > 1$, the KNF model has the advantage of increased molecular detail. On the other hand, the geometric detail makes it rather tedious to use, which is why many discussions of cooperativity in the literature still remain at the abstract level of Sect. 6.3.

In the KNF model the multisite substrate is described as a graph, with nodes representing the different sites, and edges representing which pairs of sites can directly interact with each other. For instance, if $n = 4$ and all pairs of sites are allowed to interact, the corresponding graph is a tetrahedron, as shown in Fig. 6.5a. In the case of the hemoglobin molecule, the graph is rather a square with four nodes and four edges, and each site can only directly interact with two neighbors.

Suppose that in the absence of interactions among sites, the on-rate of binding of the ligand E to a given site is $k_{on}E$, and the off-rate is k_{off}. Say that A and B represent unbound and bound sites respectively. If B_i is the concentration of substrate with i out of n bound sites, then at steady state

$$B_i = \binom{n}{i} c^i E^i B_0.$$

This equation was shown in (6.5) of the Adair model with independent sites. Here $c = k_{on}/k_{off}$ is the net affinity rate. In fact, $c^i E^i B_0$ is the concentration of each molecule given a *specific* subset of i out of the n ligands bound, and B_i adds together all $\binom{n}{i}$ of them.

Now, in the presence of interactions among neighboring sites, the geometry of the specific molecule matters considerably. Suppose given a square geometry and the molecule B_2 that has two adjacent bound sites. Out of the four edges, two of them are of type AB, one of type AA, and one of type BB (Fig. 6.5b).

The key assumption of the KNF model is that edges involving bound ligands lower the overall chemical energy of a molecule through their interaction, helping to increase the molecule concentration at steady state. Suppose that $K_{ab}, K_{bb} \geq 1$ represent the contribution of AB and BB edges to the lower energy of the molecule, respectively. In the case of the square molecule $B_{2,adj}$ with two adjacent bound sites, the new concentration at steady state is

$$B_{2,adj} = 4c^2 E^2 K_{ab}^2 K_{bb} B_0.$$

The factor 4 represents the number of different specific ligand configurations that have two adjacent bound sites, and the factor $K_{ab}^2 K_{bb}$ represents the increased reactivity towards this state due to internal site interactions.

The concentration of a molecule $B_{2,opp}$ with two opposite bound sites is $B_{2,opp} = 2c^2 E^2 K_{ab}^4 B_0$. Using the same logic, one can calculate the concentration of squares with one, three, and four bound sites (Fig. 6.5c) as $B_1 = 4cEK_{ab}^2 B_0$, $B_3 = 4c^3 E^3 K_{ab}^2 K_{bb}^2 B_0$, $B_4 = c^4 E^4 K_{bb}^4 B_0$. Now the total substrate B_{tot} is calculated as

$$B_{tot} = B_0 + B_1 + B_{2,opp} + B_{2,adj} + B_3 + B_4$$

$$B_{tot} = B_0[1 + 4cEK_{ab}^2 + 2c^2E^2(2K_{ab}^2 K_{bb} + K_{ab}^4) + 4c^3E^3K_{ab}^2K_{bb}^2 + c^4E^4K_{bb}^4]. \tag{6.6}$$

This way one can solve for B_0 as a function of E. Similarly $B_{2,opp}$ and the remaining states can be written in terms of E and the model parameters.

One can also compute the fraction of bound ligands in this model using (6.6) and canceling out B_0 in the numerator and denominator:

$$Y(E) = \frac{B_1 + 2B_{2,opp} + 2B_{2,adj} + 3B_3 + 4B_4}{4B_{tot}}$$

$$= \frac{cEK_{ab}^2 + c^2E^2(2K_{ab}^2K_{bb} + K_{ab}^4) + 3c^3E^3K_{ab}^2K_{bb}^2 + c^4E^4K_{bb}^4}{1 + 4cEK_{ab}^2 + 2c^2E^2(2K_{ab}^2K_{bb} + K_{ab}^4) + 4c^3E^3K_{ab}^2K_{bb}^2 + c^4E^4K_{bb}^4}.$$

Once again, large values of K_{ab} and especially of K_{bb} translate into a higher ultrasensitive behavior in the system. See Fig. 6.5d for graphs of this output in the case $n = 4$ and two different values of K_{bb}.

The KNF model is an implementation of the concept of *induced fit*, originally proposed by Koshland in the late 1950s. Binding sites are usually believed to be rigid and have just the right shape for a ligand to bind. Under the induced fit scenario, a binding site is flexible, and when unbound it does not have a very good affinity to the ligand. However, when bound, the binding site adapts its shape around the ligand, increasing its affinity. When the shape of the site changes, it also allosterically changes the shape of the neighboring sites, so that they also have a higher affinity to the ligand. See Fig. 6.5e.

In this way one can say that ligands have, like the MWC model, two different configurations, one with low affinity and one with high affinity. However the high affinity state is only found when the ligand is bound, so from a mathematical point of view one can simply refer to the bound and unbound states. Also importantly, the KNF model allows for different ligands to be in different states simultaneously, which is ruled out in the MWC model.

6.6 Generalized Hill Coefficients

For Hill functions $x^n/(K^n + x^n)$, the Hill coefficient n is a quantitative measure of the extent to which the function is ultrasensitive: the larger n, the stronger the all-or-none behavior. But if a dose response function $f(x)$ is not a Hill function, this definition cannot be used. Often biologists will measure the approximate Hill coefficient by carrying out a least squares minimization to find the best fitting Hill function, then use the resulting value for h as an estimate of ultrasensitive behavior. However this procedure is ill-posed for carrying out mathematical estimates.

A more useful formula was proposed by Goldbeter and Koshland [27] for use in more general sigmoidal functions:

$$H := \frac{\ln(81)}{\ln(EC90/EC10)},$$

where $EC10$ and $EC90$ are the inputs that produce 10 % and 90 % of the maximal response, respectively (Fig. 6.5d). The more ultrasensitive the function, the smaller the ratio $EC90/EC10 > 1$, and the larger H becomes.

What is interesting about this particular way of quantifying ultrasensitive behavior is that in the special case of Hill functions $f(x) = x^n/(K^n + x^n)$, it holds that $n = H$. In that sense H is a generalization of the concept of Hill coefficient for arbitrary sigmoidal functions. To see this, set $u = EC10$ and $v = EC90$, so that

$$\frac{u^n}{K^n + u^n} = 0.1, \quad \frac{v^n}{K^n + v^n} = 0.9.$$

Inverting both sides of the first equation,

$$10 = \frac{K^n + u^n}{u^n} = K^n u^{-n} + 1,$$

or $9 = K^h u^{-n}$. For v, the corresponding equation is $1/9 = K^n v^{-n}$. Dividing the two equations one obtains

$$81 = \frac{K^n u^{-n}}{K^n v^{-n}} = \left(\frac{v}{u}\right)^n.$$

Taking natural logarithm on both sides one obtains $n = \ln 81 / \ln(\frac{u}{v}) = H$.

6.7 Nonessential Modification Sites

After a tour of the classical models, we discuss more recent work carried out in the field by the author and his colleagues. In model MWC as well as KNF, the basis for ultrasensitive behavior is the assumption of internal allosteric interactions among the individual sites, either to force the system to switch globally between two states (MCW), or so that the modification of one site positively affects the rate of modification of its neighbors (KNF). In the new work we have focused on altering instead the definition of the *output* of the system, to encourage ultrasensitive behavior.

In the different models of oxygen transport discussed, the natural output is the fraction of bound ligand at steady state, or $\sum_{i=0}^{n} i S_i / (n S_{tot})$, as a function of the input E. However in signal transduction cascades it is not so much the amount of modification that matters, but the activity level of the multisite substrate. For such models, the standard output in the literature is simply the concentration S_n of the most modified protein, under the assumption that only a fully modified protein is considered active.

Suppose instead that out of n sites, a protein only needs k modifications in order to fully activate the protein [72]. At an intuitive level, this might be considered equivalent to having k sites and requiring all k sites for activation. But as it turns out, having so-called *nonessential sites* in the system can help increase its ultrasensitive behavior.

First we consider the case of a simple sequential system of the type discussed in (6.3) and Fig. 6.3a, setting simply $a_i = b_i = 1$. (In the original work [72] a constant phosphatase concentration F is assumed, which is equivalent to $b_i = F$ but can also be eliminated through a change of variables. See also the earlier work by Gunawardena [29].) In that case $A_i = 1$ for all i, and the new output can be calculated from (6.4) as

$$S_k + \ldots + S_n = S_{tot} \frac{E^k + \ldots + E^n}{1 + E + E^2 + \ldots + E^n} = S_{tot} \frac{E^k - E^{n+1}}{1 - E^{n+1}}.$$

Figure 6.6a describes this dose response for $n = 9$ and several decreasing values of k. By the standards of the previous sections, even the dose response with $k = n = 9$ is ultrasensitive. But as k decreases, the ultrasensitivity clearly becomes stronger.

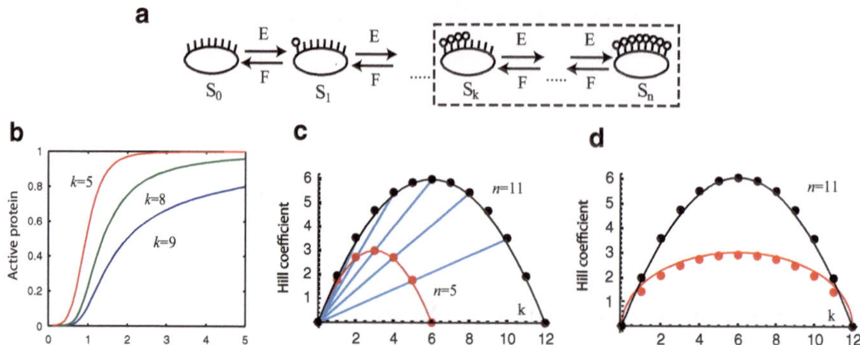

Fig. 6.6 (**a**) The output is redefined by assuming that k modifications or more are sufficient to fully activate the protein. (**b**) Dose response $S_k + \ldots + S_n$ for $n = 9$ and different values of k. (**c**) Estimated formula for the Hill coefficient (*solid lines*) along with numerical calculations (*dots*). (**d**) In the nonsequential Adair model, the Hill coefficient is proportional to the square root of that in (**a**)

One can show the ultrasensitive behavior of this dose response analytically for the special case $k = (n+1)/2$. In that case, $n + 1 = 2k$ and

$$\frac{S_k + \ldots + S_n}{S_{tot}} = \frac{E^k - E^{2k}}{1 - E^{2k}} = \frac{E^k(1 - E^k)}{(1 + E^K)(1 - E^k)} = \frac{E^k}{1 + E^k}.$$

This is once again a Hill function, with Hill coefficient k. As n increases, so does k and the ultrasensitivity increases arbitrarily.

In the paper [72] together with L. Wang and Q. Nie, we quantify the ultrasensitive behavior of the dose response for different values of n and k, using the apparent Hill coefficient defined above. We carry out an estimate to conclude that

$$H(n,k) \approx 2k\left(1 - \frac{k}{n+1}\right). \tag{6.7}$$

See Fig. 6.6b for a comparison of the estimated formula and the actual calculated values. This formula shows that in fact the largest Hill coefficient for a given n is found for $k = (n+1)/2$, and that the Hill coefficient grows linearly with n given a fixed ratio $\alpha = k/(n+1)$.

We also carried out a similar analysis in the nonsequential case using an equivalent framework to the Adair model with $K_i = 1$ and $c = 1$ [note: generalize to arbitrary c]. In that case the dose response follows from (6.5) and $S_{tot} = S_0 + \ldots S_n$:

$$S_k + \ldots + S_n = S_{tot} \frac{\sum_{i=k}^{n} \binom{n}{i} E^i}{(E+1)^n}.$$

The Hill coefficient is also estimated in that case, and surprisingly, it is essentially the square root of equation (6.7):

$$H(n,k) \approx 1.7\sqrt{k(1-\frac{k}{n+1})}.$$

In this case the maximum value of H is also reached at $(n+1)/2$. The Hill coefficient is invariant under horizontal or vertical rescaling of the dose response. In particular the same result will be obtained if, say, $c \neq 1$ in the Adair model.

6.8 General Activity Gradients

In more recent work by Ryerson and the author, we set out to further generalize the dose response under more natural assumptions. Suppose that S is a multisite substrate with n sites, and that its activity as a signaling molecule increases gradually as its sites are increasingly modified. This relaxes the previous assumption that the protein is either completely active or completely inactive for different number of modifications. To formalize this idea we define the *activity gradient*, a function $h(x) : [0,1] \to [0,1]$ representing the level of activity of a substrate molecule that has a fraction x of its sites modified.

As it turns out, the two different cases treated in the previous section now have a very different behavior. Given a fixed activity gradient $h(x)$, in the sequential case the ultrasensitivity increases arbitrarily for increasing values of n. But in the nonsequential case, the dose response $f(E)$ converges uniformly to a fixed function as $n \to \infty$. In particular, its ultrasensitive behavior is bounded by the ultrasensitivity of this bounded function. This framework can have multiple applications, for example to the regulation of pheromone signaling [Strickfaden, 65] or DNA expression [Vignali et al. 71].

6.8.1 *The Sequential Case*

Consider again the sequential model in (6.3) and Fig. 6.3a, setting $a_i = b_i = 1$. Given the activity gradient $h(x)$, suppose that activity of S_i is $h(\frac{i}{n+1})$. The dose response at steady state can be written as

$$f(E) = \sum_{i=0}^{n} h\left(\frac{i}{n+1}\right) S_i = S_{tot} \frac{h(\frac{0}{n+1}) + h(\frac{1}{n+1})E + h(\frac{2}{n+1})E^2 + \ldots + h(\frac{n}{n+1})E^n}{1 + E + \ldots + E^n}. \tag{6.8}$$

Figure 6.7a displays this dose response for the (non-ultrasensitive) activation gradient $h(x) = x/(1+x)$. Notice that as n increases, this function is increasingly

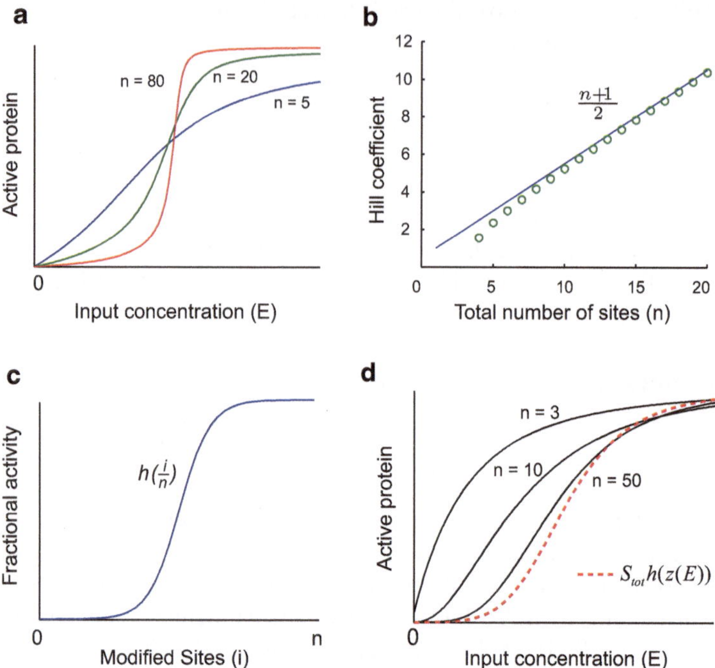

Fig. 6.7 (**a**) Dose response (6.8) in the sequential case using the activation gradient $h(x) = x/(1+x)$, for different values of n, illustrating that it becomes arbitrarily ultrasensitive as $n \to \infty$. (**b**) Estimated and exact Hill coefficient of the dose responses in (**a**). (**c**) Ultrasensitive activation gradient used for the nonsequential model (6.9). (**c**) Dose responses of (6.9) for increasing n, showing convergence towards $S_{tot}h(z(E))$

ultrasensitive. In fact, the Hill coefficient of the response appears to increase linearly, as shown in Fig. 6.7b. The following lemma calculates the estimate $H \approx const.(n+1)$ for this system.

Lemma 6.1. *Suppose that $h(x)$ is a piecewise continuous increasing function, and that $h(0) \leq 0.1h(1)$. Then the Hill coefficient of equation (6.8) satisfies $H \approx c(n+1)$, where the constant $c > 0$ does not depend on n.*

Proof. The dose response can be approximated by

$$f(E) \approx S_{tot} \frac{\int_0^{n+1} h(\frac{x}{n+1}) E^x dx}{\int_0^{n+1} E^x dx} = S_{tot} \frac{\int_0^1 h(y) E^{(n+1)y} dy}{\int_0^1 E^{(n+1)y} dy},$$

after a change of variables $y = \frac{x}{n+1}$. Defining the function

$$g(\alpha) = \frac{\int_0^1 h(y) e^{\alpha y} dy}{\int_0^1 e^{\alpha y} dy},$$

it holds

$$f(E) \approx S_{tot} g((n+1)\ln E).$$

The values $p := EC10_g$ and $q = EC90_g$ are well defined, since $g(\alpha)$ is a continuous, monotone function approaching $h(0)$ for $\alpha \to -\infty$ and $h(1)$ for $\alpha \to \infty$, and $h(0) \leq 0.1h(1)$.

Then $e^{\frac{p}{n+1}}$ is approximately the EC10 of the function $f(E)$ since

$$f(e^{\frac{p}{n+1}}) \approx S_{tot} g((n+1)\ln e^{\frac{p}{n+1}}) = S_{tot} g(p) = 0.1 g_{max} S_{tot} = 0.1 f_{max}$$

Likewise, $e^{\frac{q}{n+1}}$ is the EC90 of f. The Hill Coefficient of f is therefore

$$H_f = \frac{\ln 81}{\ln \frac{EC90_f}{EC10_f}} \approx \frac{\ln 81}{\ln \frac{e^{\frac{q}{n+1}}}{e^{\frac{p}{n+1}}}} = \frac{\ln 81}{q-p}(n+1).$$

□

6.8.2 The Nonsequential Case

Under the Adair model with $K_i = 1$, suppose that the activity of a substrate S_i is $h(i/n)$. Then the dose response is defined as the total protein activity at steady state, or

$$f(E) = \sum_{i=0}^{n} h(i/n) S_i.$$

This expression generalizes both the output used for the transport system used in the classical hemoglobin models, and the output $S_k + \ldots + S_n$ used in the previous section. In the former case this follows from setting $h(x) = x$, and in the latter case this corresponds to a Heaviside function $h(x)$.

Now, recall that the Adair model replicates the behavior of a nonsequential system with n individual sites that are modified at a rate of $k_{on}E$, and de-modified at a rate of k_{off}. One can define the function $z(E)$ as the overall fraction of modified sites at steady state. It is easy to calculate that

$$z(E) = \frac{cE}{cE+1},$$

where $c = k_{on}/k_{off}$. One can then write the steady state protein S_i in terms of $z = z(E)$:

$$S_i = S_{tot} \frac{\binom{n}{i} c^i E^i}{(cE+1)^n} = S_{tot} \binom{n}{i} z^i (1-z)^{n-i}.$$

This leads to an expression for $f(E)$ that can be approximated in surprisingly simple terms:

$$\frac{f(E)}{S_{tot}} = \sum_{i=0}^{n} h(i/n) \binom{n}{i} z^i (1-z)^{n-i} \approx h(z) \tag{6.9}$$

The middle term above is the so-called Bernstein polynomial of the function $h(x)$. For continuous $h(x)$, the left hand side has been shown to converge towards $h(z)$ uniformly on z as $n \to \infty$ [35]. Even for piecewise continuous functions $h(x)$, the left hand side converges pointwise towards $h(z)$, which is useful in the case of Heaviside functions. Either way we have the following interesting formula,

$$f(E) \approx S_{tot} h(z(E)).$$

If the function $h(x)$ is itself ultrasensitive, then $f(E)$ can also be an ultrasensitive function. In the independent Adair case $h(x) = x$, we have already shown that $f(E)/S_{tot} = z(E)$, which is consistent with this equation. In this way, one can think of the use of ultrasensitive activation functions as an alternative to the use of $K_i > 1$ in the Adair model, or to the corrections proposed in the MWC and KNF models.

See Fig. 6.7d for the dose response in this model for increasing values of n, using the activation function in Fig. 6.7c.

6.9 Other Forms of Ultrasensitivity

Although the focus of this review is on multisite mechanisms for ultrasensitivity, other mechanisms have been discussed in the literature that can create ultrasensitive responses. In some cases they are similar enough that the same system can be interpreted by more than one such mechanism, however at the conceptual level they are different enough to be distinguished.

6.9.1 Signaling Cascades

The reader may have noticed that the reaction described in Fig. 6.1a has multiple steps, i.e. the receptor complex activates Ras, which activates Raf, which in turn activates Mek, and Mek activates MAPK. Why not have the receptor activate MAPK directly and eliminate the other proteins? Such long cascades are actually the rule rather than the exception in signal transduction. For example, the last three steps here are known as a MAP kinase cascade, and almost identical cascades can be found regulating widely diverse signals in mammals, plants, yeast, etc. [58]. There are actually many reasons to have multiple cascades rather than a single regulatory step. For instance, they can help amplify an originally weak signal, and they can increase opportunities for feedback and cross-talk from other pathways [8].

Signaling cascades can also increase the ultrasensitive behavior of a response [Hooshangi 37, Markevich 48]. If the output of an initial step is the input of a second step, then at steady state in the simplest case the overall dose response is the composition of both responses. If two mildly ultrasensitive functions are composed, the result can be a more strongly ultrasensitive function. Notice that the two last proteins of the reaction in Fig. 6.1a have two phosphorylation sites each. Each of them constitutes a multisite system on its own, and the composition of several such systems can lead to a much stronger ultrasensitive response.

In actual signaling cascades, one cannot simply compose the dose responses of the individual steps to obtain the overall dose response. This is due to a retroactivity effect, in which the output molecules of an upstream component are also affected by the downstream molecules that interact with it. Huang and Ferrell investigated the behavior of MAP kinase cascades as a whole and discovered that they can be strongly ultrasensitive, even when the individual steps would not predict such behavior [38]. See also more recent work by Sarkar and colleagues on the properties of synthetically engineered cascades [55]. Also see the paper by del Vecchio, Ninfa, and Sontag [17], a seminal work that has produced a growing literature by del Vecchio and others [Vecchio 16, Ossareh 56, Vecchio 69].

6.9.2 Zero-order Ultrasensitivity

In 1981, Goldbeter and Koshland proposed an influential framework for ultrasensitive dose responses using a single protein modification site [27] under saturating conditions. In order to understand it, it is useful to describe some basic enzyme biochemistry.

The process by which an enzyme modifies a substrate and converts it into a product is usually modeled using the so-called Michaelis-Menten reactions [51]

$$S + E \leftrightarrow C \rightarrow P + E,$$

where S, E, P represent substrate, enzyme, and product respectively. The molecule C represents a transient complex formed by the substrate and the enzyme. The overall rate of conversion of substrate into product can be estimated as

$$\frac{dP}{dt} \approx kE_{tot}\frac{S}{K_m + S}. \tag{6.10}$$

Here k is the rate parameter of the reaction $C \rightarrow P + E$, E_{tot} is the total enzyme concentration, and K_m is determined solely from the reaction parameters in the system. The derivation generally uses the assumption $E_{tot} \ll S_{tot}$, although other assumptions can lead to the same result [40]. This is one of the most basic formulas in biochemistry, and it is the origin for many of the terms found in mathematical models of protein interactions.

Suppose that the substrate concentration S in the environment is much larger than K_m. Then the rate of flow of S into P can be approximated as $kE_{tot}\frac{S}{S} = kE_{tot}$, which is independent of S. In the enzyme biochemistry community this reaction is described as *zero-order*, since in general an n-th order reaction $nS \rightarrow P$ would have a rate proportional to S^n, and this reaction seems to fit that description only with $n = 0$.

Goldbeter and Koshland [27] proposed a situation in which one enzyme E modifies a substrate S, and another enzyme F eliminates this modification, assuming a very large substrate concentration of S relative to the K_m of both reactions. Then the rates of modification and de-modification are roughly independent of S, P and the net flow of S into P is approximately $k_1 E_{tot} - k_2 F_{tot}$. If F_{tot} is left constant and E_{tot} is used as an input, then very minor differences in E_{tot} can result in either a net positive or a net negative flow, resulting in a very large or very small steady state concentration of P. In this way the dose response may become highly ultrasensitive with respect to E_{tot}. (Of course, the flow is not exactly constant, otherwise S or P would become negative. Rather when one of them becomes very small the assumptions break down and the system settles into steady state.)

Because of its simplicity, this mechanism has many potential applications. It has been used in experimental studies to investigate glycogen metabolism [49] and morphogen gradients in embryonic development [50], among others. From a theoretical point of view, recent work has updated this system by taking into account e.g. stochastic effects [6] and the reversibility of the product-formation reaction in (6.10) [73].

6.9.3 Protein Relocalization

Another series of papers points to ultrasensitive behavior through mechanisms using multiple compartments and the sequestration or relocalization of proteins. For instance, Liu et al. consider a multisite protein system similar to those described in previous sections, together with an additional protein that acts as a scaffold [47]. The scaffold protein passively binds and unbinds the substrate, effectively relocalizing it to a different position inside the cell. The relocalization of the substrate affects the rates in which the enzymes can interact with it. Liu et al. show that under certain conditions the presence of the scaffold can significantly improve the ultrasensitive behavior of the dose response.

Perhaps the earliest sequestration mechanism was proposed by Ferrell in 1996 [24], in a system involving the competition of multiple different substrates for the attention of the same enzyme E. The substrate S can be activated e.g. through single-site modification by E. But other substrates are deployed as decoys to bind to the enzyme more tightly than S itself, thus relocalizing it and making it inaccessible to S. The result is that the concentration of E needs to be high enough to bind to the decoy substrates as well as S, leading to ultrasensitive behavior. This mechanism was experimentally tested in 2007 in the context of regulatory proteins in the cell cycle of frog eggs [42]. The enzyme Cdk1 alters the activity of the substrate Wee1 through phosphorylation in specific sites. But other Wee1 phosphorylation sites, as well as other proteins in the cytoplasm, appear to bind Cdk1 and prevent it from activating Wee1.

In 2009, Buchler and Cross built a synthetic circuit inside a cell, also involving protein sequestration by an inhibitor protein [10]. They were able to measure Hill coefficients as high as 12 in this synthetic system and to replicate model predictions on this experimental system.

6.9.4 Positive Feedback and Bistability

A more general way of approaching ultrasensitive behavior is through the use of positive feedback interactions (although positive feedback underlies many of the above examples). Such systems could for instance be bistable for a certain range of the input values, leading to hysteresis [4, Ozbudak 57]. A hysteretic dose response can be considered highly ultrasensitive, in the sense that a very small increase in the input beyond the bifurcation point can bring about a very large increase in the output.

Using bistability to create ultrasensitive responses presents a type of chicken and egg problem. Positive feedback interactions are by themselves not sufficient to create bistability, and usually some kind of ultrasensitive nonlinearity is also required in at least one of the feedback interactions. The mechanisms described in this chapter can precisely be used to create the nonlinear interactions necessary for bistability.

That said, positive feedback interactions can be used to enhance dose responses even when they do not lead to bistability. For instance, in a model of DNA packaging regulation, Sneppen et al. assume a positive feedback interaction between histone proteins and their respective enzymes, in order to obtain ultrasensitive dose responses [64]. When a concrete biological system does not fit the framework of any of the mechanisms above, one can also describe the dose responses in this way.

6.9.5 Allovalency and Entropic Multisite Models

Another mechanism involving the spatial distribution of proteins is known as allovalency and was proposed by Tyers and colleagues in 2003 [43,46]. It describes the binding of a multisite, disordered protein to a single binding site. An important aspect of this model is that it distinguishes between three different locations of the protein with respect to its binding site: it can be bound, proximal, and free (i.e. far apart). A thermodynamic argument is used to calculate the resulting linear transition rates between these three states depending on the number n of sites, which leads to a fair degree of ultrasensitivity for large n.

Disordered multisite proteins of course were also highly relevant in the discussion of independent ultrasensitive behavior in previous sections, and they are predicted to be found commonly in nature [39]. Another thermodynamic derivation of ultrasensitive behavior was developed by Lenz and Swain [45], using nonlinear effects in the entropy configurations of disordered multisite systems. This mechanism results in a highly ultrasensitive function $h(x)$ (using our notation) and is therefore complementary to the results described for independent multisite systems.

6.10 Discussion

There has been a lively debate in the quantitative biology community on possible ways to construct switch-like responses at the molecular level. This question is clearly of interest to experimental biologists trying to understand design principles in cellular physiology. Contributions have been made for many years, and they have been fueled more recently by advances in quantitative measurements that can potentially distinguish between different models. Clearly, the models that will have the strongest impact are the ones that actually take place in biological systems and contribute to their conceptual understanding.

Either way, there is a clear need in biochemistry to create such switch-like or sigmoidal behavior, not only for traditional signal transduction systems but also in order to create nonlinear interactions that allow nontrivial qualitative behavior such as bistability or oscillations. For instance, the oscillatory behavior of the FitzHugh-Nagumo model (Chap. 3) is possible because of a cubic term that is likely implemented biochemically using a bistable switch. The bistable switch in turn is probably made possible through the combination of a positive feedback and a sigmoidal nonlinearity.

As described in the introduction, signal transduction systems are intrinsically nonautonomous since their behavior depends entirely on the concentration of the outside input. Traditionally the outside concentration is held constant, but new experimental techniques such as microfluidics are allowing to carry out time-dependent input experiments. Recent work by Sontag et al. is already revealing some exciting properties of time-dependent signals such as fold-change detection [63]. At the same time, the stochastic properties of ultrasensitive networks have also been explored, showing for instance that signaling cascades have the ability to attenuate noise [67].

While the MCW and KNF models are extremely influential in theoretical biology, biochemistry textbooks tend to provide only an intuitive description since the mathematical details are beyond the scope of the text. A good description of these models is the book by Cornish-Bowden [13], but it still falls short of sufficient mathematical precision. I tried to strike a balance here between maintaining an informal tone for readability, and sufficient mathematical detail and generality for use as a reference.

Acknowledgements I would like to thank my colleagues Qing Nie and Liming Wang as well as my student Shane Ryerson for their participation in the described research, and Jeremy Gunawardena for his mentoring and for introducing me into this field.

References

1. C. Ajo-Franklin, D. Drubin, J. Eskin, E. Gee, D. Landgraf, I. Phillips, P. Silver, Rational design of memory in eukaryotic cells. Genes Dev. **21**, 2271–2276 (2007)
2. B. Alberts, A. Johnson, J. Lewis, M. Raff, K. Roberts, P. Walter, *Molecular Biology of the Cell* (Garland Science, New York, 2002)
3. D. Anderson, G. Craciun, T. Kurtz, Product-form stationary distributions for deficiency zero chemical reaction networks. Bull. Math. Biol. **72**(8), 1947–1970 (2010)
4. D. Angeli, J. Ferrell, E. Sontag, Detection of multistability, bifurcations, and hysteresis in a large class of biological positive-feedback systems. Proc. Natl. Acad. Sci. USA **101**(7), 1822–1827 (2004)
5. D. Beard, H. Qian, *Chemical Biophysics: Quantitative Analysis of Cellular Systems* (Cambridge University Press, Cambridge, 2008)
6. O.G. Berg, J. Paulsson, M. Ehrenberg, Fluctuations and quality of control in biological cells: zero-order ultrasensitivity reinvestigated. Biophys. J. **79**(3), 1228–1236 (2000)
7. N. Blutgen, S. Legewie, H. Herzel, B. Kholodenko, Mechanisms generating ultrasensitivity, bistability, and oscillations in signal transduction, in *Introduction to Systems Biology*, ed. by S. Choi. (Springer, Berlin, 2007), pp. 282–299
8. N. Bluthgen, H. Herzel, Map-kinase-cascade: switch, amplifier, or feedback controller? in *2nd Workshop on Computation of Biochemical Pathways and Genetic Networks* (Logos-Verlag, Berlin, 2001), pp. 55–62
9. A. Briegel, X. Li, A. Bilwes, K. Hughes, G. Jensen, B. Crane, Bacterial chemoreceptor arrays are hexagonally packed trimers of receptor dimers networked by rings of kinase and coupling proteins. Proc. Natl. Acad. Sci. USA **109**(10), 3766–3771 (2012)
10. N. Buchler, F. Cross, Protein sequestration generates a flexible ultrasensitive response in a genetic network. Mol. Syst. Biol. **5**(272), 1–7 (2009)
11. D. Burz, R. Rivera-Pomar, H. Jaeckle, S. Hanes, Cooperative DNA-binding by Bicoid provides a mechanism for threshold-dependent gene activation in the drosophila embryo. EMBO J. **17**(20), 5998–6009 (1998)
12. J.P. Changeux, J. Thiery, Y. Tung, On the cooperativity of biological membranes. Proc. Natl. Acad. Sci. USA **57**, 335–341 (1966)
13. A. Cornish-Bowden, *Fundamentals of Enzyme Kinetics* (Butterworth, London, 1979)
14. C. Craciun, Y. Tang, M. Feinberg, Understanding bistability in complex enzyme-driven reaction networks. Proc. Natl. Acad. Sci. USA **103**(23), 8697–8702 (2006)
15. V. Danos, J. Feret, W. Fontana, R. Harmer, J. Krivine, Abstracting the differential semantics of rule-based models: exact and automated model reduction, in *Annual IEEE Symposium on Logic in Computer Science* (Edinburgh, 2010)
16. D. Del Vecchio, E. Sontag, Engineering principles in bio-molecular systems: from retroactivity to modularity. Eur. J. Control (Special Issue) **15**(3–4) (2009)
17. D. Del Vecchio, A. Ninfa, E. Sontag, Modular cell biology: retroactivity and insulation. Nat. Mol. Syst. Biol. **4**, 161 (2008)
18. R. Deshaies, J. Ferrell, Multisite phosphorylation and the countdown to S phase. Cell **107**, 819–822 (2001)
19. T. Duke, D. Bray, Heightened sensitivity of a lattice of membrane receptors. Proc. Natl. Acad. Sci. USA **96**, 10104–10108 (1999)
20. T. Duke, N. Le Novere, D. Bray, Conformational spread in a ring of proteins: a stochastic approach to allostery. J. Mol. Biol. **308**, 541–553 (2001)

21. G. Enciso, E. Sontag, Monotone systems under positive feedback: multistability and a reduction theorem. Syst. Control Lett. **51**(2), 185–202 (2005)
22. M. Feinberg, Lectures on chemical reaction networks (1979), Notes of lectures given at the Mathematics Research Center of the University of Wisconsin, available at http://www.che.eng.ohio-state.edu/~FEINBERG/LecturesOnReactionNetworks/
23. Feinberg, M.: *Chemical Oscillations, Multiple Equilibria, and Reaction Network Structure* (Academic, New York, 1980), pp. 59–130
24. J. Ferrell, Tripping the switch fantastic: how a protein kinase cascade can convert graded inputs into switch-like outputs. Trends Biochem. Sci. **21**(12), 460–466 (1996)
25. J. Ferrell, Question & answer: cooperativity. J. Biol. **8**(53), 1–6 (2009)
26. F. Gnad, S. Ren, J. Cox, J. Olsen, B. Macek, M. Oroshi, M. Mann, PHOSIDA (phosphorylation site database): managment, structural and evolutionary investigation, and prediction of phosphosites. Genome Biol. **8**, R250 (2007)
27. A. Goldbeter, D. Koshland, An amplified sensitivity arising from covalent modification in biological systems. Proc. Natl. Acad. Sci. USA **78**, 6840–6844 (1981)
28. V. Gotea, A. Visel, J. Westlund, M. Nobrega, L. Pennachio, I. Ocharenko, Homotypic clusters of transcription factor binding sites are a key component of human promoters and enhancers. Genome Res. **20**(5) (2010)
29. J. Gunawardena, Multisite protein phosphorylation makes a good threshold but can be a poor switch. Proc. Natl. Acad. Sci. USA **102**, 14617–14622 (2005)
30. R. Harmer, V. Danos, J. Feret, J. Krivine, W. Fontana, Intrinsic information carriers in combinatorial dynamical ssytems. Chaos **20**, 037108 (2010)
31. S. Harvey, A. Charlet, W. Haas, S. Gygi, D. Kellogg, Cdk1-dependent regulation of the mitotic inhibitor Wee1. Cell **122**, 407–420 (2005)
32. S. Harvey, G. Enciso, N. Dephoure, S. Gygi, J. Gunawardena, D. Kellogg, A phosphatase threshold sets the level of Cdk1 activity in early mitosis in budding yeast. Mol. Biol. Cell **22**(19), 3595–3608 (2012)
33. Y. Henis, J. Hancock, I. Prior, Ras acylation, compartmentalization and signaling nanoclusters. Mol. Membr. Biol. **26**(1), 80–92 (2009)
34. K. Hertel, K. Lynch, T. Maniatis, Common themes in the function of transcription and splicing enhancers. Curr. Opin. Cell Biol. **9**, 350–357 (1997)
35. F. Herzog, J. Hill, The Bernstein polynomials for discontinuous functions. Am. J. Math. **68**(1), 109–124 (1946)
36. A. Hill, The possible effects of the aggregation of the molecules of haemoglobin on its dissociation curves. Proc. Physiol. Soc. **40**(Suppl.), iv–vii (1910)
37. S. Hooshangi, S. Thiberge, R. Weiss, Ultrasensitivity and noise propagation in a synthetic transcriptional cascade. Proc. Natl. Acad. Sci. USA **102**(10), 3581–3586 (2005)
38. C.Y. Huang, J. Ferrell, Ultrasensitivity in the mitogen-activated protein kinase cascade. Proc. Natl. Acad. Sci. USA **93**, 10078–10083 (1996)
39. L. Iakoucheva, P. Radivojac, C. Brown, T. O'Connor, J. Sikes, Z. Obradovic, A. Dunker, The importance of intrinsic disorder for protein phosphorylation. Nucleic Acid Res. **32**, 1037–1049 (2004)
40. J. Keener, J. Sneyd, *Mathematical Physiology I: Cellular Physiology* (Springer, Berlin, 2008)
41. G. Kegeles, The Hill coefficient for a Monod-Wyman-Changeux allosteric system. FEBS Lett. **103**(1), 5–6 (1979)
42. S. Kim, J. Ferrell, Substrate competition as a source of ultrasensitivity in the inactivation of Wee1. Cell **128**, 1133–1145 (2007)
43. P. Klein, T. Pawson, M. Tyers, Mathematical modeling suggests cooperative interactions between a disordered polyvalent ligand and a single receptor site. Curr. Biol. **13**, 1669–1678 (2003)
44. D. Koshland, G. Nemethy, D. Filmer, Comparison of experimental binding data and theoretical models in proteins containing subunits. Biochemistry **5**(1), 365–385 (1966)
45. P. Lenz, P. Swain, An entropic mechanism to generate highly cooperative and specific binding from protein phoshporylations. Curr. Biol. **16**, 2150–2155 (2006)

46. A. Levchenko, Allovalency: a case of molecular entanglement. Curr. Biol. **13**, R876–R878 (2003)
47. X. Liu, L. Bardwell, Q. Nie, A combination of multisite phosphorylation and substrate sequestration produces switch-like responses. Biophys. J. **98**(8), 1396–1407 (2010)
48. N. Markevich, J. Hoek, B. Kholodenko, Signaling switches and bistability arising from multisite phosphorylation in protein kinase cascades. J. Cell Biol. **164**(3), 353–359 (2004)
49. M. Meinke, J. Bishop, R. Edstrom, Zero-order ultrasensitivity in the regulation of glycogen phosphorylase. Proc. Natl. Acad. Sci USA **83**(9), 2865–2868 (1986)
50. G. Melen, S. Levy, N. Barkai, B. Shilo, Threshold responses to morphogen gradients by zero-order ultrasensitivity. Mol. Syst. Biol. **1**(0028), 1–11 (2005)
51. L. Menten, M. Michaelis, Die kinetik der invertinwirkung. Biochem. Z. **49**, 333–369 (1913)
52. J. Monod, J. Wyman, J.P. Changeux, On the nature of allosteric transitions: a plausible model. J. Mol. Biol. **12**, 88–118 (1965)
53. Y. Ohashi, J. Brickman, E. Furman, B. Middleton, M. Carey, Modulating the potency of an activator in a yeast in vitro transcription system. Mol. Cell. Biol. **14**(4), 2731–2739 (1994)
54. S. Orlicky, X. Tang, A. Willems, M. Tyers, F. Sicheri, Structural basis for phosphodependent substrate selection and orientation by the SCFCdc4 ubiquitin ligase. Cell **112**, 243–256 (2003)
55. E. O'Shaughnessy, S. Palani, J. Collins, C. Sarkar, Tunable signal processing in synthetic map kinase cascades. Cell **144**, 119–131 (2011)
56. H.R. Ossareh, A.C. Ventura, S.D. Merajver, D. del Vecchio, Long signaling cascades tend to attenuate retroactivity. Biophys. J. **100**(7), 1617–1626 (2011)
57. E. Ozbudak, M. Thattai, H. Lim, B. Shraiman, A. van Oudenaarden, Multistability in the lactose utilization network of Escherichia coli. Nature **427**, 737–740 (2004)
58. G. Pearson, F. Robinson, T.B. Gibson, B. Xu, M. Karandikar, K. Berman, M. Cobb, Mitogen-activated protein (MAP) kinase pathways: regulation and physiological functions. Endocr. Rev. **22**(2), 153–183 (2001)
59. F. Rossi, A. Kringstein, A. Spicher, O. Guicherit, H. Blau, Transcriptional control: rheostat converted to on/off switch. Mol. Cell **6**, 723–728 (200)
60. Z. Serber, J. Ferrell, Tuning bulk electrostatics to regulate protein function. Cell **128**, 441–444 (2007)
61. G. Shinar, M. Feinberg, Structural sources of robustness in biochemical reaction networks. Science **327**, 1389–1391 (2010)
62. A. Shiu, Algebraic methods for biochemical reaction network theory. Ph.D. thesis, University of California, Berkeley, 2010
63. O. Shoval, L. Goentoro, Y. Hart, A. Mayo, E. Sontag, U. Alon, Fold change detection and scalar symmetry of sensory input fields. Proc. Natl. Acad. Sci. USA **107**, 15995–16000 (2010)
64. K. Sneppen, M. Micheelsen, I. Dodd, Ultrasensitive gene regulation by positive feedback loops in nucleosome modification. Mol. Syst. Biol. **182** (2008)
65. S. Strickfaden, M.J. Winters, G. Ben-Ari, R. Lamson, M. Tyers, P. Pryciak, A mechanism for cell-cycle regulation of MAP kinase signaling in a yeast differentiation pathway. Cell **128**, 519–531 (2007)
66. S. Strogatz, *Nonlinear Dynamics and Chaos: With Applications to Physics, Biology, Chemistry, and Engineering* (Perseus Book Publishing, Cambridge, 1994)
67. M. Thattai, A. van Oudenaarden, Attenuation of noise in ultrasensitive signaling cascades. Biophys. J. **82**(6), 2943–2950 (2002)
68. M. Thomson, J. Gunawardena, Unlimited multistability in multisite phosphorylation systems. Nature **460**, 274–277 (2009)
69. D. del Vecchio, E. Sontag, Engineering principles in bio-molecular systems: from retroactivity to modularity. Eur. J. Control (Special Issue) **15**(3–4) (2009)
70. R. Verma, R. Annan, M. Huddleston, S. Carr, G. Reynard, R. Deshaies, Phosphorylation of Sic1p by G1 Cdk required for its degradation and entry into S phase. Science **278**, 455–460 (1997)
71. M. Vignali, D. Steger, K. Neely, J. Workman, Distribution of acetylated histones resulting from Gal4-VP16 recruitment of SAGA and NuA4 complexes. EMBO J. **19**(11), 2629–2640 (2000)

72. L. Wang, Q. Nie, G. Enciso, Nonessential sites improve phosphorylation switch. Biophys. Lett. **99**(10), 41–43 (2010)
73. Y. Xu, J. Gunawardena, Realistic enzymology for post-translational modification: zero-order ultrasensitivity revisited. J. Theor. Biol. **311**, 139–152 (2012)
74. C. Zhang, S. Kim, The effect of dynamic receptor clustering on the sensitivity of biochemical signaling. Pac. Symp. Biocomput. **5**, 350–361 (2000)

Chapter 7
Mathematical Concepts in Pharmacokinetics and Pharmacodynamics with Application to Tumor Growth

Gilbert Koch and Johannes Schropp

Abstract Mathematical modeling plays an important and increasing role in drug development. The objective of this chapter is to present the concept of pharmacokinetic (PK) and pharmacodynamic (PD) modeling applied in the pharmaceutical industry. We will introduce typically PK and PD models and present the underlying pharmacological and biological interpretation. It turns out that any PKPD model is a nonautonomous dynamical system driven by the drug concentration. We state a theoretical result describing the general relationship between two widely used models, namely, transit compartments and lifespan models. Further, we develop a PKPD model for tumor growth and anticancer effects based on the present model figures and apply the model to measured data.

Keywords Lifespan models • Mathematical modeling • Pharmacodynamics • Pharmacokinetics • Transit compartments • Tumor growth

7.1 Introduction to Pharmacokinetic/Pharmacodynamic Concepts

The development of new drugs is time-consuming (12–15 years) and costly. A study from 2003 [9] reports costs of approximately US$ 800 million to bring a drug to the market. It is further estimated that around 90 % of compounds (drug candidates) will

G. Koch (✉)
Department of Pharmaceutical Sciences, School of Pharmacy and Pharmaceutical Sciences, State University of New York at Buffalo, 403 Kapoor Hall, Buffalo, NY 14214, USA
e-mail: gilbert.koch@web.de

J. Schropp
Universität Konstanz, FB Mathematik und Statistik, Postfach 195, D-78457 Konstanz, Deutschland
e-mail: johannes.schropp@uni-konstanz.de

P.E. Kloeden and C. Pötzsche (eds.), *Nonautonomous Dynamical Systems in the Life Sciences*, Lecture Notes in Mathematics 2102, DOI 10.1007/978-3-319-03080-7_7,

fail during the drug development process [23]. Hence, the pharmaceutical industry is in search of new tools to support drug discovery and development. It is stated by the U.S. Food and Drug Administration that computational modeling and simulation is a useful tool to improve the efficiency in developing safe and effective drugs, see e.g. [16].

The development of a drug is usually divided into three categories. Firstly, numerous compounds are developed and screened in vitro. Secondly, promising compounds are tested for an effect in animals (in vivo). Here, the interest is also in prediction of an appropriate dose for first in man studies. Finally, the drug is tested in humans (phase I–III).

In the drug discovery and development process, so-called pharmacokinetic/pharmacodynamic experiments are conducted, which consists of two parts: The first part, called pharmacokinetics, deals with the time course of the drug concentration in blood. The interest is on absorption, distribution, metabolism and excretion of the drug in the body. The disease or more general the biological/pharmacological effect is not considered. Roughly said, one observes what the body does to the drug. The second part is the pharmacodynamics which "can be defined as the study of the time course of the biological effects of drugs, the relationship of the effects to drug exposure and the mechanism of drug action", see [14]. That means one observes what the drug does to the body. Combining pharmacokinetics (PK) and pharmacodynamics (PD) gives an overall picture of the pharmacological effect/response, where it is assumed that the drug concentration is the driving force. In this work, the pharmacological effect is understood as the measurable therapeutic effect of the drug on a disease. In [4] it is stated: "Appropriate linking of pharmacokinetic and pharmacodynamic information provides a rational basis to understand the impact of different dosage regimens on the time course of pharmacological response." Furthermore, it is believed "that by better understanding of the relationship between PK and PD one can shed light on situations where one or the other needs to be optimized in drug discovery and development", see [40]. Typically, "PKPD modeling is widely used as the theoretical basis for optimization of the dosing regimen ... of drugs in Phase II", see [6]. Finally, it is stated in [34] about PKPD modeling: "When these insights are obtained in early development they can be used in translational approach to better predict efficacy and safety in the later stages of clinical development."

From the mathematical point of view, linking of PK and PD leads to nonautonomous differential equations driven by the drug action.

Ideally, PKPD models are based on fundamental biological and pharmacological principles to mimic the underlying mechanisms of disease development and drug response. Models fulfilling these requirements are called (semi-) mechanistic. Therefore, the development of such models is in general performed in an interdisciplinary collaboration between mathematicians, biologists, pharmacologists etc.

It is written in [6]: "Not surprisingly, PKPD modeling has developed from an empirical and descriptive approach into a scientific discipline based on the (patho-) physiological mechanisms behind PKPD relationships. It is now well accepted that mechanism-based PKPD models have much improved properties for extrapolation and prediction."

Following Mager et al. [32] and Danhof et al. [7] a PKPD model consists of four parts:

(i) Modeling of pharmacokinetics to describe the drug concentration.
(ii) Modeling the binding of drug molecules at the receptor/target to describe the effect concentration relationship.
(iii) Transduction modeling, describing a cascade of processes that govern the time course of pharmacological response after drug-induced target activation.
(iv) Modeling of the disease.

An important task of a PKPD model is to describe several dosing schedules simultaneously (at least the placebo group and one dosing group) by one set of model parameter obtained from an optimization process. In a PKPD model with an estimated set of parameter only the dosing schedule is allowed to vary. Hence, a PKPD model build without existing data is mostly useless in practice. Based on a PKPD model with an appropriate amount of data, different dosing schedules could be simulated and physiological model parameter could be inter-specifically (animals to human) scaled to support e.g. first in man dose finding in early drug development. A PKPD model could also be extended with a population approach to investigate clinical data (phase I–III), see e.g. [2]. However, this additional statistical approach will not be treated in this chapter.

7.2 Pharmacokinetic Models

7.2.1 *Introduction*

The pharmacokinetics (PK) describes the behavior of an administered drug in the body over time. In detail, the PK characterizes the absorption, distribution, metabolism and excretion (called ADME concept, see e.g. [14]) of a drug.

First pharmacokinetic models representing the circulatory system were published by the Swedish physiologist T. Teorell [39] in 1937. The German pediatrist F.H. Dost is deemed to be the founder of the term pharmacokinetics, see [41]. In his famous books "Der Blutspiegel" from 1953 [10] and "Grundlagen der Pharmakokinetik" from 1968 [11], he presented a broad overview and analysis of drug behavior in time based on linear differential equations.

In pharmacokinetic experiments the drug concentration in blood over time is measured. In order to develop a PK model the body is typically divided into several parts. In this work we focus on the widely used two-compartment model approach dividing the body in a heavily with blood supplied part and the rest. Such a model is based on linear differential equations and is from the modeling point of view an empirical approach to describe the drug concentration. However, the time course of most drugs could be well described by such models.

Further note, that the PK model is the driving force in a full PKPD model and therefore, a handy and descriptive representation of the solution is necessary from the computational point of view.

7.2.2 *Two-Compartment Models*

In this work we focus on two-compartment pharmacokinetic models representing the body based on linear differential equations. We consider oral absorption and intravenous administration of a drug. In practice, for drug concentration measurements, blood samples have to be taken from the patients and therefore, the availability of data is limited due to ethical constraints. It turned out in application that two compartments are sufficient to appropriate describe the time course in blood for most drugs. For a more detailed overview of pharmacokinetic models see e.g. the book from Gabrielsson [14].

A two-compartment model consists of two physiological meaningful parts (see e.g. [28]):

- The central compartment is identified with the blood and organs heavily supplied with blood like liver or kidney.
- The peripheral compartment describes for example tissue or more generally, the part of the body which is not heavily supplied with blood.

The compartments are connected among each other in both directions and therefore, a distribution between central and peripheral compartment takes place.

Main assumption in pharmacokinetics:

- The drug is completely eliminated (metabolism and excretion) from the body through the central (blood) compartment.

In case of oral administration of a drug (p.o.), absorption through the gastrointestinal tract takes place. Therefore, the distribution in the blood is not immediate and also only a part of the drug will reach the blood circulation (called bioavailability). In contrast, in case of intravenous dosing (i.v.) the drug is directly applied to the blood circulation and it is assumed that the drug is immediately distributed in the body.

With the formulated assumptions a two-compartment model for oral drug administration (p.o.) at time $t = 0$ reads

$$x_1' = -k_{10}x_1 - k_{12}x_1 + k_{21}x_2 + k_{31}x_3, \qquad x_1(0) = 0 \tag{7.1}$$

$$x_2' = k_{12}x_1 - k_{21}x_2, \qquad x_2(0) = 0 \tag{7.2}$$

$$x_3' = -k_{31}x_3, \qquad x_3(0) = f \cdot dose \tag{7.3}$$

where $0 < f \leq 1$ is a fraction parameter representing the amount of drug which effectively reaches the blood. We set without loss of generality $f \equiv 1$. The blood compartment is described by (7.1), the peripheral compartment by (7.2) and (7.3) describes the absorption. Note that Eq. (7.3) is not a part of the body. It is understood as additional hypothetical compartment necessary to describe the absorption. The model (7.1)–(7.3) has the parameters

$$\theta = (k_{10}, k_{12}, k_{21}, k_{31})$$

and the variable *dose*.

The parameter $k_{10} > 0$ describes the elimination rate from the body, $k_{12}, k_{21} > 0$ stand for the distribution between central and peripheral compartment and $k_{31} > 0$ is the absorption rate. In case of i.v. administration $k_{31} = 0$ (no absorption) and $x_1(0) = dose$ and therefore, the model reduces to (7.1)–(7.2).

In practice, the drug is measured as concentration in blood plasma. Therefore, the parameter volume of distribution $V_1 > 0$ of the central compartment $x_1(t)$ is introduced to obtain the drug concentration

$$c(t) = \frac{x_1(t)}{V_1}. \tag{7.4}$$

In this work, $c(t)$ will always denote the drug concentration in blood.

The representation of the two-compartment model based on ordinary differential equation is unhandy in application because in a full PKPD model the drug concentration has to be evaluated many times. Also for multiple dosing the representation is not appropriate. In order to reduce the computational effort the analytical solution of the blood compartment is presented in the next section.

7.2.3 Single Dosage

Applying the Laplace transform to (7.1)–(7.3) gives for the blood compartment in concentration terms for p.o. administration

$$\begin{aligned} c^{p.o.}(t) &= \frac{dose k_{31}(k_{21}-\alpha)}{V_1(k_{31}-\alpha)(\beta-\alpha)}\exp(-\alpha t) + \frac{dose k_{31}(k_{21}-\beta)}{V_1(k_{31}-\beta)(\alpha-\beta)}\exp(-\beta t) \\ &\quad + \frac{dose k_{31}(k_{21}-k_{31})}{V_1(k_{31}-\beta)(k_{31}-\alpha)}\exp(-k_{31}t) \\ &= dose A_{po}\exp(-\alpha t) + dose B_{po}\exp(-\beta t) \\ &\quad - dose(A_{po}+B_{po})\exp(-k_{31}t) \end{aligned} \tag{7.5}$$

where

$$\alpha, \beta = \frac{1}{2}\left(k_{12} + k_{21} + k_{10} \pm \sqrt{(k_{12} + k_{21} + k_{10})^2 - 4k_{21}k_{10}}\right).$$

The final parameterization of the two-compartment p.o. model with a given *dose* reads

$$\theta = \left(A_{po}, B_{po}, \alpha, \beta, k_{31}\right) \tag{7.6}$$

and is called macro constant parameterization. Typically, (7.5)–(7.6) is used to fit data. However, this parameterization is not physiological interpretable. Following the clearance concept (see e.g. [14]), one obtains the physiological parameterization (see e.g. [18]) standing in a one-to-one correspondence to (7.6)

$$\theta = (Cl, Cl_d, V_1, V_2, k_{31})$$

where $Cl = k_{10}V_1$ is the hepatitic clearance, $Cl_d = k_{12}V_1 = k_{21}V_2$ the intercompartmental clearance and V_2 the volume of distribution of the peripheral compartment.

Finally, we give a short comment on classical allometric (inter-species) scaling of physiological parameters like clearance or volume of distribution. First, to perform a scaling, the underlying pharmacokinetic mechanism for the different species has to be similar. Second, it is commonly believed that clearance or volume of distribution depend on the body weight w, see [33]. A typical allometric model for scaling a physiological parameter p is based on a power law and reads

$$p(w) = aw^b \tag{7.7}$$

where $a, b > 0$ are allometric parameters, see [14,33] or [42]. It is suggested that at least 4 to 5 species are necessary to predict from mouse to human. A typical chain is mouse, rat, rabbit, monkey and finally human.

7.2.4 *Multiple Dosage*

The next step to describe the pharmacokinetics of a drug is to handle multiple dosing events, that means, a drug is administered several times to the body. Hence, one has also to account the remaining drug concentration in the body from a previous dosage.

A drug is often designed for equidistant administration, i.e. every day, every second day, every week and so on. This makes the application of drugs more secure for patients and therefore, increases the success on the market.

Let $\tau > 0$ be the length of the dosing interval, $m \in \mathbb{N}$ the maximal number of doses and $j \in \{1, \ldots, m\}$ denote the actual number of dosage. Using the superposition principle one obtains the multiple dosing formula for p.o. administration represented by a composed function. The drug concentration at time t reads

$$c^{p.o.}(t) = \begin{cases} c_j^{p.o.}(\xi), & t = j\tau + \xi, \quad \xi \in [0, \tau] \quad j \in \{1, \ldots, m-1\} \\ c_j^{p.o.}(\xi), & t = j\tau + \xi, \quad \xi \geq 0 \qquad j = m \end{cases}$$

with

$$\begin{aligned} c_j^{p.o.}(\xi) = {} & dose A_{po} \frac{1 - \exp(-\tau j \alpha)}{1 - \exp(-\tau \alpha)} \exp(-\alpha \xi) \\ & + dose B_{po} \frac{1 - \exp(-\tau j \beta)}{1 - \exp(-\tau \beta)} \exp(-\beta \xi) \\ & - dose(A_{po} + B_{po}) \frac{1 - \exp(-\tau j k_{31})}{1 - \exp(-\tau k_{31})} \exp(-k_{31} \xi) \end{aligned} \tag{7.8}$$

see e.g. the book from Gibaldi [15]. We remark that for multiple p.o. administration, $c^{p.o.}(t)$ is a continuous function whereas in case of i.v., $c^{i.v.}(t)$ is not continuous at the dosing time points.

7.2.5 Discussion and Outlook

The pharmacokinetics describes the behavior of a drug in the body over time. Two-compartment models are widely used in industry and academics to describe the drug concentration in blood empirically because the time course of most drugs is reflected quite well. Such models have an analytical representation and mainly serve as input (driving force) in a full nonautonomous PKPD model. Also note that in experiments often a sparse PK data situation exists because only a limited number of blood samples can be taken from the animals or patients.

A mechanistic description of pharmacokinetic processes (ADME) to predict the kinetics of drugs in the whole body is provided by physiologically based pharmacokinetic models (PBPK), see [19]. Such models are composed of several compartments representing relevant organs (like kidney, liver, lung, gut, etc.) and tissues described by weight or volume and blood perfusion rates. PBPK models admit a mechanistic understanding of the drug's kinetics in the body and its implication to toxicological assessment. It is commonly stated that these models are superior when estimating human pharmacokinetic parameters based on animal data in contrast to classical allometric approaches based on empirical compartment PK models, see [19]. In addition, these models allow to differentiate for the prediction in PK between children and adults, see [1]. Nevertheless, in this work we skip a detailed description of PBPK because in a full PKPD model the drug concentration is usually described by empirical models.

7.3 Pharmacodynamic Models

7.3.1 Introduction

The pharmacodynamics (PD) "can be defined as the study of the time course of the biological effects of drugs, the relationship of the effects to drug exposure and the mechanism of drug action", see [14].

Of major importance in PD is the binding of the drug at the receptor (target) because "receptors are the most important targets for therapeutic drugs", see [31]. For that purpose we introduce effect-concentration models (compare [14]). Such models are typically used as subunits in larger systems describing the pharmacological effect/response on a disease provoked by the binding of the drug at the receptor/target.

Hence, the next step is "the process of target activation into pharmacological response. Typically, binding of a drug to its target activates a cascade of electrophysiological and/or biochemical events resulting in the observable biological response", see [6]. For that we consider models with a zero order inflow and first order outflow and also focus on cascades of these models, so-called transit compartments.

Further, we present lifespan models to describe the lifespan of subjects in a population, e.g. typically used to describe maturation of cells. Such models have a zero order in- and outflow term, an explicit lifespan parameter and a description of the past. Finally, we show an important relationship between transit compartments and lifespan models.

From the mathematical point of view, the resulting pharmacodynamic models are differential equations. In the following we understand the existence of the solution in two different terms. If the right hand side is continuous (p.o. case) the existence of the solution is understood in terms of Picard–Lindelöf. If the right hand side is piecewise continuous in time (i.v. case) we understand the existence in the sense of Carathéodory, see [5].

7.3.2 Effect-Concentration Models

In Sect. 7.2, pharmacokinetic models describing the time course of the drug concentration $c(t)$ were introduced. Now we consider models that put the drug concentration in relationship to an effect denoted by

$$e(\sigma, c(t)) \tag{7.9}$$

where we call σ the drug-related parameter. The only requirement on (7.9) is

$$e(\sigma, c) \geq 0 \quad \text{and} \quad e(\sigma, 0) = 0 \,.$$

The simplest approach for an effect of a drug is a linear term

$$e(k_{pot}, c(t)) = k_{pot} c(t) \tag{7.10}$$

where $k_{pot} > 0$ describes the drug potency. Such a parameter could be used to rank different compounds among each other in preclinical screening. The approach (7.10) is also useful if only few dosing groups are available for a simultaneous fit. For more dosing groups this approach is only locally true because the effect of a drug is in the majority of cases only linear in a small range of different doses.

The classical drug-receptor binding theory states that the amount of binding possibilities at the receptor is limited. Therefore, the effect of a drug will saturate and more drug will not lead to more effect. The most common nonlinear model to relate drug concentration and effect is

$$e(\sigma, c(t)) = \frac{E_{max} c(t)^h}{EC_{50}^h + c(t)^h} \tag{7.11}$$

with $\sigma = (E_{max}, EC_{50}, h)$, see [31]. $E_{max} > 0$ is the maximal effect, $EC_{50} > 0$ is the concentration needed to produce the half-maximal effect and $h > 0$ is the Hill coefficient. Equation (7.11) is one of the basic principles in PKPD and called the E_{max} model, see also Chap. 6.

7.3.3 Indirect Response Models

In pharmacodynamics, one is often faced with a so-called indirect drug response, that means, the drug stimulates or inhibits factors which control the response, see [8]. Further, one often assumes that the system describing the pharmacological action is in a so-called baseline condition. For example, think of heart rate, blood pressure, biomarkers etc. . The aim is to describe a perturbation of the baseline by a drug $c(t)$. Moreover, if the perturbation vanishes, it is pharmacological assumed that the response runs back into its baseline.

The basic equation of an indirect response model (IDR) with constant inflow $k_{in} > 0$ and outflow $k_{out} > 0$ reads

$$x' = k_{in} - k_{out} x \,, \qquad x(0) = x^0 \geq 0 \,. \tag{7.12}$$

For the baseline condition the initial value is set equal to the steady state $x^* = \lim_{t \to \infty} x(t)$,

$$x^0 = x^* = \frac{k_{in}}{k_{out}} \,.$$

In nonautonomous indirect response models, the drug effect is often described by an E_{max} term modulating the in- and outflow. Depending on which rate is stimulated or inhibited, one obtains four possible models, see originally Jusko and coworkers [8] or summarized in [14], presented here in compact form

$$x' = k_{in} \cdot \left\{ \left(1 - \frac{I_{max} c(t)^h}{IC_{50}^h + c(t)^h}\right), \left(1 + \frac{S_{max} c(t)^h}{SC_{50}^h + c(t)^h}\right) \right\} - k_{out} \cdot \left\{ \left(1 - \frac{I_{max} c(t)^h}{IC_{50}^h + c(t)^h}\right), \left(1 + \frac{S_{max} c(t)^h}{SC_{50}^h + c(t)^h}\right) \right\} \cdot x \tag{7.13}$$

with the initial value

$$x(0) = \frac{k_{in}}{k_{out}}$$

where $0 < I_{max} \leq 1$. IDRs of the form (7.13) are one of the most popular models in PKPD and are extensively studied and applied in the last 20 years.

7.3.4 General Inflow–Outflow Models

Consider a state $x : \mathbb{R}_{\geq 0} \rightarrow \mathbb{R}_{\geq 0}$ controlled by two processes, namely, an inflow into the state and an outflow from this state. A reasonable realization is by a zero-order inflow and a first-order outflow. Let $k_{in} : \mathbb{R}_{\geq 0} \rightarrow \mathbb{R}_{\geq 0}$ and $k_{out} : \mathbb{R}_{\geq 0} \rightarrow \mathbb{R}_{\geq 0}$ be piecewise continuous and bounded functions with

$$\lim_{t\to\infty} k_{in}(t) = k_{in}^* \geq 0 \quad \text{and} \quad \lim_{t\to\infty} k_{out}(t) = k_{out}^* > 0$$

describing inflow and outflow, respectively. We call

$$x'(t) = k_{in}(t) - k_{out}(t)x(t), \qquad x(0) = x^0 \geq 0 \tag{7.14}$$

an inflow–outflow model (IOM). Model (7.14) has an asymptotically stable stationary point

$$x^* = \lim_{t\to\infty} x(t) = \frac{k_{in}^*}{k_{out}^*}. \tag{7.15}$$

Note that indirect response models are a special case of IOMs.

7.3.5 Transit Compartment Models

Widely used models in PKPD are transit compartment models (TCMs). Such models are chains of inflow–outflow models where the inflow in the j-compartment is just the outflow of the $j-1$. The corresponding equations read

$$x_1' = k_{in}(t) - kx_1, \qquad x_1(0) = x_1^0 \geq 0 \tag{7.16}$$

$$x_2' = kx_1 - kx_2, \qquad x_2(0) = x_2^0 \geq 0 \tag{7.17}$$

$$\vdots \qquad \vdots$$

$$x_n' = kx_{n-1} - kx_n, \qquad x_n(0) = x_n^0 \geq 0 \tag{7.18}$$

where $k_{in} : \mathbb{R}_{\geq 0} \to \mathbb{R}_{\geq 0}$ is a piecewise continuous and bounded function. For example, $k_{in}(t)$ could describe the PK and therefore, in case of i.v. discontinuities exist. The transit rate between the compartments is $k > 0$. Roughly said, the states $x_2(t), \ldots, x_n(t)$ can be viewed as delayed versions of $x_1(t)$.

The application of (7.16)–(7.18) is versatile in PKPD modeling. TCMs can be motivated by signal transduction processes, see [38], and therefore, mimic biological signal pathways. But TCMs are also often used to just produce delays, see [30] (delayed drug course) or [12] (delayed cytokine growth). Hence, the states $x_i(t)$ often lose their pharmacological interpretation and the TCM concept is downgraded to a help technique. Historically, Sheiner was the first in 1979, see [36], who suggested to apply a TCM with $n = 1$ to describe a delay between pharmacokinetics and effect which is also called an effect compartment.

TCMs are also applied to describe populations, see e.g [37]. When looking at a TCM one sees that one could assign a mean residence/transit time of $\frac{1}{k}$ for an individual to stay in the i-th compartment, $i \in 1, \ldots, n$, see e.g. [38]. In this sense, a TCM could be reinterpreted as a model describing an age structured population and $x_i(t)$ describes the number of individuals with age a_i, where $a_i \in (\frac{i-1}{k}, \frac{i}{k}]$. Hence, spoken in terms of population, the $x_1(t), \ldots, x_n(t)$ describe the age distribution of a total population

$$y_n(t) = x_1(t) + \cdots + x_n(t)\,.$$

Therefore, the secondary parameter

$$T = \frac{n}{k}$$

describes the mean transit/residence time needed for an object created by k_{in} to pass through all states $x_i(t)$ for $i = 1, \ldots, n$.

However, in most cases it is obvious that the choice of the number of compartments n is somehow arbitrary. In application, n is often chosen in such a way that the final PKPD model fits the data best. For example, Savic and Karlsson [35] used a TCM to describe an absorption lag which is often seen in PK p.o. data because "some time passes before drug appears in the systemic circulation." They calculated the optimal number of compartments based on fitting results for different drugs.

A reasonable extension of (7.16)–(7.18) is to use a time-variant transit rate satisfying $kg(t) > 0$. This leads to

$$x_1' = k_{in}(t) - kg(t)x_1, \qquad x_1(0) = x_1^0 \geq 0 \tag{7.19}$$

$$x_2' = kg(t)x_1 - kg(t)x_2, \qquad x_2(0) = x_2^0 \geq 0 \tag{7.20}$$

$$\vdots \qquad\qquad \vdots$$

$$x_n' = kg(t)x_{n-1} - kg(t)x_n, \qquad x_n(0) = x_n^0 \geq 0 \tag{7.21}$$

where $g : \mathbb{R} \to \mathbb{R}_{>0}$ is a piecewise and bounded function with finitely many discontinuity points. We call (7.19)–(7.21) a generalized TCM and will have a closer look at it in Sect. 7.3.7.

7.3.6 Lifespan Models

Another class of pharmacodynamic models are lifespan models introduced by Krzyzanski and Jusko in 1999 [26] to PKPD. Generally, such models describe populations where the individuals have a certain lifespan. Krzyzanski and Jusko applied this approach to hematological cell populations in the context of indirect response models.

Let $y : \mathbb{R}_{\geq 0} \to \mathbb{R}_{\geq 0}$ be a state controlled by production (birth of individuals) and loss (death of individuals). The general form of a lifespan model (LSM) is

$$y'(t) = k_{in}(t) - k_{out}(t), \quad y(0) = y^0 \tag{7.22}$$

where k_{in} and k_{out} are piecewise continuous and bounded functions.

In this chapter we consider two different cases. First, we present LSMs with a constant lifespan, that means every individual in the population has the same lifespan $T > 0$. This approach is first of all an idealized situation. However, this assumption is reasonable in most applications due to the data situation. Second, we additionally consider distributed lifespans.

7.3.6.1 Constant Lifespan

Assuming a constant lifespan T, the outflow from state y at time t is equal to the inflow at time $t - T$ and we obtain the relation

$$k_{out}(t) = k_{in}(t - T) \quad \text{for } t \geq 0\,.$$

Hence, the LSM for constant lifespan reads

$$y'(t) = k_{in}(t) - k_{in}(t - T), \quad y(0) = y^0 . \tag{7.23}$$

In applications one has seldom the freedom of choosing the initial value $y(0) = y^0$ arbitrarily. For example, in populations the initial value y^0 has to be set in such a way that it describes the amount of individuals already born and also died in the interval $[-T, 0]$. Therefore, we obtain

$$y^0 = \int_{-T}^{0} k_{in}(s)\, ds . \tag{7.24}$$

The solution of (7.23)–(7.24) reads

$$y(t) = \int_{t-T}^{t} k_{in}(s)\, ds \quad \text{for } t \geq 0 .$$

An important situation in application is a constant production in the past (e.g. in context of cell production)

$$k_{in}(s) = k_{in}^{-} \quad \text{for } s \leq 0 .$$

Then the initial value (7.24) is

$$y^0 = T k_{in}^{-} .$$

7.3.6.2 Distributed Lifespan

Let X be a random variable with a probability density function $l : \mathbb{R} \to \mathbb{R}_{\geq 0}$ where $l(s) = 0$ for $s < 0$ describes the lifespan of individuals and $T = \mathbb{E}[X]$. The outflow term then reads

$$k_{out}(t) = \int_{0}^{\infty} k_{in}(t - \tau) l(\tau) d\tau = (k_{in} * l)(t)$$

see e.g. [25], [27]. The LSM is

$$y'(t) = k_{in}(t) - (k_{in} * l)(t), \quad y(0) = y^0 . \tag{7.25}$$

Again the initial value y^0 has to be chosen in such a way that it describes the amount of individuals already born and died. One obtains

$$y^0 = \int_0^\infty \int_{-\tau}^0 k_{in}(s)\, ds\; l(\tau)\, d\tau \tag{7.26}$$

see [20]. The solution of (7.25)–(7.26) reads

$$y(t) = \int_0^\infty \int_{t-\tau}^t k_{in}(s)\, ds\; l(\tau)\, d\tau \quad \text{for } t \geq 0\,.$$

For constant past $k_{in}(s) = k_{in}^-$ for $s \leq 0$ we have $y^0 = Tk_{in}^-$.

7.3.7 *General Relationship Between Transit Compartments and Lifespan Models*

In this section we present an important relationship between transit compartments and lifespan models with constant lifespan. Roughly said, if the number of compartments tends to infinity and the parameter

$$T = \frac{n}{k}$$

is fixed, then in the limit the sum of all compartments is a lifespan model with constant lifespan $T > 0$.

An initial result was presented by Krzyzanski in 2011, see [24]. He investigated equal initial values for the generalized TCM (7.19)–(7.21) and constant past for the LSM (7.23)–(7.24).

Here, we consider (7.19)–(7.21) with arbitrary initial values $x_1^0 \geq 0, \ldots, x_n^0 \geq 0$ and look for the corresponding generalized LSM with arbitrary past. This generalization covers more pharmacological situations. An important role plays

$$\tau(t) = \int_0^t g(s)\, ds\,, \quad t \in \mathbb{R}\,.$$

Note that τ is a strongly increasing function with $\tau(0) = 0$ and inverse τ^{-1}. Furthermore, $\tau(t)$ could be interpreted as a time-transformation.

Theorem 7.1. *Consider the generalized transit compartment model*

$$x_1' = k_{in}(t) - kg(t)x_1, \qquad x_1(0) = x_1^0 \geq 0 \tag{7.27}$$

$$x_2' = kg(t)x_1 - kg(t)x_2, \qquad x_2(0) = x_2^0 \geq 0 \tag{7.28}$$

$$\vdots \qquad\qquad\qquad\qquad \vdots$$

$$x_n' = kg(t)x_{n-1} - kg(t)x_n, \qquad x_n(0) = x_n^0 \geq 0, \tag{7.29}$$

where $k_{in} : \mathbb{R} \to \mathbb{R}_{\geq 0}$ *and* $g : \mathbb{R} \to \mathbb{R}_{>0}$ *are piecewise continuous and bounded functions with finitely many discontinuity points and* $k > 0$*. Again let* $\tau(t) = \int_0^t g(s)\,ds$ *and let* $h : [0,1] \to \mathbb{R}_{\geq 0}$ *be an arbitrary piecewise continuous function with* $h(0) = k_{in}(0)$*. Assume that the initial values of (7.27)–(7.29) satisfy*

$$x_i(0) = \frac{1}{k} h\left(\frac{i}{n}\right) \quad \textit{for } i = 1, \ldots, n\,. \tag{7.30}$$

Let

$$T = \frac{n}{k} > 0$$

be an arbitrary but fixed value. Further consider the total population based on (7.27)–(7.29)

$$y_n(t) = x_1(t) + \cdots + x_n(t)\,.$$

Then the limit

$$y(t) = \lim_{n \to \infty} y_n(t) \quad \textit{for } t \geq 0 \tag{7.31}$$

fulfills the lifespan model

$$y' = k_{in}(t) - g(t)\frac{k_{in}(z)}{g(z)}, \qquad y(0) = y^0 \tag{7.32}$$

$$z' = \frac{g(t)}{g(z)}, \qquad z(0) = \tau^{-1}(-T) \tag{7.33}$$

provided the input function k_{in} *satisfies*

$$\frac{k_{in}\left(\tau^{-1}(t)\right)}{g\left(\tau^{-1}(t)\right)} = h\left(-\frac{t}{T}\right) \quad \textit{for } -T \leq t \leq 0\,. \tag{7.34}$$

The initial value of (7.32) reads

$$y^0 = T \int_0^1 h(s)\,ds\,. \tag{7.35}$$

Proof. The matrix notation of (7.27)–(7.30) is

$$x' = g(t)Ax + k_{in}(t)e^1 \,, \qquad x(0) = \frac{1}{k}\hat{x}^0 \tag{7.36}$$

with $\hat{x}_i^0 = h\left(\frac{i}{n}\right), i = 1, \ldots, n$ and

$$A = \begin{pmatrix} -k & & & \\ k & -k & & \\ & \ddots & \ddots & \\ & & k & -k \end{pmatrix} \in \mathbb{R}^{n,n} \,.$$

Consider in addition

$$u' = Au + \frac{k_{in}(\tau^{-1}(t))}{g(\tau^{-1}(t))}e^1 \,, \qquad u(0) = \frac{1}{k}\hat{x}^0 \tag{7.37}$$

where the time dependency of the transit rate is shifted into the inflow term. Note that $u_1(t), \ldots, u_n(t)$ describe a TCM with constant transit rate k and inflow

$$\tilde{k}_{in}(t) = \frac{k_{in}(\tau^{-1}(t))}{g(\tau^{-1}(t))} \,. \tag{7.38}$$

It is obvious that the solutions of (7.36) and (7.37) are linked via

$$x(t) = u(\tau(t)) \,. \tag{7.39}$$

Next we use (7.39) and obtain

$$y_n'(t) = x_1'(t) + \cdots + x_n'(t) = k_{in}(t) - kg(t)u_n(\tau(t)) \,.$$

Because (7.37) is a TCM with constant transit rate k and inflow $\tilde{k}_{in}(t)$ (see (7.38)), we can apply the convergence result from [21]. This yields

$$\lim_{n\to\infty} ku_n(s) = \tilde{k}_{in}(s-T) = \frac{k_{in}(\tau^{-1}(s-T))}{g(\tau^{-1}(s-T))} \quad \textit{for } s \in \mathbb{R}$$

provided that (7.34) holds. Hence, the equation for the limit (7.31) reads

$$y'(t) = k_{in}(t) - g(t)\lim_{n\to\infty} ku_n(\tau(t)) = k_{in}(t) - g(t)\frac{k_{in}(\tau^{-1}(\tau(t)-T))}{g(\tau^{-1}(\tau(t)-T))}$$

with the initial value

$$y^0 = \lim_{n\to\infty} \sum_{i=1}^{n} x_i(0) = \lim_{n\to\infty} \sum_{i=1}^{n} \frac{T}{n} h\left(\frac{i}{n}\right) = T \int_0^1 h(s)\, ds\,.$$

Furthermore,

$$z(t) = \tau^{-1}(\tau(t) - T)$$

satisfies

$$z'(t) = \frac{1}{\tau'\left(\tau^{-1}\left(\tau(t) - T\right)\right)} \tau'(t) = \frac{g(t)}{g(z(t))}\,, \quad z(0) = \tau^{-1}(-T)\,.$$

Summarizing, we obtain the stated result. □

Remark 7.1. (a) In case of $g \equiv 1$ in (7.27)–(7.29), the lifespan model (7.32)–(7.33) reduces to

$$y'(t) = k_{in}(t) - k_{in}(t - T)\,, \quad y(0) = y^0\,, \qquad z(t) = t - T$$

which is well known from [21].

(b) The assumption $g : \mathbb{R} \to \mathbb{R}_{>0}$ is pharmacological reasonable. For example, g could describe a stimulation or inhibition term depending on the drug concentration as applied in (7.13).

(c) The solution of (7.32)–(7.33) reads

$$y(t) = \int_{z(t)}^{t} k_{in}(s)\, ds\,, \quad z(t) = \tau^{-1}\left(\tau(t) - T\right)\,.$$

7.3.8 Discussion and Outlook

Typical (semi-) mechanistic pharmacodynamic models describing the pharmacological effect applied in academics and industry were presented. We introduced models to describe the effect-concentration relationship, stated inflow/outflow models typically applied to describe perturbations of a baseline and finally, presented lifespan models for populations. In Theorem 7.1 we presented an important relationship between general transit compartments and lifespan models.

In the next section we will develop a model for a disease progression (tumor growth) and the effect of drug on the disease. For that we apply an effect concentration term and mimic the dying of proliferating cells by either transit compartments or lifespan models. For another application of PKPD see Chap. 8 in this volume.

For further reading about PKPD modeling we recommend the books from Gabrielsson and Weiner [14], from Bonate [2] for a more statistical oriented data analysis and also from Macheras and Iliadis [31] for a more general biological/mathematical point of view. Finally, several excellent review articles about PKPD modeling were published in the last years where we like to highlight the manuscripts from Danhof et al. [6, 7] or Mager et al. [32].

7.4 Pharmacokinetic/Pharmacodynamic Tumor Growth Model for Anticancer Effects

In this section we develop a PKPD model to describe tumor growth and the anticancer effects of a drug along the guideline (i)–(iv) listed in Sect. 7.1. We firstly model the disease development (iv) without drug action, here called unperturbed tumor growth. Then we present the modeling of the drug effect on the disease (compare (ii)–(iii)) called perturbed tumor growth. Finally, we include the pharmacokinetics of a specific drug into the model, see (i).

7.4.1 Introduction and Experimental Setup

It is generally stated that the work of Laird [29] "Dynamics of tumor growth" published in 1964 initiated the mathematical modeling of tumor growth. Laird applied the Gompertz equation (here presented in the original formulation)

$$\frac{W}{W_0} = e^{\frac{A}{\alpha}(1-e^{-\alpha t})}$$

to describe unperturbed (no drug administration) tumor growth. W denotes the tumor size in time, W_0 is the initial tumor size and A, α are growth related parameters. This model realizes a sigmoid growth behavior and therefore, describes the three significant phases of tumor growth. First, a tumor grows exponentially, after a while the tumor growth becomes linear due to limits of nutrient supply and finally, the tumor growth saturates. Laird applied the Gompertz equation to data from mice, rats and rabbits.

In the book from Wheldon [43] it is stated that the saturation property of tumors could seldom be measured in patients because the host dies in the majority of cases before saturation begins. Also in preclinics, the experiments have to be terminated when a critical tumor size is reached due to ethical constraints and according to the animal welfare law. Hence, in this work we present a tumor growth model without saturation and focus on the first two tumor growth phases, namely exponential growth followed by linear growth.

We consider experiments performed in xenograft mice. Such mice are applied as a model for human tumor growth. It is stated by Bonate in [3]: "Most every drug approved in cancer was first tested in a xenograft model to determine its anticancer activity". Xenograft mice develop human solid tumors based on implantation of human cancer cells. The tumor grows in the flank of the mice and its volume is measured by an electronic caliber and recalculated to weight based on tissue consistency assumptions. Roughly said, the tumor size could be measured "from the outside" without stressing the animals in contrast to PK where blood samples have to be taken. Therefore, in general more data is available in PD in contrast to PK.

However, we also mention two disadvantages of xenografts formulated by Bonate, see [3]: "First, these are human tumors grown in mice and so the mice must be immunocompromised for the tumor growth in order to prevent a severe transplant reaction from occurring in the host animal. Second, since these tumors are implanted in the flank, they do not mimic tumors of other origins, e.g. a lung cancer tumor grown in the flank may not representative for a lung cancer tumor in the lung."

7.4.2 *Unperturbed Tumor Growth*

The growth of a tumor without an anticancer drug is called unperturbed growth. The aim of this section is to model this behavior with a realistic right hand side of the differential equation

$$w' = f(w), \quad w(0) = w_0 \tag{7.40}$$

where $w_0 > 0$ is the inoculated tumor weight, more precisely, the amount of implanted human tumor cells into the xenograft mouse. The tumor weight is denoted by $w(t)$.

In 2004, Simeoni et al. [37] presented a model consisting of an exponential and a linear growth phase in order to describe the tumor growth in xenograft mice in time by the function

$$g_s(w) = \begin{cases} \lambda_0 w, & w \le w_{th} \\ \lambda_1, & w > w_{th} \end{cases}, \qquad w_{th} = \frac{\lambda_1}{\lambda_0} \tag{7.41}$$

for (7.40). In (7.41), the parameter $\lambda_0 > 0$ describes the exponential growth rate and $\lambda_1 > 0$ the linear growth rate. If the weight w reaches a threshold w_{th}, then the exponential growth switches immediately to linear growth in (7.41). This produces a fast transition between the exponential and linear phase in $w(t)$. It is suggested by Simeoni to apply the approximation

$$g_a(w) = \frac{\lambda_0 w}{\left[1 + \left(\frac{\lambda_0}{\lambda_1} w\right)^{20}\right]^{\frac{1}{20}}}$$

for (7.41) in practice.

Another growth function for (7.40)

$$g(w) = \frac{2\lambda_0 \lambda_1 w}{\lambda_1 + 2\lambda_0 w} \tag{7.42}$$

was presented in [22] which is based on the Michaelis–Menten approach and produces a longer transition between these two essentially different growth phases. The parameter in (7.42) have the same meaning as in Simeoni's model, see [22] for argumentation and derivation.

In this work we use the disease progression model

$$w' = \frac{2\lambda_0 \lambda_1 w}{\lambda_1 + 2\lambda_0 w}, \qquad w(0) = w_0 \tag{7.43}$$

for unperturbed tumor growth $w(t)$ with the three parameter

$$\theta = (\lambda_0, \lambda_1, w_0)\,.$$

In Fig. 7.1, measurements from four different human tumor cell lines in xenograft mice, namely RKO (cancer of the colon), PC3 (prostate cancer), MDA (breast cancer) and A459 (lung cancer) were fitted with (7.43).

7.4.3 Perturbed Tumor Growth Based on Transit Compartments

The next step towards a PKPD tumor growth model is to include the pharmacokinetics of a drug, or more precisely, the perturbation of the tumor growth by an anticancer agent. It is generally observed that the anti-cancer effect is delayed due to the drug concentration. Hence, the attacked tumor cells could be considered as a population with a lifespan. Simeoni and co-workers applied a transit compartment model and assumed that proliferating cells attacked by the drug will pass through different damaging stages until the cells finally and irrevocably die, see [37].

We apply the linear effect-concentration term

$$e(k_{pot}, c(t)) = k_{pot} c(t)$$

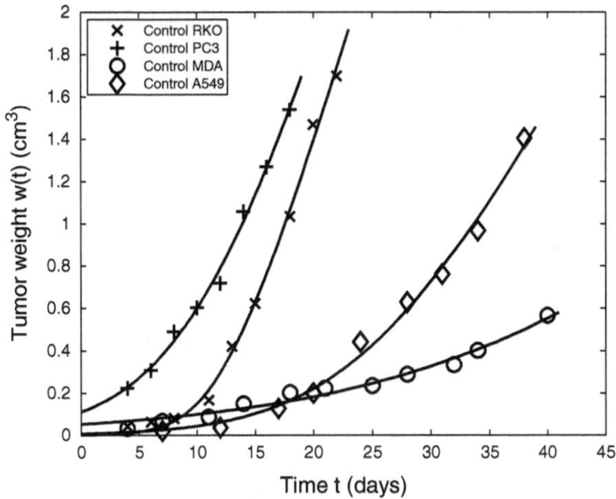

Fig. 7.1 Different human tumor cell lines (RKO, PC3, MDA and A549) in xenograft mice fitted with model (7.43)

to describe the action of the drug at the target (proliferating cells). The pharmacokinetics is denoted by $c(t)$ and $k_{pot} > 0$ describes the potency parameter of a drug. The PK $c(t)$ is a two-compartment model with either p.o. or i.v. administration. In our performed experiments we have two dosing groups, namely, a placebo and a drug administration group. Therefore, the linear effect term is an appropriate choice.

In a first approach we also apply a transit compartment model to describe the different stages of dying non-proliferating tumor cells initiated by the drug action. We denote by $p(t)$ the amount of proliferating tumor cells and by $d_1(t), \ldots, d_n(t)$ the different stages of dying tumor cells attacked by an anticancer agent. Since, the non-proliferating cells $d_1, \ldots, d_n$ still add to total tumor mass, the total tumor w is the sum of proliferating tumor cells p and non-proliferating tumor cells $d_1, \ldots, d_n$. Only proliferating cells that are not affected by drug action contribute to the tumor growth. Therefore, the growth function $g(w)$ of the total tumor consisting of proliferating and non-proliferating cells is slowed down by the factor $\frac{p}{w}$.

The PKPD model with transit compartments reads

$$p' = \frac{2\lambda_0\lambda_1 p}{\lambda_1 + 2\lambda_0 p}\frac{p}{w(t)} - k_{pot}c(t)p, \qquad p(0) = w_0 \tag{7.44}$$

$$d_1' = k_{pot}c(t)p - kd_1, \qquad d_1(0) = 0 \tag{7.45}$$

$$d_2' = kd_1 - kd_2, \qquad d_2(0) = 0 \tag{7.46}$$

$$\vdots \qquad \qquad \vdots$$

$$d_n' = kd_{n-1} - kd_n, \qquad d_n(0) = 0 \tag{7.47}$$

$$w(t) = p(t) + d_1(t) + \cdots + d_n(t) \tag{7.48}$$

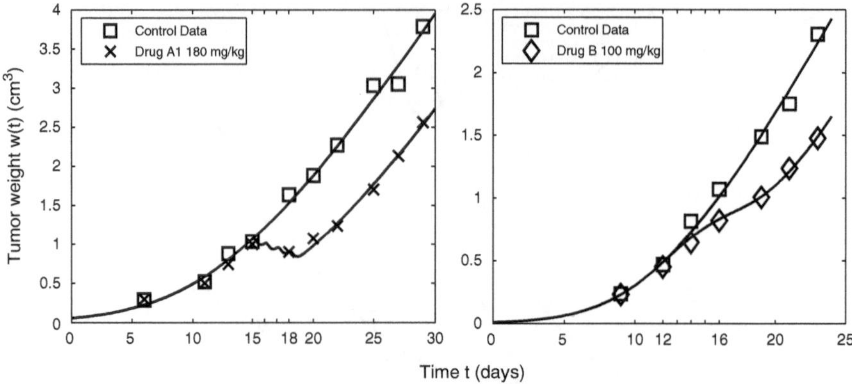

Fig. 7.2 In every plot the unperturbed and perturbed tumor growth data was simultaneously fitted with (7.44)–(7.48) and $n = 3$. In the *left panel* the drug A1 was administered at day 15, 16, 17 and 18 and in the *right panel* the drug B was administered at day 12, 13, 14, 15 and 16

with the model parameter

$$\theta = (\lambda_0, \lambda_1, w_0, k_{pot}, k).$$

The total tumor weight is denoted by $w(t)$. The average lifespan of attacked tumor cells is computed after a fitting process by

$$T = \frac{n}{k}. \tag{7.49}$$

In Fig. 7.2 we present two simultaneous fits of unperturbed and perturbed data with (7.44)–(7.48) and $n = 3$.

7.4.4 *Perturbed Tumor Growth Based on the Lifespan Approach*

In this section we apply Theorem 7.1 to the tumor growth model based on transit compartments. From a schematic point of view the model (7.44)–(7.48) can be regarded as a system with a TCM represented by (7.16)–(7.18) with input

$$k_{in}(t) = e(\sigma, c(t))p(t). \tag{7.50}$$

On the way to a description of the pharmacological process with an LSM we set

$$d(t) = d_1(t) + \cdots + d_n(t)$$

representing the totality of cells attacked by the anticancer agent and replace the TCM (7.45)–(7.47) by a LSM for the population $d(t)$. Using (7.50) this leads to

$$d'(t) = k_{in}(t) - k_{in}(t-T) = e(\sigma, c(t))p(t) - e(\sigma, c(t-T))p(t-T)$$

completed by the initial condition $d(0) = 0$ and the past

$$e(\sigma, c(s))p(s) = 0\,, \quad -T \leq s < 0\,. \tag{7.51}$$

In applications, (7.51) is fulfilled because no drug is administered before inoculation of the tumor cells.

Then the reformulation of (7.44)–(7.48) in the lifespan model context reads

$$p'(t) = \frac{2\lambda_0\lambda_1 p(t)}{\lambda_1 + 2\lambda_0 p(t)}\frac{p(t)}{w(t)} - e(\sigma, c(t))p(t), \qquad p(0) = w_0 \tag{7.52}$$

$$d'(t) = e(\sigma, c(t))p(t) - e(\sigma, c(t-T))p(t-T), \qquad d(0) = 0 \tag{7.53}$$

$$w(t) = p(t) + d(t)\,. \tag{7.54}$$

In the LSM formulation (7.51)–(7.54) we have exactly two differential equations, one for the proliferating cells $p(t)$ and one governing the population of the attacked tumor cells $d(t)$. Note that it is not necessary to provide information about $p(s)$ for $-T \leq s < 0$ due to (7.51). The parameters are

$$\theta = (\lambda_0, \lambda_1, w_0, k_{pot}, T)$$

where T is the lifespan of the dying tumor cells which is now fitted directly from the data.

The sum of squares and parameter estimates of (7.51)–(7.54) and (7.44)–(7.48) are similar. The new formulation (7.51)–(7.54) is also from the modeling point of view a serious alternative to the classical formulation. Here the number of dying tumor stages is reduced to exactly one stage for the total population of cells attacked by the anticancer agent. This coincides with the situation in practice, where the choice of the number of compartments n is more or less arbitrary because the different stages could not be measured.

7.4.5 *Discussion and Outlook*

It is estimated that every third European develops cancer once in life time. Hence, mathematical modeling of tumor growth data is an important task to support drug development. The PKPD model structure presented by Simeoni et al. in 2004 [37] is one of the most applied tumor growth models in the last years.

In this work we focused on administration of one single drug. However, an important topic in anticancer drug development is the combination of different drugs and the search of synergistic effects in order to maximize the pharmacological effect. Based on a synergistic combination of drug effects the dosage could be reduced to minimize the side effects in patients. Hence, a new direction in tumor growth modeling is the development of realistic and mechanistic models for drug combination approaches. In [22] an approach which explicitly quantifies the synergy by a parameter and also describes combination therapy data was presented. The model could be used to rank different combination therapies. Nevertheless, this modeling field is subject of active research, see e.g. [17] for preclinical and [13] for clinical phase. To our knowledge no widely accepted mechanistic PKPD tumor growth combination therapy model is developed yet.

Acknowledgements The present project is supported by the National Research Fund, Luxembourg, and cofunded under the Marie Curie Actions of the European Commission (FP7-COFUND). The authors like to thank Dr. Antje Walz and Dr. Thomas Wagner for their valuable comments and remarks.

References

1. J.S. Barrett, O. Della Casa Alberighi, S. Läer, B. Meibohm, Physiologically based pharmacokinetic (PBPK) modeling in children. Clin. Pharmacol. Ther. **92**, 40–49 (2012)
2. P.L. Bonate, *Pharmacokinetic-Pharmacodynamic Modeling and Simulation* (Springer, London, 2006)
3. P.L. Bonate, D.R. Howard, *Pharmacokinetics in Drug Development: Advances and Applications*, vol. 3 (Springer, Berlin, 2011)
4. M.E. Burton, L.M. Shaw, J.J. Schentag, W.E. Evans, *Applied Pharmacokinetics & Pharmacodynamics: Principles of Therapeutic Drug Monitoring* (Lippincott Williams & Wilkins, Philadelphia, 2006)
5. E.A. Coddington, N. Levinson, *Theory of Ordinary Differential Equations* (McGraw-Hill, New York, 1955)
6. M. Danhof, J. de Jongh, E.C.M. De Lange, O. Della Pasqua, B.A. Ploeger, R.A. Voskuyl, Mechanism-based pharmacokinetic-pharmacodynamic modeling: Biophase distribution, receptor theory, and dynamical systems analysis. Annu. Rev. Pharmacol. Toxicol. **47**, 357–400 (2007)
7. M. Danhof, E.C.M. de Lange, O.E. Della Pasqua, B.A. Ploeger, R.A. Voskuyl, Mechanism-based pharmacokinetic-pharmacodynamic (PK-PD) modeling in translational drug research. Trends Pharmacol. Sci. **29**, 186–191 (2008)
8. N.L. Dayneka, V. Garg, W.J. Jusko, Comparison of four basic models of indirect pharmacodynamic responses. J. Pharmacokinet. Biopharm. **21**, 457–478 (1993)
9. J.A. DiMasi, R.W. Hansen, H.G. Grabowski, The price of innovation: New estimates of drug development costs. J. Health Econ. **22**, 151–185 (2003)
10. F.H. Dost, *Der Blutspiegel: Kinetik der Konzentrationsabläufe in der Kreislaufflüssigkeit* (Georg Thieme, Leipzig, 1953)
11. F.H. Dost, *Grundlagen der Pharmakokinetik* (Georg Thieme, Stuttgart, 1968)
12. J.C. Earp, D.C. DuBois, D.S. Molano, N.A. Pyszczynski, C.E. Keller, R.R. Almon, W.J. Jusko, Modeling corticosteroid effects in a rat model of rheumatoid arthritis I: Mechanistic disease

progression model for the time course of collagen-induced arthritis in Lewis rats. J. Pharmacol. Exp. Ther. **326**, 532–545 (2008)
13. N. Frances, L. Claret, R. Bruno, A. Iliadis, Tumor growth modeling from clinical trials reveals synergistic anticancer effect of the capecitabine and docetaxel combination in metastatic breast cancer. Cancer Chemother. Pharmacol. **68**, 1413–1419 (2011)
14. J. Gabrielsson, D. Weiner, *Pharmacokinetic and pharmacodynamic data analysis: Concepts and applications* (Swedish Pharmaceutical Press, Stockholm, 2006)
15. M. Gibaldi, D. Perrier, *Pharmacokinetics Second Edition Revised and Expanded* (Marcel Dekker, New York, 1982)
16. J.V.S. Gobburu, P.J. Marroum, Utilisation of pharmacokinetic-pharmacodynamic modelling and simulation in regulatory decision-making. Clin. Pharmacokinet. **40**, 883–892 (2001)
17. K. Goteti, C.E. Garner, L. Utley, J. Dai, S. Ashwell, D.T. Moustakas, M. Gönen, G. Schwartz, S.E. Kern, S. Zabludoff, P.J. Brassil, Preclinical pharmacokinetic/pharmacodynamic models to predict synergistic effects of co-administered anti-cancer agents. Cancer Chemother. Pharmacol. **66**, 245–254 (2010)
18. S.A. Hill, Pharmacokinetics of drug infusions. Continuing Educ. Anaesth. Crit. Care Pain **4**, 76–80 (2004)
19. H.M. Jones, I.B. Gardner, K.J. Watson, Modelling and PBPK simulation in drug discovery. Astron. Astrophys. Suppl. **11**, 155–166 (2009)
20. G. Koch, *Modeling of Pharmacokinetics and Pharmacodynamics with Application to Cancer and Arthritis*. KOPS (Das Institutional Repository der Universität Konstanz, 2012). http://nbn-resolving.de/urn:nbn:de:bsz:352-194726
21. G. Koch, J. Schropp, General relationship between transit compartments and lifespan models. J. Pharmacokinet. Pharmacodyn. **39**, 343–355 (2012)
22. G. Koch, A. Walz, G. Lahu, J. Schropp, Modeling of tumor growth and anticancer effects of combination therapy. J. Pharmacokinet. Pharmacodyn. **36**, 179–197 (2009)
23. I. Kola, J. Landis, Can the pharmaceutical industry reduce the attrition rates? Nat. Rev. Drug Discov. **3**, 711–715 (2004)
24. W. Krzyzanski, Interpretation of transit compartments pharmacodynamic models as lifespan based indirect response models. J. Pharmacokinet. Pharmacodyn. **38**, 179–204 (2011)
25. W. Krzyzanski, J.J. Perez Ruixo, Lifespan based indirect response models. J. Pharmacokinet. Pharmacodyn. **39**, 109–123 (2012)
26. W. Krzyzanski, R. Ramakrishnan, W.J. Jusko, Basic pharmacodynamic models for agents that alter production of natural cells. J. Pharmacokinet. Biopharm. **27**, 467–489 (1999)
27. W. Krzyzanski, S. Woo, W.J. Jusko, Pharmacodynamic models for agents that alter production of natural cells with various distributions of lifespans. J. Pharmacokinet. Pharmacodyn. **33**, 125–166 (2006)
28. Y. Kwon, *Handbook of Essential Pharmacokinetics, Pharmacodynamics and Drug Metabolism for Industrial Scientists* (Springer, New York, 2001)
29. A.K. Laird, Dynamics of tumor growth. Br. J. Cancer **18**, 490–502 (1964)
30. E.D. Lobo, J.P. Balthasar, Pharmacodynamic modeling of chemotherapeutic effects: Application of a transit compartment model to characterize methotrexate effects in vitro. AAPS PharmSci. **4**, 212–222 (2002)
31. P. Macheras, A. Iliadis, in *Modeling in Biopharmaceutics, Pharmacokinetics, and Pharmacodynamics*. Interdisciplinary Applied Mathematics, vol. 30 (Springer, Berlin, 2006)
32. D.E. Mager, E. Wyska, W.J. Jusko, Diversity of mechanism-based pharmacodynamic models. Drug Metab. Dispos. **31**, 510–519 (2003)
33. J. Mordenti, S.A. Chen, J.A. Moore, B.L. Ferraiolo, J.D. Green, Interspecies scaling of clearance and volume of distribution data for five therapeutic proteins. Pharm. Res. **8**, 1351–1359 (1991)
34. B.A. Ploeger, P.H. van der Graaf, M. Danhof, Incorporating receptor theory in mechanism-based pharmacokinetic-pharmacodynamic (PK-PD) modeling. Drug Metab. Pharmacokinet. **24**, 3–15 (2009)

35. R.M. Savic, D.M. Jonker, T. Kerbusch, M.O. Karlsson, Implementation of a transit compartment model for describing drug absorption in pharmacokinetic studies. J. Pharmacokinet. Pharmacodyn. **34**, 711–726 (2007)
36. L.B. Sheiner, D.R. Stanski, S. Vozeh, R.D. Miller, J. Ham, Simultaneous modeling of pharmacokinetics and pharmacodynamics: Application to d-tubocurarine. Clin. Pharmacol. Ther. **25**, 358–371 (1979)
37. M. Simeoni, P. Magni, C. Cammia, G. De Nicolao, V. Croci, E. Pesenti, M. Germani, I. Poggesi, M. Rocchetti, Predictive pharmacokinetic-pharmacodynamic modeling of tumor growth kinetics in xenograft models after administration of anticancer agents. Cancer Res. **64**, 1094–1101 (2004)
38. Y. Sun, W.J. Jusko, Transit compartments versus gamma distribution function to model signal transduction processes in pharmacodynamics. J. Pharm. Sci. **87**, 732–737 (1998)
39. T. Teorell, Kinetics of distribution of substances administered to the body I. The extravascular modes of administration. Archs. Int. Pharmacodyn. Ther. **57**, 205–225 (1937)
40. P.H. van der Graaf, J. Gabrielsson, Pharmacokinetic-pharmacodynamic reasoning in drug discovery and early development. Future Med. Chem. **1**, 1371–1374 (2009)
41. J.G. Wagner, History of pharmacokinetics. Pharmacol. Ther. **12**, 537–562 (1981)
42. G.B. West, J.H. Brown, The origin of allometric scaling laws in biology from genomes to ecosystems: Towards a quantitative unifying theory of biological structure and organization. J. Exp. Biol. **208**, 1575–1592 (2005)
43. T.E. Wheldon, *Mathematical Models in Cancer Research* (Adam Hilger, Bristol/Philadelphia, 1988)

Chapter 8
Viral Kinetic Modeling of Chronic Hepatitis C and B Infection

Eva Herrmann and Yusuke Asai

Abstract Chronic infection with hepatitis C or hepatitis B virus are important world-wide health problems leading to long-term damage of the liver. There are, however, treatment options which can lead to viral eradication in hepatitis C or long-term viral suppression in hepatitis B in some patients. Nevertheless, there is still room for improvement. Mathematical compartment models based on ordinary differential equation systems have successfully been applied to improve antiviral treatment. Here, we illustrate how mathematical and statistical analysis of such models influenced clinical research and give an overview on the most important models for hepatitis C and hepatitis B viral kinetics.

Keywords Models in medicine • Parameter estimation • Ordinary differential equations • PKPD models

8.1 Basic Viral Kinetic Models

There are various mathematical approaches to model acute or chronic viral infections. The scale of these models ranges from modeling viral infection and/or replication on cell level (see, e.g., as a recent example, [18]) up to epidemiological models on the world-wide spread of viral infections which may even account for modern air traffic data, see, e.g., [6]. Nevertheless, an accepted tool to optimize treatment for some important acute and chronic infections bases on modeling the viral dynamics inside a single infected patient which is referred to as viral kinetics. This is especially true for chronic infections with the human immunodeficiency

E. Herrmann (✉) · Y. Asai
Department of Medicine, Institute of Biostatistics and Mathematical Modeling, Goethe University Frankfurt, Deutschland
e-mail: herrmann@med.uni-frankfurt.de; asai@med.uni-frankfurt.de

P.E. Kloeden and C. Pötzsche (eds.), *Nonautonomous Dynamical Systems in the Life Sciences*, Lecture Notes in Mathematics 2102, DOI 10.1007/978-3-319-03080-7_8,

virus (HIV), hepatitis B virus (HBV) and hepatitis C virus (HCV). In the following, we will focus on such models and their impact on clinical research.

The most important basic viral kinetic model is described by a three-dimensional, nonlinear and autonomous ordinary differential equation system (ODS) given by

$$\dot{V} = pI - cV \tag{8.1}$$

$$\dot{I} = \beta TV - \delta I \tag{8.2}$$

$$\dot{T} = s - dT - \beta TV. \tag{8.3}$$

It was proposed by [20] but some earlier approaches on defining compartment models in viral infections had already been published.

The compartments here are a compartment of circulating virus V, productively infected cells I and non-infected target cells T. The parameters of the model are the production rate of infected cells p, the clearance rate of circulating virus c, the *de-novo* infection rate β, the death rate of infected cells δ, a production rate of target cells s as well as a death rate of target cells d.

A detailed description of this model and early applications and generalizations when model HIV and HBV viral kinetics can be found in [21]. This model and slightly modified versions has found many applications to model HIV, HBV and HCV viral kinetics.

The model is easiest to interpret if infected cells continuously produce and release virus in contrast to releasing a certain amount of virus while dying. Furthermore, the amount of virus that infects cells is typically negligible in comparison to the amount of virus that is cleared. Therefore, typically, a term of the form $-\beta TV$ is ignored in Eq. (8.1).

This basic model can be seen as an adaptation of the epidemiological basic SIR model (see Chap. 1 of Kloeden and Pötzsche in this volume) which describes the population dynamics of infections and goes back to W.O. Kermack and A.G. McKendrick in the twenties of the last century [14], see also [16, Sects. 10.1–10.3] for an overview. It also still forms the basis of highly advanced models for describing global spread of emergent diseases, e.g. the models accounting for international air transport data and allowing predictions in [6].

The basic viral kinetic model was initially proposed to model chronic viral infection. Such chronic infections can run over months and years without obvious damage to the infected patient besides physiological stress. Therefore, as long as there is a stable and untreated chronic infection, the compartment model can be assumed to be in a steady state. This is a special technical feature of such modeling approaches and allow to reduce the number of parameters describing the model characteristics by steady state assumptions. Obviously, if V^*, I^* and T^* describe the steady state levels of the corresponding compartments during such a chronic phase, we obtain

$$p = c\frac{V^*}{I^*}, \quad \beta = \delta\frac{I^*}{V^*T^*}, \quad \text{and} \quad s = dT^* + \delta I^*.$$

Then we can normalize the ODS and obtain

$$\dot{x} = cy - cx \tag{8.4}$$

$$\dot{y} = \delta\tau xz - \delta y \tag{8.5}$$

$$\dot{z} = s^* - dz - \delta\tau xz \tag{8.6}$$

with $x = \frac{V}{V^*}$, $y = \frac{I}{I^*}$ and $z = \frac{T}{T^*}$.

The normalized parameter $s^* = \frac{s}{T^*} = \frac{d}{\tau} + \delta$ characterizes the production or regeneration rate of the normalized z compartment whereas the additional parameter $\tau = \frac{I^*}{T^*}$ characterizes the status of infection. High levels of τ indicate a high proportion of infected cells and, therefore, a more active infection.

In particular, when modeling HCV and HBV kinetics and fitting the model to clinical data, the normalized version of the differential equation system by Eqs. (8.4)–(8.6) has the advantage, that there is no need to specify the true amount of infected and uninfected cells I^* and T^*. Indeed, typically, we just monitor the viral load in blood described by $V(t) = V^* \cdot x(t)$ but it is not possible to monitor the dynamics of the cell compartments or the total amount of cells or even the steady state level of infected or uninfected liver cells.

Thus, the remaining parameters describing the dynamics of the V compartment only are V^*, c, δ, d and τ.

8.2 Assessing Treatment Effects from HCV Kinetics

The basic viral kinetic model was first proposed to model chronic viral infection. Starting from Eqs. (8.4)–(8.6), several viral kinetic models were developed and used to analyze treatment effects.

Note, that chronic infection with hepatitis C is an important health problem and affects around 150 million people worldwide. Although the virus was identified in the late 1980s, treatment of chronic HCV infection even started before the virus was characterized. In those days, chronic infection was typically characterized as non-A non-B hepatitis.

Development of more effective treatments was successfully done by a combination of different approaches: Just exploring the efficacy of several general antiviral treatments, designing specific inhibitors of HCV proteins but also optimizing and quantifying treatment effects by mathematical modeling approaches.

8.2.1 Modeling Treatment Effects of Interferon

First treatment schedules comprise treatment with standard interferons alone or in combination with ribavirin. These general antivirals lead to a sustained virological response in 20 %, nearly none, and around 40 % of patients when treated with interferon, ribavirin and a combination of both, respectively.

A benchmarking publication of A.U. Neumann et al. in 1998 [19] used the basic viral kinetic model and assumed different possible treatment effects for fitting data in a dose-escalating trial. In contrast to earlier publications, they assumed a single but only partial treatment effect on several potential pathways: viral production, *de novo* infection, infected cell loss and viral clearance, respectively. Mathematically, such an effect can be introduced by a factor $1 - \varepsilon \in [0, 1]$ in the term of the ODS where a blocking should be modeled or as an inflation factor $M > 1$ when an inflation should be modeled. For example, modeling a partial blocking of viral production with an efficiency factor of $\varepsilon \in [0, 1]$ would then change Eq. (8.4) of the normalized ODS (8.4)–(8.6) to

$$\dot{x}(t) = (1 - \varepsilon)cy - cx \tag{8.7}$$

and uses the steady state initial values as $x(0) = y(0) = 1$ and $z(0) = \frac{1}{\tau}$.

Of course, treatment effect should be stronger, i.e., ε would be greater in the higher dosing regimes.

A partial effect on *de novo* infection and infected cell loss, respectively, would lead to viral dynamics which differ from the observed ones as then viral decline would start slowly and become faster afterwards, see the top panels in Fig. 8.1.

In contrast, a partial effect on viral production lead to a biphasic decline with a steep first phase and a slower second phase of decline where the extend of the first phase is dose-dependent, see Fig. 8.1. This fits very well to observed clinical data [19]. Furthermore, in contrast to an effect on the viral clearance rate, the decay rate of the first phase is nearly unchanged.

8.2.2 Estimation of Kinetic Parameters

The individual viral kinetic function can be fitted in the framework of nonlinear parametric regression.

Even so the basic model is relatively simple, estimation of kinetic parameters from sequential quantifications of viral load in blood is still challenging for the following reasons.

- We can typically only observe data from one compartment, in particular the parameters d and τ do not have much influence on short-term viral kinetics. Both parameters mainly describe if the dynamics is not completely biphasic but slows down after some time, see Fig. 8.2.

- Nevertheless, it is relatively easy to assess two of the clinical important parameters: the efficiency factor ε and the infected cell loss δ as the first characterizes the amount of decay during the first phase and the second is the

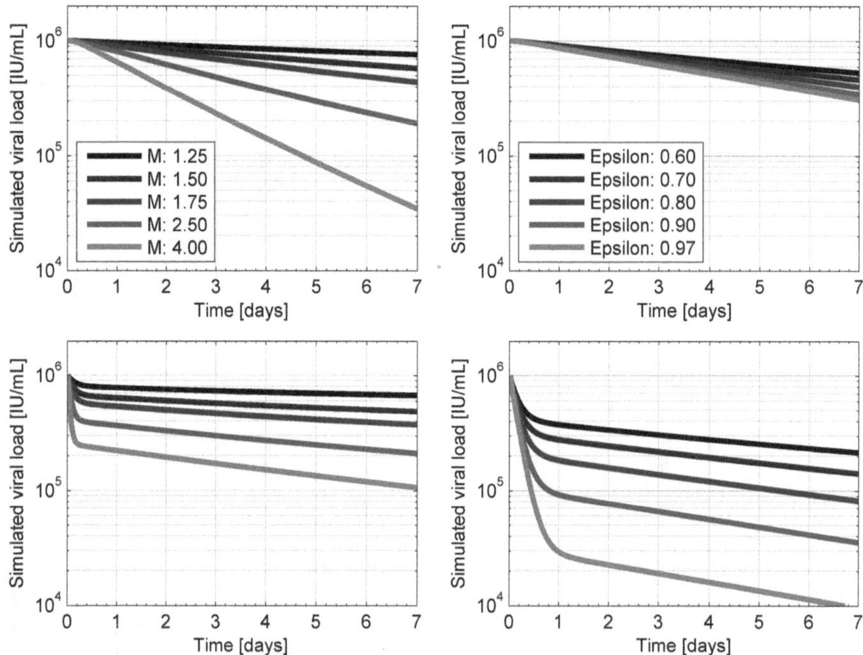

Fig. 8.1 Modeling HCV kinetics for 7 days with a dose-dependent treatment effect on infected cell loss (*left above panel*), viral *de novo* infection (*right above panel*), clearance of virus (*left below panel*, same inflations as the panel above) and viral production (*right below panel*, same efficiency factors as the panel above)

main contributor to the second phase decline. Unfortunately, both parameters are highly variable between patients and different patient groups.

- It has been proven to be suitable to fit log values of viral quantifications to logarithmically transformed model function of the viral load compartment. Furthermore, logarithmic transformations for the viral kinetic parameters V_0, c, δ, d and τ and probit transformations for the efficacy parameter ε should be used.

- If there is an effective treatment, viral load may not be observable soon during therapy. Detection and quantification limits (lower as well as upper limits) are problematic within a least squares approach but can be solved with a more advanced maximum likelihood approach. A description of the approach can be found in Guedj et al. [10]. The approach can easily be extended to the situation where one has to account for a combination of such limits.

- Furthermore, fitting viral kinetic parameters cannot be done with standard routines and needs an iterated approach of maximizing the likelihood while solving the nonlinear differential equation system numerically at each iteration step.

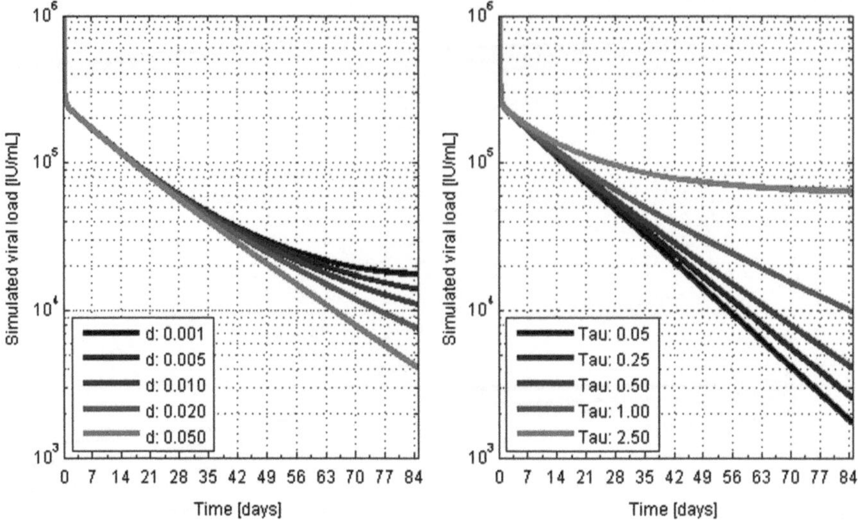

Fig. 8.2 HCV kinetic model functions for 12 weeks using different values of the parameter d on the left and of the parameter τ on the right. Note that, in contrast to Fig. 8.1, simulations do not assume a treatment effect on the varied parameter but of blocking 80 % of viral production

Table 8.1 Typical (mean or median) viral kinetic parameters estimated from clinical trials in patients chronically infected with HCV

Treatment	ε	c (per day)	δ (per day)	Patients	Reference
Interferon monotherapy (different doses)	0.81–0.96	6.2	0.14	23	[19]
Interferon monotherapy (different genotypes, formulations)	0.64–0.88	2.12–3.90	0.22–0.88	16–17	[36]
Interferon plus ribavirin (HCV genotype 1 only, different phases)	0.67	4.7	0.05–0.55	10	[12]
Interferon plus ribavirin	0.92	8.0	0.14	31	[9]
Interferon plus ribavirin (HCV genotype 1 only)	0.77	8.0	0.35	30	[34]

Typical values of estimated viral kinetic parameters can be found in Table 8.1. Note that patient groups and treatments slightly differ.

Parameter estimates from the biphasic viral kinetic model also reflected very well easy versus difficult to treat patients groups (e.g., mono-infected patients vs. patients coinfected with HBV and/or HIV, white Americans vs. African Americans) with slower or faster viral declines already in the first weeks of therapy, see, e.g., [13] for an overview on the impact of such standard host factors on viral kinetics and [7] for additional comments on the association between HCV kinetics and IL28B polymorphism. This important host factor was detected in 2009. It can explain at least part of the observed differences between human races.

Overall, patient groups with lower response rates to antiviral treatment for 24–48 weeks consistently showed slower viral kinetics in the first weeks described by lower efficiency factors ε or slower infected cell loss δ. This coincidence and the fact that the basic viral kinetic model is still quite easy to understand helped to establish the use of viral kinetic models for clinical research.

Interestingly, as early viral kinetics already reflects differences in treatment response of the most important viral and host factors, their additional predictive values decrease if viral kinetics is already known. Therefore, it may suffice to adapt individualized treatment optimization according to viral kinetics instead of accounting for a variety of different factors. Nevertheless, individualization of treatment is complicated but see, e.g., [27] for a successful sophisticated approach.

8.2.3 Optimizing Interferon Dosing Regimes Using PK-PD Models

The viral kinetic analysis of Neumann et al. [19] and others illustrates a high activity during chronic infection in spite of the equilibrium. Each day, a large amount of virus is cleared and newly produced. Furthermore, they quantified the main important rates of viral infection.

Even turnover rates of infected liver cells are much larger than was previously thought. Therefore, doubts occur if the trice weekly injections with standard interferon can be optimal because of the short half-life of standard interferon of only a few hours. Modeling results as the relatively short duration of the viral replication cycle as well as the short infected cell half-life strongly support the development of long-acting interferons. Pegylation of interferon was successfully developed. and lead to peginterferons which need only to be applied once weekly and still have a more constant drug level than standard interferons. This property does also translate in improved sustained virological response rates. The varying drug profiles even of pegylated interferons also cause doubts if the basic model with a constant efficacy is indeed appropriate. Therefore, full PK-PD models for interferon were developed. General forms of PK-PD models are discussed in detail in Chap. 7 of Koch and Schropp in this volume.

Note that interferon is given by injections and that pegylated interferons have a relatively slow PK profile. Therefore, a simple Bateman-function suffices to describe the PK of interferon. The drug concentration as a function of time after a single injection at time t_1 can then be modeled by

$$C(t) = \frac{D}{V_d(k_1 - k_2)}(\exp(-k_1(t - t_1)) - \exp(-k_2(t - t_1)))$$

for $t \geq t_1$ and $C(t) = 0$ for $t \leq t_1$. Here, D is the dose of interferon, V_d the volume of distribution, and k_1 and k_2 describe the drug absorption and degradation rates, respectively.

As we have weekly injections during therapy, let $T = \{t_1, \ldots, t_n\}$ describe the set of dosing times with t_1 as starting time of treatment, we can assume a PK function of

$$C(t) = \sum_{t_i \in T, t_i < t} \frac{D}{V_d(k_1 - k_2)} \big(\exp(-k_1(t - t_i)) - \exp(-k_2(t - t_i))\big) \tag{8.8}$$

for all $t \geq 0$.

The Hill function (see Chap. 6 of Enciso in this volume) can be used to model the treatment efficacy as function of the drug concentration

$$\varepsilon(t) = \frac{C(t)^h}{C(t)^h + IC_{50}} \tag{8.9}$$

where h is the Hill parameter and IC_{50} gives the drug concentration leading to a 50 % blocking of viral production.

Therefore, the efficacy is now time-dependent and smaller at the end of each dosing period. The efficacy function can easily be inserted in Eq. (8.7) of the ODS. The additional parameters k_1, k_2 and D/V_d can be fitted to serial observations of the individual pharmacokinetics. The parameters h and IC_{50} can be assessed from the viral quantifications instead of fitting a constant drug efficacy ε.

A first publication of this modeling approach was described in [23], see also [22] for an overview. The advantage of such an approach is that we can explore the expected treatment response by varying dose or dosing schedule.

This model approach also allows the comparison of the two available peginterferons which differ in their pharmacokinetics. It is also possible to explain the response to an induction therapy. From the basic viral kinetic model, clinicians learned that the extent of the first phase is a marker of drug efficacy. Therefore, it was explored if a higher dose in the first few weeks of treatment would improve treatment response. Unfortunately, this could not be confirmed in clinical trials. Analogously, it is also obvious from the PK-PD model that a higher efficacy of an increased initial dose (e.g., double dose) will soon nearly be lost after switching back to standard dose. See Fig. 8.3 for an illustration of the model predictions and [3] for a clinical trial with relative frequent observations during an induction dosing scheme which behave as predicted.

8.2.4 HCV Kinetic Models Including Cell Proliferation

Clinical data with detailed data from antiviral treatment and using interferons in combination with ribavirin, still show some systematic deviations from the basic model or the PK-PD model. This is especially true in cases with relatively low efficacy (e.g., in coinfected patients or in African American patients). One can observe a flat intermediate phase between the fast first phase of viral decay and

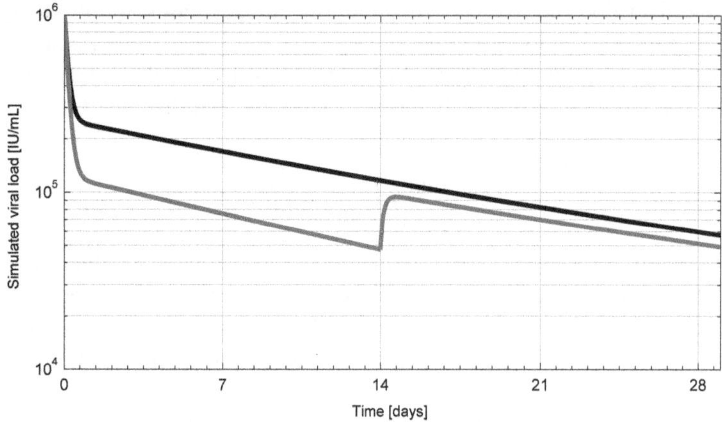

Fig. 8.3 Simulated viral kinetic curve of an induction dose (double dosing regime) during the first 14 days (*gray line*) in comparison to a standard dose (*black line*) for 28 days

the final viral decay phase more frequently as can be expected from just random variations in the quantifications. There are different approaches to explain this effect in a more descriptive effect [12] or a mechanistic way [8].

The mechanistic model explanation of Dahari et al. [8] includes proliferation terms of infected and non-infected cells in the model. As in [9], the treatment effect of ribavirin is assumed to be independent of the interferon treatment effect and is modeled by rendering a fixed proportion $\rho \in (0, 1)$ of newly produced virus as being noninfectious. The ODS then includes a new compartment N of noninfectious virus and is given by

$$\dot{V} = (1-\varepsilon)(1-\rho)pI - cV \tag{8.10}$$

$$\dot{N} = (1-\varepsilon)\rho pI - cN \tag{8.11}$$

$$\dot{I} = \beta TV + p_I I\left(1 - \frac{T+I}{T_{max}}\right) - \delta I \tag{8.12}$$

$$\dot{T} = s + p_T T\left(1 - \frac{T+I}{T_{max}}\right) - dT - \beta TV, \tag{8.13}$$

where p_T and p_I denote the proliferation rates of non-infected and infected cells, respectively.

In [28], this model was used to fit and derive parameter estimates with a population data approach to a large data base and predict long-term treatment response.

Nevertheless, this model can have unexpected dynamics. If there is a low efficacy of interferon, i.e., ε is small, the total number of observed cells $T + I$ can dramatically increase. This does not really reflect the true situation where the size of the liver does not change. In some low efficacy cases, the model does not predict a virus eradication for long enough treatment but viremia converges to a during treatment steady state level. This coincides well with clinical data of slow or partially responding patients. To be more explicit, this will happen if

$$(1-\varepsilon)(1-\rho) < \frac{\delta c}{\beta T_{max} p}. \tag{8.14}$$

The new steady state level can be calculated as a function of the other viral kinetic parameters. Surprisingly, this level can even be greater than the steady state level before treatment [31]. This is not really plausible and seems to indicate a deficiency of the model. It can be overcome by simplifying the model equation of target cells in the system.

The liver has in general a strong regeneration property that lead to a quite fast regeneration, e.g., after resection or even a decay of liver volume if a too large liver is transplanted in animal experiments. Therefore, it is reasonable just to model the target cell compartment according to an easy dynamics towards a fixed number of liver cells T_{max}. In this model variant, Eq. (8.13) will be replaced by

$$\dot{T} = \gamma T\left(1 - \frac{T+I}{T_{max}}\right). \tag{8.15}$$

This model based on Eqs. (8.10) to (8.12) and (8.15), which may also include a time-dependent interferon efficacy, is flexible enough to fit a broad variety of clinical data and allows reasonable clinical interpretations (see, e.g., [17]).

8.2.5 Modeling Quasispecies Dynamics and Resistance

New treatment options comprise direct-acting inhibitors of viral proteins as, e.g., HCV protease and polymerase. These drugs are typically highly effective but can lead to very fast resistance already during the first week of monotherapy in nearly all patients.

A mechanistic model of such resistance development is described in [1]. It differentiates between different viral strains $V_0, \ldots, V_k$ as well as the respective infected cell compartments $I_0, \ldots, I_k$. Here, V_0 and I_0 denote the wild type virus and the infected cells infected by wild type virus, respectively.

Additional parameters are the mutation rate between the different viral strains $m_{i,j}, i = 0, \ldots, k, j = 0, \ldots, k, i \neq j$ and $m_i i = 1 - \sum_{j=0,\ldots,k; j \neq i} m_{i,j}$ for all $i = 0, \ldots, k$. Furthermore, there are fitness parameters $f_1, \ldots, f_k$ which describe the deficiency of the mutated viral strains in the replication cycle compared with wild type virus (set to $f_0 = 1$) in the absence of treatment.

Of course, each viral strain also has a different drug efficacy $(\varepsilon_0, \ldots, \varepsilon_k)$ to model resistance.

The model in [1] has another interesting feature. It assumes that infected cell loss will increase somewhat during highly efficient treatment as this seems to be reflected in such data. It may be explained by the infected cell loss due to intra-cell viral eradication as additional treatment effect. Then the model equations can be summarized as

$$\dot{V}_i = \sum_{j=0}^{k}(1-\varepsilon_j)m_{i,j}pI_j - cV_i, \quad i = 0,\dots,k \tag{8.16}$$

$$\dot{I}_i = \beta T V_i - (\delta_0 - \delta_1 \log_{10}(1-\varepsilon_i))\, I_i, \quad i = 0,\dots,k \tag{8.17}$$

$$\dot{T} = s - dT - \beta T \sum_{i=0}^{k} V_i. \tag{8.18}$$

In [1], the authors were able to fit this complex model to detailed data during and after short-term monotherapy with an HCV protease inhibitor. Besides serial quantifications of overall viral load, the data set comprises serial sequence data, which monitors the relative frequency of wild type and 7 mutant strains which were already known to be resistant to this protease inhibitor from *in vitro* experiments.

There is also another sophisticated approach to explain the relatively large infected cell loss δ observed during highly effective antiviral treatment using also a mechanistic approach for intracellular viral degradation and, therefore, infected cell cure [11] without accounting for full quasispecies dynamics. They base on the model including proliferation rates but do not use a compartment of noninfectious virus as they do not model combination treatment with ribavirin. Instead, they use two additional compartments: compartment U which models the intra-cellular replication units and compartment R, which describes the intracellular RNA. Using appropriate rate for proliferation of the replication units p_U, intracellular viral replication α and intracellular degradation of replication units and viral RNA γ and σ, respectively, they derive at the following differential equation system

$$\dot{V} = pIR - cV \tag{8.19}$$

$$\dot{R} = (1-\varepsilon)\alpha U - \sigma R \tag{8.20}$$

$$\dot{U} = p_U R \left(1 - \frac{U}{U_{max}}\right) - \gamma U \tag{8.21}$$

$$\dot{I} = \beta T V - \delta I \tag{8.22}$$

$$\dot{T} = p_T T \left(1 - \frac{T+I}{T_{max}}\right) - dT - \beta T V. \tag{8.23}$$

They also propose an extension of this model including two viral compartments, a wild type virus as well as one general compartment of resistant strains. Of course, the same differentiation has to be made for intracellular RNA and replication units, but not necessarily for infected cells. Therefore, the straight-forward modification lead to an ordinary differential equation system with 8 differential equations. These models were, however, only evaluated qualitatively and have not yet be used for fitting clinical data.

A similar approach as in [1] was used in [33] for less detailed data for a related drug. In this analyses, resistance parameters were obtained from in vitro data whereas fitness parameters of the single mutated strains were estimated by a sophisticated statistical approach from relatively sparse clinical data.

Another detailed theoretical and qualitative analysis of such models was presented in [26]. Using variants similar to that from Eqs. (8.16)–(8.18). Besides such a full mutant model, a simplified version with only two virus compartments (wild type and one compartment of resistant virus) was assessed. Furthermore, also models including proliferation terms were analyzed. They focus on the velocity of resistance development in comparison with clinical data.

See also [5] for further future challenges in hepatitis C viral kinetic models in the context of new drug developments.

8.2.6 Stochastic Models

In contrast to some approaches of modeling *in vitro* viral kinetics or intracellular viral kinetics as well as in modeling the acute phase of viral infection (see, e.g., [25]), there are only few approaches to include further stochastic terms in the viral kinetic model equations when modeling HCV viral kinetics in chronically infected patients. One reason may be that the residual variance when fitting deterministic models is already within the limit of the variance for the quantification assays. Therefore, parameters directly influencing viremia as drug efficacy and viral eradication do not seem to include much random variation. Nevertheless, the situation may be different for parameters with more indirect effects as, especially, infected cell loss δ. This parameter reflects the individual patient immune response. Here, random variations are highly reasonable. In [2], an approach for including such a variation within fixed bounds ($\delta \pm \delta_1$ with $\delta > \delta_1$) may be included in a random ordinary differential equation system and compares explicit numerical algorithms to simulate viral kinetics. The system base on the basic viral kinetic model described by Eqs. (8.1)–(8.3) and is given by

$$\begin{aligned}
\dot{V} &= (1-\varepsilon)pI - cV \\
\dot{I} &= \beta TV - \left(\delta + \delta_1 \frac{2}{\pi} \arctan W_t\right) I \\
\dot{T} &= s - dT - \beta TV,
\end{aligned}$$

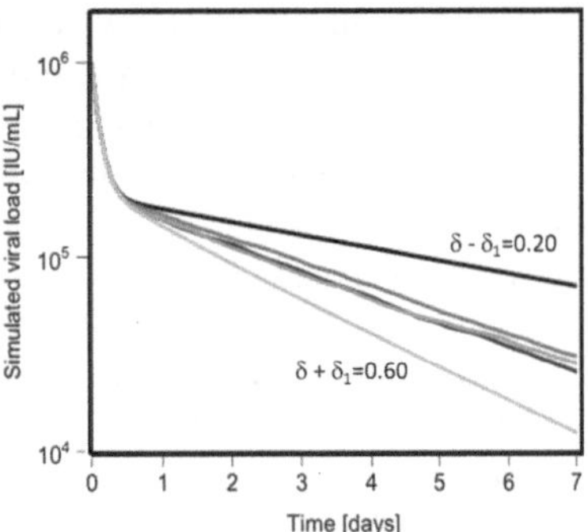

Fig. 8.4 Sample paths of the random ordinary differential equation system described in [2] in comparison with the deterministic viral kinetic model functions using $\delta \pm \delta_1$

where W_t denotes a Wiener process. In Fig. 8.4, sample paths of this model are compared with deterministic model functions.

A very different situation and a different model is used in [4] which may be seen as a first model addressing a topic with increasing clinical relevance. A major problem with chronic infection is the increased risk for hepatocellular carcinoma. Chakrabarty and Murray [4] use a standard viral kinetic model supplemented by a differential equation for a single immune response compartment to assess steady state levels. Later, a stochastic model basing on birth and death poisson processes for the development of hepatocellular carcinoma is used. Here, the risk of cancer is simulated with respect to years from infection. The birth process depends on immune response and viremia.

Much more modeling approaches exist for modeling infection with HIV. The approaches comprise stochastic differential equations and stochastic processes for acute and chronic infection and resistance development. A recent overview as well as a thorough analysis of some stochastic differential equation systems is given in [35].

8.3 Modeling HBV Kinetics

Clinically, chronic infection with HBV has some important differences in comparison with chronic HCV infection. First of all, in contrast to HCV, an effective vaccine is available which can prevent HBV infection. Nevertheless, there are still around 300 million of patients chronically infected with HBV, especially in Asia, Northern Africa, and South America. After infecting liver cells, HBV as a DNA virus implements itself inside the nucleus and initiates production of viral proteins and viral replication. Chronic HBV is characterized by different phases of disease including an immune tolerant phase with minor liver injury and no necessity of antiviral treatment as well as immune clearance phases with risk of worsening liver disease.

There are mainly two HBV core proteins which help to diagnose and characterize HBV infection. The most basic protein is the HBV s-antigen (HBsAg) which serves as marker of ongoing chronic infection. Furthermore, HBV e-antigen (HBeAg) may be lost during therapy and the loss of HBeAg as well as the detection of anti-HBeAg serves as marker of successful treatment and typically characterizes less active chronic phases.

Sometimes HBeAg is also lost by mutation. Therefore, HBeAg negative patients prior to antiviral treatment are typically just patients infected with a different type of HBV. Hence, it is not surprising that also treatment response can differ in these patients.

In general, HBV kinetics is less intensively studied than HCV kinetics. This may simply reflect the less active clinical research in these patients during the last decade.

Several treatment options are, however, available in patients chronically infected with HBV.

- Patients during the immune tolerate phase may not need treatment at all.

- Patients with more active disease can be treated by some nucleoside or nucleotide analogs as lamivudine or tenofovir. These treatments are typically well tolerated and effective in most patients and lead to a viral decay under the detection limit. If treatment is stopped, however, nearly all patients show a rapid viral rebound, so these kinds of drugs are considered for long-term treatment. Sometimes, after long-term treatment, resistance can occur. Therefore, patients are typically monitored every 3 months. For viral kinetic analyses, such clinical data is too sparse to allow a detailed modeling approach as described in Sect. 8.2.5.

  Unfortunately, even so viral production can be suppressed effectively, still a large amount of viral proteins is produced and there is the possibility that they may already cause long-term clinical complications as hepatocellular carcinoma.

- As in HCV infected patients, interferon may be used in patients chronically infected with HBV. Due to side-effects, treatment is typically limited to 12–48 weeks. Viral decay is typically much slower than during treatment with nucleoside or nucleotide analogs but in around 10 % of patients even suppression of HBsAg under detection limit and seroconversion to positive detection of anti-HBsAg can be observed. This is the best marker of successful treatment and may really prevent an increased risk of liver damage.

8.3.1 Basic Modeling Approaches

Standard models of viral kinetics can also be used for treatment of patients chronically infected with HBV. Similarly to HCV, all available treatment do suppress production of complete virus. Therefore, from monitoring viral load, models as described by Eqs. (8.5)–(8.7) as well as models including proliferation terms can be used to assess antiviral effectivity.

An up-to-date model without aiming to model quasispecies development and resistance is

$$\dot{V} = (1 - \varepsilon)pI - cV \tag{8.24}$$

$$\dot{I} = (1 - \eta)\beta TV - \delta I \tag{8.25}$$

$$\dot{T} = \gamma T\left(1 - \frac{T + I}{T_{max}}\right). \tag{8.26}$$

where $\eta \in [0, 1]$ is an additional efficacy parameter on viral infection, see, e.g., [32]. Interestingly, viral decay as well as the development of resistance is much slower in HBV infection, see, e.g., [29]. Viral decay during effective treatment is around 5–10 times slower. When HCV has a fast first-phase of around 1 day, it takes around 1 week in HBV.

Some years ago, a high variability during viral decay was observed and discussed. But as viral quantification assays became more accurate, this effect vanishes and therefore it may be mainly explained by a deficiency of the earlier laboratory methods.

A detailed comparison of viral kinetic parameters is given in [24]. They show that mean half-life of free virions was about 13.1 ± 1.1 h and 25.2 ± 1.7 h in HBeAg-negative and HBeAg-positive patients, respectively. This corresponds to viral clearance rates c of 1.3 and 0.7 per day only. Also, half-life of infected cells was 12.1 ± 1.4 days and 16.0 ± 1.7 days in HBeAg-negative and HBeAg-positive patients, respectively. This corresponds to infected cell loss rate δ of 0.06 and 0.04 per day. Compare Table 8.1 for a comparison with the respective rates in HCV infection.

Overall, they illustrate that viral kinetics is faster in HBeAg-negative patients when compared with those in HBeAg-positive patients. This indicates also a more active infection during the untreated chronic state in HBeAg-negative patients.

8.3.2 Combination Treatments and PK-PD

Even so, monotherapy is used typically when treating chronic HBV, combination treatment, especially combination of a nucleoside or nucleotide analog with interferon may be considered. There are a few clinical trials which analyze this combination also with viral kinetic models. Here, a PK-PD model for interferon should be used as was pointed out in [30].

The PK of interferon can be assessed as previously described and again Eq. (8.9) can be used to model interferon treatment efficacy. If there is a combination treatment with a nucleoside or nucleotide analog, a further efficacy parameter ε_{nuc}

can be included. Then Eq. (8.24) in the ordinary differential equation system (8.24)–(8.26) can be replaced by

$$\dot{V} = (1 - \varepsilon(t))(1 - \varepsilon_{nuc})pI - cV$$

for all instants t during the combination treatment phase is used.

This modeling of the treatment effect can be easily extended to model sequential therapy. In [15], a treatment schedule of first using 8 weeks of interferon monotherapy, followed by 24 weeks of combination treatment of interferon and lamivudine followed by 28 weeks of lamivudine. This approach enables to compare treatment effects of different treatments in the same patient and allows a thorough assessment of the advantage of combination treatment. Indeed, it was demonstrated that viral decay of the combination treatment was significantly faster. Furthermore, such a combination treatment allows to effectively reduce viremia from treatment with lamivudine and to have also the chance for a decay of HBsAg.

As immune response to HBV infection is not homogeneous and might be positively influenced by new treatment approaches, a future challenge of HBV kinetic modeling may be the inclusion of further immune compartments as well as of HBsAg or also HBeAg. Serial quantifications of HBsAg and HBeAg are now possible at a sufficiently high accuracy, but it is still challenging to obtain them. Therefore, it may be very interesting to define and fit long-term models including the dynamics of these compartments.

References

1. B.S. Adiwijaya, E. Herrmann, B. Hare, T. Kieffer, C. Lin, A.D. Kwong, V. Garg, J.C. Randle, C. Sarrazin, S. Zeuzem, P.R. Caron, A multi-variant, viral dynamic model of genotype 1 HCV to assess the in vivo evolution of protease-inhibitor resistant variants. PLoS Comput. Biol. **6**, e1000745 (2010)
2. Y. Asai, E. Herrmann, P.E. Kloeden, Stable integration of stiff random ordinary differential equations. Stoch. Anal. Appl. **31**, 293–313 (2013)
3. F.C. Bekkering, J.T. Brouwer, B.E. Hansen, S.W. Schalm, Hepatitis C viral kinetics in difficult to treat patients receiving high dose interferon and ribavirin. J. Hepatol. **34**, 435–440 (2001)
4. S.P. Chakrabarty, J.M. Murray, Modelling hepatitis C virus infection and the development of hepatocellular carcinoma. J. Theor. Biol. **305**, 24–29 (2012)
5. A. Chatterjee, J. Guedj, A.S. Perelson, Mathematical modelling of HCV infection: what can it teach us in the era of direct-acting antiviral agents? Antivir. Ther. **17**, 1171–1182 (2012)
6. V. Colizza, A. Barrat, M. Barthélemy, A. Vespignani, The modeling of global epidemics: stochastic dynamics and predictibility. Bull. Math. Biol. **68**, 1893–1921 (2006)
7. H. Dahari, J. Guedj, A.S. Perelson, T.J. Layden, Hepatitis C viral kinetics in the era of direct acting antiviral agents and IL28B. Curr. Hepat. Rep. **10**, 214–227 (2011)
8. H. Dahari, E. Shudo, R.M. Ribeiro, A.S. Perelson, Mathematical modeling of HCV infection and treatment. Meth. Mol. Biol. **510**, 439–453 (2009)
9. N.M. Dixit, J.E. Layden-Almer, T.J. Layden, A.S. Perelson, Modelling how ribavirin improves interferon response rates in hepatitis C virus infection. Nature **432**, 922–924 (2004)

10. J. Guedj, R. Thiébaut, D. Commenges, Maximum likelihood estimation in dynamical models of HIV. Biometrics **63**, 1198–1206 (2007)
11. J. Guedj, A.U. Neumann, Understanding hepatitis C viral dynamics with direct-acting antiviral agents due to the interplay between intracellular replication and cellular infection dynamics. J. Theor. Biol. **267**, 330–340 (2010)
12. E. Herrmann, J.H. Lee, G. Marinos, M. Modi, S. Zeuzem, Effect of ribavirin on hepatitis C viral kinetics in patients treated with pegylated interferon. Hepatology **37**, 1351–1358 (2003)
13. E. Herrmann, C. Sarrazin, Hepatitis C viral kinetics. J. Gastroenterol. Hepatol. **19**, S133–S137 (2004)
14. W.O. Kermack, A.G. McKendrick, Contributions to the mathematical theory of epidemics I, Reprint from the Proc. Roy. Soc. Lond. Ser. A **115**, 700–721 (1927), in Bull. Math. Biol. **53**, 33–55 (1991)
15. U. Mihm, H.L. Chan, S. Zeuzem, A.M. Chim, A.Y. Hui, V.W. Wong, J.J. Sung, E. Herrmann, Virodynamic predictors of response to pegylated interferon and lamivudine combination treatment of hepatitis B e antigen-positive chronic hepatitis B. Antivir. Ther. **13**, 1029–1037 (2008)
16. J.D. Murray, *Mathematical Biology: I. An Introduction*, 3rd edn. Interdisciplinary Applied Mathematics, vol. 17 (Springer, Berlin, 2001)
17. S. Naggie, A. Osinusi, A. Katsounas, R. Lempicki, E. Herrmann, A.J. Thompson, P.J. Clark, K. Patel, A.J. Muir, J.G. McHutchison, J.F. Schlaak, M. Trippler, B. Shivakumar, H. Masur, M.A. Polis, S. Kottilil, Dysregulation of innate immunity in hepatitis C virus genotype 1 IL28B-unfavorable genotype patients: impaired viral kinetics and therapeutic response. Hepatology **56**, 444–454 (2012)
18. J. Nakabayshi, A compartmentalization model for hepatitis C virus replication: An appropriate distribution of HCV RNA for effective replication. J. Theor. Biol. **300**, 110–117 (2012)
19. A.U. Neumann, N.P. Lam, H. Dahari, D.R. Gretch, T.E. Wiley, T.J. Layden, A.S. Perelson, Hepatitis C viral dynamics in vivo and the antiviral efficacy of interferon-alpha therapy. Science **282**, 103–107 (1998)
20. M.A. Nowak, S. Bonhoeffer, A.M. Hill, R. Boehme, H.C. Thomas, H. McDade, Viral dynamics in hepatitis B virus infection. Proc. Natl. Acad. Sci. USA **93**, 4398–4402 (1996)
21. M.A. Nowak, R.M. May, *Virus Dynamics* (Oxford University Press, New York, 2000)
22. A.S. Perelson, E. Herrmann, F. Micol, S. Zeuzem, New kinetic models for the hepatitis C virus. Hepatology **42**, 749–754 (2005)
23. K.A. Powers, N.M. Dixit, R.M. Ribeiro, P. Golia, A.H. Talal, A.S. Perelson, Modeling viral and drug kinetics: hepatitis C virus treatment with pegylated interferon alfa-2b. Sem. Liver Dis. **23**(Suppl 1), 13–18 (2003)
24. R.M. Ribeiro, G. Germanidis, K.A. Powers, B. Pellegrin, P. Nikolaidis, A.S. Perelson, J.M. Pawlotsky, Hepatitis B virus kinetics under therapy sheds light on differences between e-antigen positive and negative infection. J. Infect. Dis. **202**, 1309–1318 (2010)
25. R.M. Ribeiro, H. Li, S. Wang, M.B. Stoddard, G.H. Learn, B.T. Korber, T. Bhattacharya, J. Guedj, E.H. Parrish, B.H. Hahn, G.M. Shaw, A.S. Perelson, Quantifying the diversification of hepatitis C virus (HCV) during primary infection: estimates of the in vivo mutation rate. Plos Pathogens **8**, e1002881 (2012)
26. L. Rong, R.M. Ribeiro, A.S. Perelson, Modeling quasispecies and drug resistance in hepatitis C patients treated with a protease inhibitor. Bull. Math. Biol. **74**, 1789–1817 (2012)
27. C. Sarrazin, S. Schwendy, B. Möller, N. Dikopoulos, P. Buggisch, J. Encke, G. Teuber, T. Goeser, R. Thimme, H. Klinker, W.O. Boecher, E. Schulte-Frohlinde, R. Prinzing, E: Herrmann, S. Zeuzem, T. Berg, Improved responses to pegylated interferon alfa-2b and ribavirin by individualizing treatment for 24–72 weeks. Gastroenterology **141**, 1656–1664 (2011)
28. E. Snoeck, P. Chanu, M. Lavielle, P. Jacqmin, E.N. Jonsson, K. Jorga, T. Goggin, J. Grippo, N.L. Jumbe, N. Frey, A comprehensive hepatitis C viral kinetic model explaining cure. Clin. Pharmacol. Therapeut. **87**, 706–713 (2010)

29. V. Soriano, A.S. Perelson, F. Zoulim, Why are there different dynamics in the selection of drug resistance in HIV and hepatitis B and C viruses? J. Antimicrob. Chemother. **62**, 1–4 (2008)
30. V.A. Sypsa, K. Mimidis, N.C. Tassopoulos, D. Chrysagis, T. Vassiliadis, A. Moulakakis, M. Raptopoulou, C. Haida, A. Hatzakis, A viral kinetic study using pegylated interferon alfa-2b and/or lamivudine in patients with chronic hepatitis B/HBeAg negative. Hepatology **42**, 77–85 (2005)
31. M. Soschinsky, Plausibility of predicted steady states in simulations with HCV kinetic models and prognostic significance of first phase viral kinetics. Medical Doctoral Thesis, Goethe University Frankfurt, Department of Medicine, 2013
32. D.J. Suh, S.H. Um, E. Herrmann, J.H. Kim, Y.S. Lee, H.J. Lee, M.S. Lee, Y.S. Lee, W. Bao, P. Lopez, H.C. Lee, C. Avila, S. Zeuzem, Early viral kinetics of telbivudine and entecavir: results of a 12-week randomized exploratory study with patients with HBeAg-positive chronic hepatitis B. Antimicrob. Agents Chemother. **54**, 1242–1247 (2010)
33. S. Susser, C. Welsch, Y. Wang, M. Zettler, F.S. Domingues, U. Karey, E. Hughes, R. Ralston, X. Tong, E. Herrmann, S. Zeuzem, C. Sarrazin, Characterization of resistance to the protease inhibitor boceprevir in hepatitis C virus-infected patients. Hepatology **50**, 1709–1718 (2011)
34. K.H. Tang, E. Herrmann, H. Cooksley, N. Tatman, S. Chokshi, R. Williams, S. Zeuzem, N.V. Naoumov, Relationship between early HCV kinetics and T-cell reactivity in chronic hepatitis C genotype 1 during peginterferon and ribavirin therapy. J. Hepatol. **43**, 776–782 (2005)
35. S.W. Vidurupola, L.J.S. Allen, Basic stochastic models for viral infection within a host. Math. Biosci. Eng. **9**, 915–935 (2012)
36. S. Zeuzem, E. Herrmann, J.H. Lee, J. Fricke, A.U. Neumann, M. Modi, G. Colucci, W.K. Roth, Viral kinetics in patients with chronic hepatitis C treated with standard or peginterferon alpha2a. Gastroenterology **120**, 1438–1447 (2001)

Chapter 9
Some Classes of Stochastic Differential Equations as an Alternative Modeling Approach to Biomedical Problems

Christina Surulescu and Nicolae Surulescu

Abstract Stochastic differential equations (SDEs) provide an appropriate framework for modeling biomedical problems, since they allow detailed a priori biochemical knowledge to be accounted for and at the same time are able to describe the noise in the systems under investigation and in the data without excessively complicating the settings. We present three application paradigms related to an intracellular signaling pathway, to radio-oncological treatments, and to cell dispersal.

Keywords Cell dispersal • Intracellular signaling pathways • Nonparametric estimation • Stochastic differential equations • Stochastic processes • Tumor control probability

9.1 Introduction

In the last decades differential equations have become the main ingredient of many mathematical models. Typically, in the framework of ordinary differential equations (ODEs) such a model takes the following form:

$$\dot{\mathbf{x}} = \mathbf{F}(t, \mathbf{x}, \mathbf{a}), \tag{9.1}$$

C. Surulescu (✉)
Felix Klein Center of Mathematics, University of Kaiserslautern, Paul-Ehrlich-Str. 31, D-67663 Kaiserslautern, Deutschland
e-mail: surulescu@mathematik.uni-kl.de

N. Surulescu
Institute for Mathematical Statistics, University of Münster, Einsteinstr. 62, D-48149 Münster, Deutschland
e-mail: nicolae.surulescu@uni-muenster.de

P.E. Kloeden and C. Pötzsche (eds.), *Nonautonomous Dynamical Systems in the Life Sciences*, Lecture Notes in Mathematics 2102, DOI 10.1007/978-3-319-03080-7_9,

where $\mathbf{x}$ denotes some vectorial state variable and $\mathbf{a}$ is a vector of parameters. In more complex situations several variables (e.g., time and space) are conditioning the dynamics of the quantity of interest and the models involve partial derivatives, thus leading to partial differential equations (PDEs):

$$\frac{\partial u}{\partial t}(t, \mathbf{x}) = Lu(t, \mathbf{x}, \mathbf{a}) \tag{9.2}$$

with L denoting a partial differential operator whose precise form depends on the process(es) to be modelled and usually accounts for diffusion and/or transport. The PDE model class can be further extended to include integral operators, in which case one has to deal with a partial integro-differential equation (PIDE). In this work we focus on models relying on ODEs and P(I)DEs as starting points.

However, the deterministic settings e.g., with ordinary differential equations (ODEs) cannot accommodate the random effects which are often encountered in biological systems. The sources of stochasticity are manifold, depending on the concrete problem under consideration. For instance, stochastic variation is an inherent property of any particle interactions, since not every species involved in the reaction kinetics is present in such abundant quantities that the corresponding temporal variations are continuous and deterministic (*intrinsic noise*, caused by probability events among the small numbers of molecules in the cell). Moreover, the environmental conditions, cell-to-cell differences, and the phenotype of an organism can also trigger randomness (*extrinsic noise*). For a very informative discussion on the role and nature of intrinsic/extrinsic noise we refer e.g., to the review article by Qian [67].

When starting from an ODE setting, these considerations lead to a more realistic model of the following form:

$$\dot{\mathbf{x}} = \mathbf{F}(t, \mathbf{x}, \mathbf{a}) + \text{“noise”}, \tag{9.3}$$

where “noise” denotes the stochastic part capturing the previously mentioned random effects in the system. Hence, the noise is a kind of “black box”, nevertheless it needs to have a certain mathematical structure. This is a nontrivial issue and one of the reasons which have prevented the more widespread use of such modeling approaches.

Stochastic differential equations (SDEs) [44, 60] and random differential equations[1] [72] provide an adequate tool for handling this issue. In the present work we focus on the potential of some classes of SDE in modeling several problems from biology and medicine.

[1]Random ODEs are ODEs that include random variables or stochastic processes in their coefficients. Unlike SDEs they can be handled pathwise using deterministic rather than stochastic calculus, see e.g., [37].

In the framework of Itô calculus the noise can be modeled with the aid of Brownian motions, leading to mathematical expressions of the form

$$d\mathbf{x}(t) = \mathbf{F}(t, \mathbf{x}(t), \mathbf{a})dt + \boldsymbol{\Sigma}(t, \mathbf{x}(t), \mathbf{a})\, d\mathbf{W}_t, \tag{9.4}$$

or in the more general form of a nonlocal SDE [42, 43, 50]

$$d\mathbf{x}(t) = \mathbf{F}(t, \mathbf{x}(t), \mathbb{E}(\mathbf{x}(t)), \mathbf{a})dt + \boldsymbol{\Sigma}(t, \mathbf{x}(t), \mathbb{E}(\mathbf{x}(t)), \mathbf{a})\, d\mathbf{W}_t, \tag{9.5}$$

where $\mathbf{W}_t$ is a (multivariate) Brownian motion and $\boldsymbol{\Sigma}$ denotes the diffusion matrix. This description is however formal, since the trajectories of the Brownian motion are nowhere differentiable in the usual sense, but a rigorous characterization can be given with the concept of stochastic integration. For a general theory of systems of the type (9.4) including existence and uniqueness results we refer e.g., to [44,52,60] and for (9.5) to [42, 43, 50]. In [81] such nonlocal SDEs have been deduced as asymptotic limits of SDEs characterizing the dynamics of the membrane potentials for a neuron network (under the assumption of a large enough number of involved neurons), upon relying on a probabilistic approach.

Of course, the noise in the systems mentioned above can be modelled more generally by using extensions of the Brownian motion like Lévy processes or fractional Brownian motions. However, the detailed description of the corresponding models involving SDEs driven by such processes would go beyond the scope of this chapter.

Here we present three applications of modeling with SDEs or SDE-like processes; they are related to intracellular signaling pathway, to radio-oncological treatments, and to cell dispersal, respectively. For the last of these problems (to be addressed in Sect. 9.4) the modeling via SDEs or simply via stochastic processes[2] offers an alternative to the PIDE approach enabling to numerically handle complex, more realistic (even multiscale) situations which cannot be treated in the PDE framework.

The system set up in Sect. 9.3 will allow to describe the evolution of the number of cancer cells affected by irradiation, individually for each patient, which is a modeling novelty. A particular class of SDEs in this context will be specified below and relies on generalizations of the classical *geometric Brownian motion* given by

$$d\mathbf{x}(t) = \alpha\mathbf{x}(t)dt + \beta\mathbf{x}(t)dW_t, \quad t \geq 0, \quad \alpha, \beta > 0. \tag{9.6}$$

This offers a new modeling framework for *tumor control probability* (TCP) problems, which is able to account for much more effects, inaccessible by the usual ODE models (e.g., the evolution of the illness is treatment and patient specific).

For some other biological problems the past dynamics are relevant for the evolution of a system, for instance in the maturation of one or several populations

[2] All these models can actually be put in the framework of SDEs, some of which are however driven by jump processes and not by Brownian motions.

(see e.g. [58]), in pharmacokinetics and pharmacodynamics [7], immune responses [53], gene expression or feedback control in signal transduction networks. Delay differential equations (DDEs) are the classical framework to account for such phenomena explicitly and the mathematical equipment for their analysis and numerics is well developed [1, 20, 25, 31, 47]. In this context the special class of stochastic differential delay equation (SDDE) models can be used in order to accommodate randomness which cannot be described with deterministic delay equations. In the SDDE framework the drift and diffusion coefficients supplementary depend e.g., on $\mathbf{x}(t-\tau)$ with τ denoting the (constant) delay. However, the issues of parameter inference and numerical simulations for SDDEs where one or several states are directly depending on the time lag are highly challenging and—as far as parameter estimation is concerned—still open. Such difficulties can be eluded through an appropriate modeling of the relevant phenomena so as to keep the delay in the system, but being able to handle it deterministically, preserving the stochasticity only on the states which do not explicitly involve delay. This will be illustrated in Sect. 9.2 on a problem related to intracellular signaling. An alternative handling of the delay in the SDDE context has been proposed in [75]. We recall that model in Sect. 9.2 too: it extends previous (S)DDE settings by using a classical idea in continuous time series analysis (see e.g., [77] and the references therein) which suggests the use of deterministic time varying instead of constant coefficients. The same idea is taken up again in one of the nonlocal SDE models for the signaling dynamics.

Finally, the potential of the new settings will be addressed in Sect. 9.5.

9.2 Intracellular Signaling: The JAK–STAT Pathway

Cells continuously communicate with each other and with their environment through signals and messages. Signal transduction is concerned with the study of the biochemical information used by the cells to interact, with the methods for its detection, and with the investigation of the cellular mechanisms for transferring this information, as well as decoding and responding to signals. The understanding of these complex mechanisms is motivated by the role they play in the functioning of a biological system; defective signaling is at the origin of many diseases, like cancer, diabetes, achondroplasia a.o. [22].

Until a few decades ago the main research efforts were directed towards establishing diagrams describing the qualitative behavior of the interacting components of a signaling pathway. These static graphical schemes, however, cannot provide information about the dynamics of such a system. The latter is essential in the quest for characterizing life: the aim is to comprehend the way this dynamics emerges and how it can be controlled. Since nature is too complex to be accurately described, one needs to make drastic simplifications and the result are more or less realistic mathematical models. Most of them involve ordinary differential equations (ODEs) along with the relationship between input and output data. Life is highly

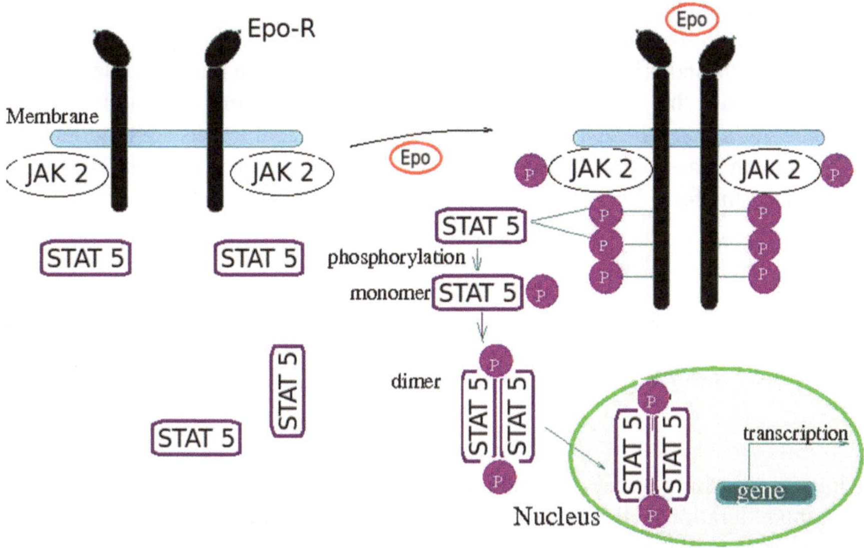

Fig. 9.1 The JAK–STAT signaling pathway (after [78])

erratic, however, and one cannot expect an assembly of deterministic equations to give a faithful characterization of processes happening in living organisms. These random effects are inherent in each biological system [56] and a current way to modeling them is the use of stochastic differential equations (SDEs) instead of ODEs. As already mentioned in Sect. 9.1, these equations are obtained by allowing for randomness in the coefficients of ODEs, which clearly provides more realistic modeling of the actual phenomenon.

The modeling ideas presented in the Introduction are now applied in the concrete context of signaling pathways, in order to extend the classical settings in this area. As an illustration we will consider the JAK–STAT signaling pathway.

The family of STATs (Signal Transducers and Activators of Transcription) comprises cytoplasmic transcription factors, which are responsible for cellular functions like growth, development, division, metabolism and apoptosis. Stimulation by extracellular signals (like cytokines and hormones) leads to transient activation of the STATs by phosphorylation through receptor-bound Janus kinases (JAKs); the activated STATs are released, the phosphorylated monomeric STATs form dimers and migrate to the nucleus to activate transcription. Figure 9.1[3] shows a cartoon of the JAK–STAT5 signaling pathway where the stimulus is the Erythropoietin (Epo) hormone binding to the Epo receptor. For more details about the JAK–STAT signalling pathways we refer to [15, 65] and the references therein.

[3]Taken from [75].

In the nonlinear deterministic models developed for the JAK–STAT signal transduction pathway in [78] the dynamics were established upon using some empirical facts and prior biochemical knowledge only, which makes the description quite uncertain; this explains the variety of existing deterministic models. More realistic settings are those where some of the equation coefficients are random.

Assuming mass-action kinetics, Timmer et al. [78] gave a mathematical transcription of the feed-forward cascade in the figure above:

$$\begin{aligned}
\dot{x}_1 &= -k_1 x_1 EpoR + 2k_4 x_4 \\
\dot{x}_2 &= k_1 x_1 EpoR - k_2 x_2^2 \\
\dot{x}_3 &= -k_3 x_3 + \frac{1}{2} k_2 x_2^2 \\
\dot{x}_4 &= k_3 x_3 - k_4 x_4
\end{aligned} \tag{9.7}$$

with corresponding initial conditions and $t \in [0, T_{max}]$, where T_{max} represents the maximum duration of the experiment. Here $EpoR$ is the amount of activated Epo-receptors, x_1 is the unphosphorylated monomeric STAT5, x_2 is the phosphorylated monomeric STAT5, x_3 stands for the phosphorylated dimeric STAT5 in the cytoplasm, while x_4 means phosphorylated dimeric STAT5 in the nucleus; k_1 to k_7 are parameters.

This description was then improved [78] by allowing for nucleocytoplasmic cycling and taking into account the sojourn time τ of STAT5 in the nucleus, modelled by a fixed time delay. This led to the following system of delay differential equations (DDEs):

$$\begin{aligned}
\dot{x}_1 &= -k_1 x_1 EpoR + 2k_4 x_3(t-\tau) \\
\dot{x}_2 &= k_1 x_1 EpoR - k_2 x_2^2 \\
\dot{x}_3 &= -k_3 x_3 + \frac{1}{2} k_2 x_2^2 \\
\dot{x}_4 &= k_3 x_3 - k_4 x_3(t-\tau).
\end{aligned} \tag{9.8}$$

again with appropriate initial conditions. Applying the modeling ideas from Sect. 9.1 we deduce from the previous DDE system an SDDE system for the JAK–STAT pathway:

$$dx_1 = \Big[- k_1 x_1 EpoR + 2k_4 x_3(t-\tau) \Big] dt + \sigma_1(t)\, dW_1(t) \tag{9.9}$$

$$dx_2 = \Big[k_1 x_1 EpoR - k_2 x_2^2 \Big] dt + \sigma_2(t)\, dW_2(t) \tag{9.10}$$

$$dx_3 = \Big[- k_3 x_3 + \frac{1}{2} k_2 x_2^2 \Big] dt + \sigma_3(t)\, dW_3(t) \tag{9.11}$$

$$dx_4 = \Big[k_3 x_3 - k_4 x_3(t-\tau) \Big] dt + \sigma_4(t)\, dW_4(t), \tag{9.12}$$

where $\sigma_i(t)$ $(i = 1,2,3,4)$ are some stochastic processes. Their precise form has to account for the specific biological problem to be modeled. An adequate choice would be to take these diffusion coefficients proportionally to the corresponding states i.e. $\sigma_i(t) = \sigma_i x_i(t)$ for some nonnegative constants σ_i, hence considering multiplicative noise. A simpler option would have been to take them constants (additive noise), however this is an unrealistic choice, since it would mean that the states have no influence on the biochemical processes which have been ignored in the deterministic description. Moreover, when the diffusion coefficients $\sigma_i(t)$ do not depend on the states x_i the solutions may become unrealistically negative. Actually, it is reasonable to assume that the amplitude of the random fluctuations is proportional to the level of the concrete state variables, which motivates our choice.

In practice, however, only combinations of the states x_1, x_2, x_3 can be observed, since individual STAT5 populations are experimentally difficult to access [78]. Thus, the measurements in the cytoplasm include the amount of tyrosine phosphorylated STAT5:

$$y_1 = k_5(x_2 + 2x_3)$$

and the total amount of STAT5:

$$y_2 = k_6(x_1 + x_2 + 2x_3),$$

with k_5 and k_6 some scaling parameters. The measurements for y_1 and y_2 have been made in an interval of 60 min on an unequally spaced grid of time points [78]. In such a context of partial observations the above system (9.9)–(9.12) is statistically hardly (if at all) tractable.

One way to avoid this difficulty has been proposed in [75], where relying on the idea of a delay chain approach [55, 68], the system (9.9)–(9.12) has been replaced with classical SDEs including a single supplementary state equation allowing for a time-varying delay:

$$dx_1 = \Big[- k_1 x_1 EpoR + 2k_4 z_1 \Big] dt + \sigma_1 x_1 dW_1(t) \tag{9.13}$$

$$dx_2 = \Big[k_1 x_1 EpoR - k_2 x_2^2 \Big] dt \tag{9.14}$$

$$dx_3 = \Big[- k_3 x_3 + \frac{1}{2} k_2 x_2^2 \Big] dt \tag{9.15}$$

$$dx_4 = \Big[k_3 x_3 - k_4 z_1 \Big] dt \tag{9.16}$$

$$dz_1 = \theta(t)[x_3 - z_1]dt, \tag{9.17}$$

with $\theta(\cdot)$ an appropriate positive continuous deterministic function, a concrete parametric form of which can be for instance $\theta(t) = \alpha/(1 - A^{\alpha} \exp(-\alpha t))$ for

$t \in [0, T_{max}]$, with $A \in (0, 1)$ a constant, and $0 < \alpha$. The initial conditions are as for all systems in this section $x_1(0) = x_1^0$, $x_i(0) = 0$ $(i = 2, 3, 4)$[4] and thus also $z_1(0) = 0$. By using this and from the last equation above, the function z_1 can be expressed explicitly w.r.t. x_3, which allows to verify easily that e.g., for $\delta = -\ln A$ and for all $t \geq 0$ we have $z_1(t) \in [\min_{s\in[-\delta,t]} x_3(s), \max_{s\in[-\delta,t]} x_3(s)]$. Since x_3 is a stochastic process with continuous trajectories it follows that for all $t \geq 0$ there exists another stochastic process $\{\tau(t)\}_{t\geq 0}$ with $\tau(t) \in [0, t+\delta]$ and such that it is the smallest one for which $z_1(t) = x_3(t - \tau(t))$ holds. Observe that $\tau(t)$ describes a time-varying delay and z_1 depends on the whole trajectory of state x_3 (up to the current time t), which renders the dependence on the past more flexible that in the setting of (9.9)–(9.12). This also makes the drift of x_1 depend on the entire history of x_3. Moreover, notice that for this new model the classical nonlinear filtering theory can be applied to solve the inference problem in the mentioned context of partial observations.[5] For further details to this model and some numerical simulations see [75].

Another way to deal with the above statistical tractability problem is to extend the model (9.8) in the sense of nonlocal SDEs of the type (9.5), see [42, 43, 50]. In this spirit we propose two extensions of the DDE system (9.8). Since the cartoon presented in Fig. 9.1 only includes the states x_1 to x_4 of the pathway and ignores all other interactions (e.g., activation of receptors and of their domains, cross talk with other pathways) it can be conceived that the process x_1 itself is not prone to participate entirely in a deterministic way to the formation of x_2, but influences its dynamics by the time-varying mean level $\mathbb{E}(x_1)$. Therefore, a first nonlocal SDE model accounting for this feature can be

$$dx_1(t) = \Big[-k_1x_1(t)EpoR + 2k_4x_3(t-\tau)\Big]dt + \sigma_1 x_1(t)dW_1(t) \tag{9.18}$$

$$dx_2(t) = \Big[k_1\mathbb{E}(x_1(t))EpoR - k_2x_2^2(t)\Big]dt \tag{9.19}$$

$$dx_3(t) = \Big[-k_3x_3(t) + \frac{1}{2}k_2x_2^2(t)\Big]dt \tag{9.20}$$

$$dx_4(t) = \Big[k_3x_3(t) - k_4x_3(t-\tau)\Big]dt, \tag{9.21}$$

where the random effects for the state variables x_2, x_3 (the last equation can be simply decoupled) are ignored and the focus is instead on the random effects for x_1.

[4] We used that $x_3(0) = 0$ and extended x_3 by this value on the interval $[-\delta, 0]$.

[5] Due to the nonlinearity of the system, the large number of parameters to be estimated, and the rather small amount of available data, however, the practical handling of this issue is still not feasible.

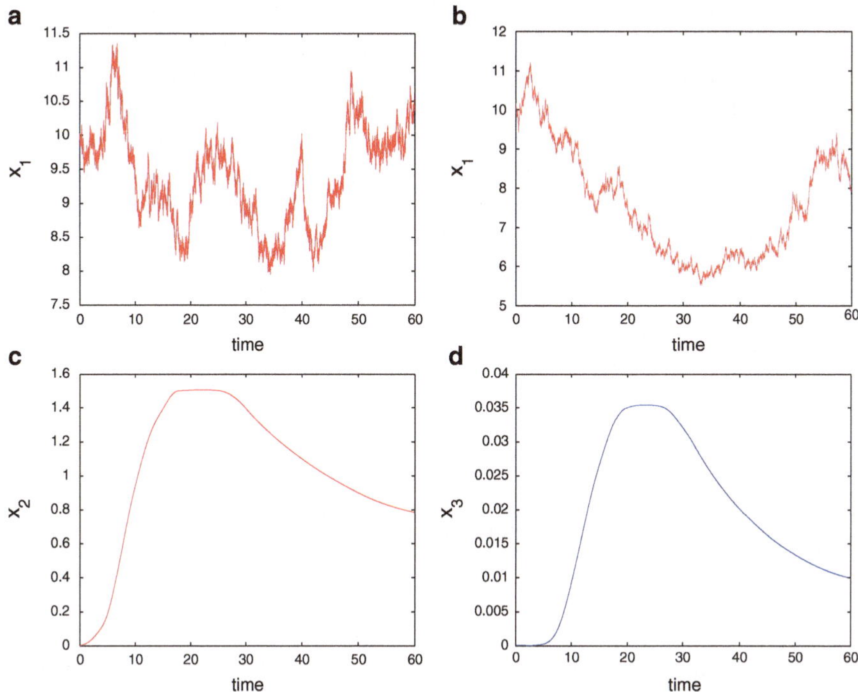

Fig. 9.2 Numerical simulations of model (9.18)–(9.21) and of model (9.23)–(9.26). (**a**) Concentration of x_1 (for model (9.18)–(9.21)). (**b**) Concentration of x_1 (for model (9.23)–(9.26)). (**c**) and (**d**) Concentration of x_2 and of x_3, respectively

Equation (9.18) is affine in x_1 and can be solved explicitly to yield

$$x_1(t) = x_1(0)\exp\Big(-(\frac{\sigma_1^2}{2} + k_1 EpoR)t + \sigma_1 W_1(t)\Big) \tag{9.22}$$

$$+ 2k_4 \int_0^t x_3(s-\tau)$$

$$\exp\Big(-(\frac{\sigma_1^2}{2} + k_1 EpoR)(t-s) + \sigma_1(W_1(t) - W_1(s))\Big)ds,$$

which is obviously positive as long as x_3 it is.

Numerical simulations for the system (9.18)–(9.21) have been performed upon using the Euler-Maruyama scheme (see e.g., [44]) with a time step of 0.001 and the following parameters: $k_1 = 0.02$, $k_2 = 0.0235$, $k_3 = 0.7494$, $k_4 = 0.7492$, $\sigma_1 = 0.0491$, $x_1(0) = 2.3$, $x_i(0) = 0$ $(i = 2, 3, 4)$, $\tau = 6\,\text{min}$. The results are illustrated in Fig. 9.2 below.

Further extensions of both ODE and SDE models can be obtained upon letting the coefficients of the respective systems depend on the states. This can be seen as having certain similarities with the idea of using random ODEs or DDEs to go beyond the classical settings. Following the deterministic modeling process for the JAK–STAT pathway it can be seen that in (9.7) it was the term involving the conversion rate k_4 of state x_4 which had to be changed to get a better model. This indicates that the rate k_4 should receive more attention when trying to achieve an enhanced modeling. In this sense we propose a further nonlocal SDE model of the type (9.18)–(9.21), but with the coefficient k_4 now depending on the state x_1 e.g., in the following way: $k_4 = k_4(x_1(t)) = \frac{\tilde{k}_4 x_1(t)}{\sqrt{\alpha + (\mathbb{E}(x_1(t)))^2}}$, with $\tilde{k}_4$ and α some positive constants. The resulting system takes the form

$$dx_1(t) = \Big[- k_1 x_1(t) EpoR + 2k_4(x_1(t)) x_3(t-\tau) \Big] dt + \sigma_1 x_1(t)\, dW_1(t) \tag{9.23}$$

$$dx_2(t) = \Big[k_1 \mathbb{E}(x_1(t)) EpoR - k_2 x_2^2(t) \Big] dt \tag{9.24}$$

$$dx_3(t) = \Big[- k_3 x_3(t) + \frac{1}{2} k_2 x_2^2(t) \Big] dt \tag{9.25}$$

$$dx_4(t) = \Big[k_3 x_3(t) - k_4(x_1(t)) x_3(t-\tau) \Big] dt, \tag{9.26}$$

whereby $\tilde{k}_4 \le k_3$. A representation similar to (9.22) above can be obtained for the solution of (9.23) as well. The theoretical aspects related to existence and positivity of solutions to the systems (9.18)–(9.21) and (9.23)–(9.26) above are future work.

The setting (9.23)–(9.26) amounts to seeing now the conversion rate of phosphorylated dimeric STAT5 in the cytoplasm into unphosphorylated monomeric STAT5 as a function depending on the expulsion rate k_4 of x_3 from the nucleus and on the state x_1 (scaled by its expectation, in order to allow capturing nonlocal information about its unknown current condition). The new rate thus implicitly accounts for the processes related to recycling of phosphorylated STAT5 dimers from the nucleus, their splitting into phosphorylated monomers, and the subsequent dephosphorylation, which are prone to random perturbations and have not been explicitly modelled in the previous versions of the system. Also, the rate $k_4 = \frac{\tilde{k}_4 x_1(t)}{\sqrt{\alpha + (\mathbb{E}(x_1(t)))^2}}$ can be interpreted as a Michaelis–Menten-like saturation[6] [57] in the production of x_1 by influence of x_3 (with the corresponding delay for nuclear expulsion). It is again the *expectation* of the stochastic process $x_1(t)$ being involved in the production of x_2: this nonlocality suggests an averaging over the influences of possible conditions of x_1 on the growth dynamics of x_2. Observe as before that the

[6] Hereby the growth limiting is realized by the expectation of the stochastic process of the relevant concentration.

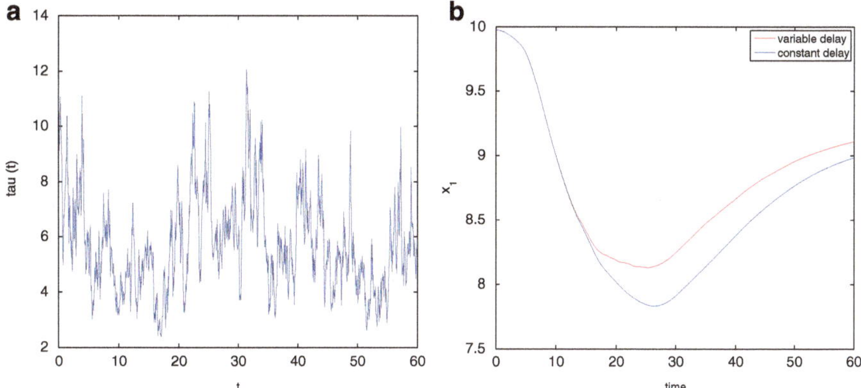

Fig. 9.3 Dynamics of variable delay $\tau(t)$ (**a**) and concentration of x_1 for the system with variable and constant delay, respectively (**b**). The mean reverting rate in (9.27) is $\alpha = 1$ and the diffusion coefficient is $\beta = 0.5$

last equation in systems (9.18)–(9.21) and (9.23)–(9.26) above is decoupled from the rest and can be solved separately, after having solved the remaining equations in the above systems.

The advantage of the new models is that the time delay now enters the equations in a deterministic way, which enables its handling even in the framework of SDEs. Moreover, observe that the DDE system proposed by Timmer et al. in [78] can be reobtained in the limit taking $\sigma_1 \to 0$ and then $\alpha \to 0$.

Numerical simulations for the system (9.23)–(9.26) have been performed as above upon using the Euler–Maruyama scheme with a time step of 0.001 and the same parameters as for system (9.18)–(9.21), with additionally $\alpha = 10^{-3}$. The results are illustrated in Fig. 9.2 above (the concentrations x_2 and x_3 do not change).

It can be observed that the trends of the states x_1 to x_3 are similar to those of the corresponding ones from the deterministic case (see [78]). More accurate shapes will be made possible after estimating the model parameters from data. This is ongoing work.

The above idea of making k_4 a stochastic process can be used, of course, for other system parameters as well. For instance, one can choose $\tau(t)$ to be a mean reverting stochastic process given by

$$d\tau(t) = \alpha(\tau_m - \tau(t))dt + \beta\tau(t)dV_t, \tag{9.27}$$

where V_t is a Brownian motion independent on $W_1(t)$ in (9.23), τ_m the mean delay, and α and β denote a positive mean reverting rate and some positive constant diffusion coefficient, respectively. Simulations of such a process are illustrated in Fig. 9.3 for $\alpha = 1$, $\beta = 0.5$, $\tau_m = 6$, $\tau(0) = 6$, and respectively in Fig. 9.4 for $\alpha = 0.3$, $\beta = 0.9$ and $\tau_m = 6$, $\tau(0) = 6$. The parameter inference for the system (coupled to (9.27)) is ongoing work, too.

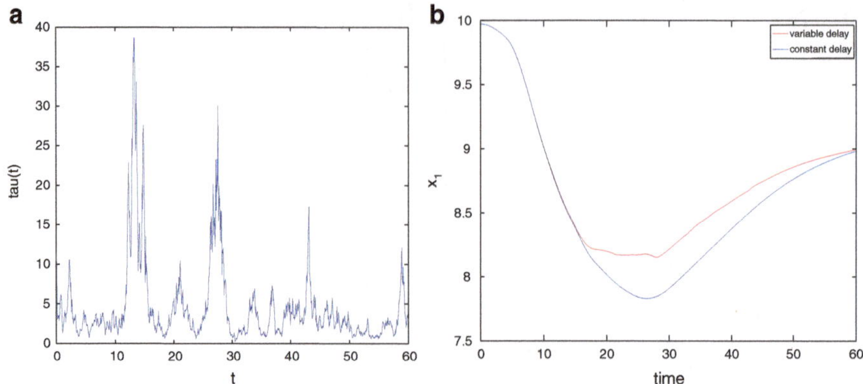

Fig. 9.4 Dynamics of variable delay $\tau(t)$ (**a**) and concentration of x_1 for the system with variable and constant delay, respectively (**b**). The mean reverting rate in (9.27) is $\alpha = 0.3$ and the diffusion coefficient is $\beta = 0.9$

9.3 Radiation Therapy

In this section[7] we provide a further example where modeling with SDEs can lead to a more realistic setting allowing to enhance the level of information that can be captured about the phenomenon of interest.

Radiation therapy is one of the most common methods for cancer treatment. It consists in using ionizing radiation to affect the viability of neoplastic cells, while trying to spare the surrounding healthy tissue. A quantity of interest for assessing the success of a treatment schedule is the tumor control probability (TCP), which gives the probability that no clonogenic cells survive the radiation treatment. It is determined by complex interactions between tumor biology, tumor microenvironment, radiation dosimetry, and patient-related variables. The complexity of these joint factors constitutes a challenge for building predictive models for routine clinical practice.

Most TCP models rely on simple statistics in connection to cell survival. A model class very popular for its straightforwardness is the one considering a discrete distribution (Poisson or binomial) for the number of cells surviving radiation treatment [11, 14, 49, 80]. These settings, however, are not able to capture in satisfactory detail the effect of the treatment schedules on cancer cell dynamics like cell repair, proliferation, sensitivity to radiation etc. Instead, cell population models have been proposed, which describe the evolution of the number of cancer cells via differential equations. The proliferation is, thereby, modeled via exponential, logistic or Gompertzian growth (see e.g., [45, 46]), possibly also accounting for the effects of the cell cycle upon dividing the cell population into several compartments.

[7]See [75].

Such standard proliferation functions, however, prescribe monotonically increasing growth and can fail to model unexpected changes in growth rates, while an approach involving stochastic differential equations (SDEs) as in [21] can handle these variations.

Furthermore, deterministic models involving differential equations are adequate for large cell populations. By radiation treatment such a population is supposed to shrink drastically, so that only a small number of cells remains. This renders the use of a deterministic model problematic, thus calling for the accommodation of stochasticity in the modeling process. To this purpose previous models consider stochastic birth and death processes (see e.g., Zaider and Minerbo [84]), however they again lead eventually to a system of ODEs, which are the mean field equations for the expected cell number and where the effect of the birth and death processes is captured via a hazard function. Subsequent models involving cell cycle dynamics [16, 34] are extensions of Zaider and Minerbo's model. Here we propose a new class of models relying on stochastic jump processes and which is more flexible than previous approaches, since it allows to account for interesting features like the probability distribution of the time it takes for patients to get rid of their cancer cells under a certain treatment schedule.

One of the classical models describing cancer population growth has the form

$$\dot{C} = (b - d - h(t))C, \tag{9.28}$$

where $C(t)$ denotes the number of clonogenic cells at time t, the constants b and d are per capita rates representing birth and death, respectively, and $h(t)$ denotes the hazard function characterizing the radiation induced cell death depending on the cumulative dose (see e.g., [34]). However, such a description with an explicit hazard function is rather artificial; below we propose instead a model where the role of the hazard function is implicitly accounted for.

Furthermore, the previous setting has to be adapted to some patient dependent model for tumor evolution and able to include random effects like e.g., patient positioning errors or organ motion. This could be realized by an appropriate stochastic perturbation of the above ODE, but the resulting SDE is still not flexible enough to capture the usual succession of irradiation treatments. Therefore it would be more realistic to describe the evolution of clonogens with the aid of some stochastic process $C_\gamma(t)$[8] such that each of its trajectories describes the tumor growth dynamics for a specific patient. Thereby, one appropriate choice can be given by:

$$C_\gamma(t) = (n_0 + \gamma)N_B(t) - \gamma, \quad t \geq 0,\ \gamma \geq 0 \tag{9.29}$$

[8]Characterizing the evolution of the number of clonogens w.r.t. the standard reference population of N_{ref} individuals (usually in literature $N_{ref} = 10^6$) after the moment of the illness detection. Hence t represents the time passed since the diagnosis of cancer has been set. A patient is considered to be cured at the first time τ_c when $C_\gamma(\tau_c) = 0$.

with $N_B(t)$ a geometric Brownian motion defined by

$$dN_B(t) = R_{bd}N_B(t)\,dt + N_B(t)\sigma_B\,dW_t, \quad t \geq 0, \quad \sigma_B > 0,\ N_B(0) = 1, \quad (9.30)$$

where n_0 is a random variable independent of the standard Brownian motion W_t representing the fraction of tumor cells at the moment where the illness has been detected. R_{bd} is some positive constant denoting the analogue of the net population growth (difference between birth and death rates) in the classical models. The positive parameter γ characterizes how curable the cancer is (if $\gamma = 0$, then $C_\gamma(t) > 0$ with probability 1, meaning that the tumor cannot be exhaustively eliminated). Here we consider γ to be a constant, however it could also be a random variable like n_0. Note that (9.29) corresponds to the case without treatment. The current setting will be extended below to a model accounting for treatment effects.

From a mathematical point of view the above process $C_\gamma(t)$ is defined on a probability space $(\Omega, \mathcal{K}, P)$, where Ω denotes e.g., the set of cancer patients being considered, under the assumption of each patient having only one tumor. However, this setting can also be extended to more complex situations where a patient can have several tumors or for an arbitrary cardinality of Ω.

If we denote by τ_c the random time indicating the moment where $C_\gamma(t)$ first becomes zero, then the tumor control probability is given by

$$TCP(t) = P(\tau_c \leq t), \qquad t \geq 0 \quad (9.31)$$

i.e. the TCP is the cumulative distribution function (cdf) of the random variable τ_c.

In order to render the distribution of τ_c more realistic we allow the diffusion coefficient σ_B of $N_B(t)$ to be a stochastic process e.g., of the form $\sigma_B = f(Y_t)$, with f some positive, smooth function and Y_t an Ornstein–Uhlenbeck (OU) process given by

$$dY_t = \alpha_Y(m_Y - Y_t)dt + \beta_Y\,d\hat{Z}_t, \quad t \geq 0, \quad (9.32)$$

where $\hat{Z}_t$ denotes another Brownian motion, possibly correlated with W_t. The positive constants α_Y and β_Y denote the rate of mean reversion of the process Y_t, respectively the diffusion coefficient, and m_Y is the long run mean value of Y_t.

In the following we assume for the sake of simplicity that all patients follow the same treatment schedule (for instance, daily irradiations with breaks on weekends), however the models can be easily adapted to more general (e.g., Poissonian) schedules.

Let us denote a time sequence $\{t_i\}_{i=1,\dots,N}$ with $0 < t_1 < t_2 < \cdots < t_N$ modeling the times when the patients are treated. Let further $(\xi_k)_{k=1,\dots,N}$ be a sequence of i.i.d. positive random variables describing the radiation effect on tumor cells (ξ_k corresponds to the treatment at the moment t_k). The mean value of these r.v. is positive and can be chosen e.g., to be a nondecreasing function of the radiation dose D.

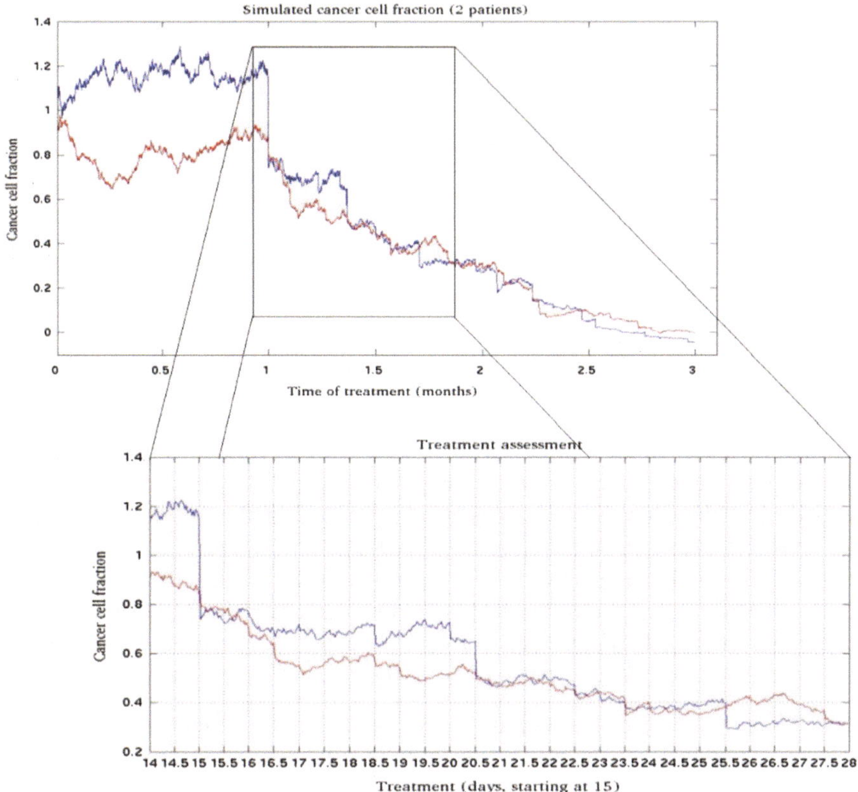

Fig. 9.5 Simulated cancer cell fraction for two patients

The process L_t describing the effect of treatment can now be given by

$$L_t = \begin{cases} 0 & , \quad 0 \leq t < t_1, \\ \sum\limits_{\{k\,:\,t_k \leq t\}} \xi_k & , \quad t \geq t_1. \end{cases} \tag{9.33}$$

Thus, in order to account for the treatment effects the previously introduced process $C_\gamma(t)$ has to be correspondingly modified. The resulting model incorporating all these features can be now summarized as follows:

$$C_\gamma(t) = (n_0 + \gamma) N_B(t) \exp(-L_t) - \gamma, \tag{9.34}$$

$$dN_B(t) = R_{bd} N_B(t)\, dt + N_B(t) f(Y_t)\, dW_t, \tag{9.35}$$

$$dY_t = \alpha_Y (m_Y - Y_t) dt + \beta_Y d\hat{Z}_t, \quad t \geq 0. \tag{9.36}$$

This model is illustrated in Fig. 9.5 with a simulation result. Where applicable, we chose the parameters according to [9, 69]. We simulated two patients receiving

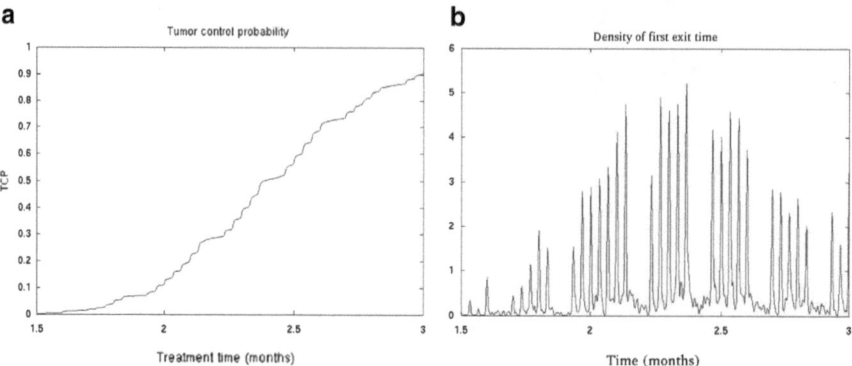

Fig. 9.6 Results for 1,000 simulated patients. (**a**) Evolution of TCP. (**b**) Density of τ_c until T_{max}

a daily treatment (except on weekends), starting with $t_1 = 2$. The fraction of initially available clonogens was chosen to be 0.01 and for the rest of the parameters we took the values $\gamma = 0.8$, $R_{bd} = 0.0491$, $m_Y = 0.15$, $\beta_Y = 2.2361$, $\alpha_Y = 60$. The simulation of the process $N_B(t)$ has been carried out with the classical Euler–Maruyama scheme (see e.g., [44]) and we chose $f = \exp$. The treatment effect was modeled by choosing for ξ_k a rescaled noncentered χ^2 distribution (i.e., $\xi_k \sim \sigma_D^2 \chi^2(1, \frac{aD+\beta D^2}{\sigma_D^2})$), where $\sigma_D = 0.1$, $D = 2\ Gy$, $a = 0.145\ Gy^{-1}$, $\beta = 0.0353\ Gy^{-2}$. The choice of the noncentral parameter is motivated by studying the connection between the mean value of ξ_k and the dose D: clearly, they are proportional, which means that this dependence is analogous to the one for the survival probability. Here we used the linear quadratic (LQ) function, which is so far the most popular one in radiation treatment (see, e.g., [59] and the references therein[9]).

Observe a certain periodicity in the decay of the clonogens, which is in accordance to the treatment schedules and captures the effects of no irradiation on weekends (the cancer cells start to recover during these breaks and are hit again on mondays etc.).

Since our setting is more complicated than the classical ones, we cannot explicitly compute the formula of the TCP for the general model (9.34)–(9.36), however, it can be assessed numerically for any time moment of interest with the aid of simulations of the kind described above.

Starting from (9.31) we can compute the TCP upon using a large enough number S of simulations for the process $C_\gamma(t)$. For instance, denoting by $n_{\tau_c}(t)$ the number of simulations for which $\tau_c \leq t$, then the estimated TCP will be given by

$$\widehat{TCP}_S(t) = \frac{n_{\tau_c}(t)}{S}, \quad t \geq 0$$

and converges in probability to $TCP(t)$ for $S \to \infty$.

[9]However, it can also have some drawbacks (see, e.g., [32, 41]).

Using this estimator and the same parameters as above we illustrate with Fig. 9.6a the evolution of the TCP for a sample size of $S = 1{,}000$ patients.

Supplementary information on the TCP can, of course, be obtained by estimating the density of the random time τ_c, but this is a nontrivial issue. In Fig. 9.6b we plot the relevant part of the nonparametric estimation of this density on the time interval of interest $t \in [0, 3]$. Hence, $TCP(t)$ can be characterized alternatively as being the area under the curve up to t.

More details on the nonparametric method used for the density estimation are given in Sect. 9.4.3.

Observe again the consequence of the weekend breaks during the treatments: they make up the periodic larger gaps in the density of the random time T, while the spikes represent the effects of the irradiation. A noticeable impact of the radiation treatment can thus be seen after more than one month of treatment for a daily schedule (excepting weekends) and with the prescribed doses given above.

9.4 Cell Dispersal

We now analyze a multiscale model for cell dispersal in the framework of partial (integro-) differential equations and present an alternative approach starting from the underlying stochastic processes for the velocity jump movement of cells under the influence of a chemoattractant signal and of the intracellular dynamics. The latter can be put in the form of SDEs driven by jump processes (see Sect. 9.5 for more comments on this issue), which motivates the title of the present section. This new approach will be handled via a nonparametric density estimation method that will be recalled shortly. Here we are mainly concerned with bacterial motion as a paradigm, however, the methods can be applied to investigate the motility of any random walker biasing its behavior according to biochemical or physical cues.

The motion of most unicellular organisms can be characterized as a random walk through physical space. For instance, flagellated bacteria like *E.coli* or *Salmonella typhimurium* alternate a smooth, rather straight swimming (runs) with a brief and abrupt reorientation tumbling, which, however, does not cause a significant change in its location [3, 4]. Their motion can be biased by environmental signals: for instance, the chemotactic behavior enables such organisms to approach advantageous locations and to avoid hostile ones in response to chemical stimuli. So far chemotaxis phenomena have been mainly investigated either from a macroscopic (Patlak–Keller–Segel and its variants, see e.g. [40, 48]) or a mesoscopic (kinetic) viewpoint, without accounting for the inner dynamics on which the entire motility related processes—and thus also the behavior of the chemotactically moving population—are relying. Among the first multiscale settings we mention that proposed by Firmani, Guerri and Preziosi [29] in the context of tumor immune system competition with medically induced activation/disactivation and those aligning to the general kinetic theory of active particles (KTAP) proposed by Bellomo et al., see e.g., [2] and the references therein. The same idea has been exploited in a more recent model by Erban and Othmer [26, 27] dealing with the multiscale aspects of chemotaxis and where the density function of bacteria obeys a Boltzmann-like equation coupled to a reaction-diffusion equation for the chemotactic signal, while the internal dynamics are described by an evolutionary system. The global existence of solutions for the resulting chemotaxis model has been shown in [28] (one-dimensional case, incomplete assumptions) and in [5] (higher dimensions), the latter with the aid of dispersion and Strichartz estimates under some borderline growth assumptions on the turning kernel. In [76] we proposed a model where the evolution of the bacterial population density is characterized with the aid of an integro-differential transport equation coupled to a reaction-diffusion equation for the chemoattractant and involving as in [5,26–28] intracellular dynamics, whose influence is stated in an explicit way. A simple and natural proof of global existence of a unique solution to this system was given for all biologically relevant dimensions and under usual assumptions on the involved turning kernel. The framework also allows for kernels which are more general than in previous works and also for weakening the assumptions made on the chemotactic signal. We shortly recall the model[10] and the idea of the proof, referring to [76] for further details.

9.4.1 *Problem Setting and Modeling Aspects in the PDE Framework*

Let $f(t, \mathbf{x}, \mathbf{v}, \mathbf{y})$ be the density function of bacteria in a $(2N+d)$-dimensional phase space ($N = 1, 2, 3$, $d \geq 1$) with coordinates $(\mathbf{x}, \mathbf{v}, \mathbf{y})$, where $\mathbf{x} \in \mathbb{R}^N$ is the position of a cell, $\mathbf{v} \in \mathbb{R}^N$ its velocity, and $\mathbf{y} \in \mathbb{R}^d$ the vector characterizing its internal state. The components y_i, $i = 1, \dots, d$ of $\mathbf{y}$ are concentrations of chemical species involved in intracellular signaling pathways controlling the motion of the cell. Thus $f(t, \mathbf{x}, \mathbf{v}, \mathbf{y})d\mathbf{x}d\mathbf{v}d\mathbf{y}$ is the number of cells at time t with position between $\mathbf{x}$ and $\mathbf{x} + d\mathbf{x}$, velocity between $\mathbf{v}$ and $\mathbf{v} + d\mathbf{v}$ and internal state between $\mathbf{y}$ and $\mathbf{y} + d\mathbf{y}$.

The macroscopic density of individuals at the position $\mathbf{x} \in \mathbb{R}^N$ and at the time t is given by

$$n(t, \mathbf{x}) = \int_V \int_Y f(t, \mathbf{x}, \mathbf{v}, \mathbf{y}) d\mathbf{v} d\mathbf{y}, \tag{9.37}$$

where $V \subset \mathbb{R}^N$ is the set of velocities and $Y \subset \mathbb{R}^d_+$ denotes the set of internal states.

[10]For a comprehensive deduction of a corresponding model in a related, but much more complex framework (cancer cell migration through a tissue network) we refer to [38, 39].

The density f of particles satisfies a Boltzmann like integro-differential equation [62], where the integral operator in the right hand side describes the turning events instead of the usual collisions [10][11]:

$$\partial_t f(t,\mathbf{x},\mathbf{v},\mathbf{y}) + \mathbf{v}\cdot\nabla_{\mathbf{x}} f(t,\mathbf{x},\mathbf{v},\mathbf{y}) + \nabla_{\mathbf{y}}\cdot(\mathbf{F}(S(t,\mathbf{x}),\mathbf{y})f) \tag{9.38}$$
$$= \int_V \lambda(\mathbf{y})K[S](\mathbf{v},\mathbf{v}',\mathbf{y})[f(t,\mathbf{x},\mathbf{v}',\mathbf{y}) - f(t,\mathbf{x},\mathbf{v},\mathbf{y})]d\mathbf{v}',$$

with the initial condition

$$f(0,\mathbf{x},\mathbf{v},\mathbf{y}) = f_0(\mathbf{x},\mathbf{v},\mathbf{y}). \tag{9.39}$$

Here $K[S](\mathbf{v},\mathbf{v}',\mathbf{y})$ stands for the turning kernel and gives the likelihood of a velocity jump from the $\mathbf{v}'$ to the $\mathbf{v}$ regime. The reorientations are modelled with a Poisson process with intensity λ, which is usually taken to be a positive constant, but should actually depend on the dynamics of intracellular signaling pathways.

The evolution of the inner dynamics is characterized by the ODE system

$$\frac{d\mathbf{y}}{dt} = \mathbf{F}(S(t,\mathbf{x}),\mathbf{y}), \qquad \mathbf{y}(0) = \mathbf{y}_0. \tag{9.40}$$

An explicit form for these equations has been given e.g., in [26, 27] for a simplified excitation–adaptation mechanism responsible for the activation of the flagellar rotor. However, in order to allow for true excitability that form has been replaced in [5] by a FitzHugh–Nagumo type system:

$$\dot{y}_1 = \frac{1}{t_e}(h(S) - q(y_1) - y_2) \qquad \text{(excitation)} \tag{9.41}$$
$$\dot{y}_2 = \frac{1}{t_a}(h(S) + y_1 - y_2) \qquad \text{(adaptation)},$$

where t_e and t_a denote the excitation, respectively adaptation time, the saturating ligand function $h(S) = \frac{S}{1+S}$ illustrating the fact that the signal is transmitted via receptors and that binding equilibrates rapidly. q is a cubic function of the form $q(u) = u(u-\gamma_1)(u-\gamma_2)$, with γ_1 and γ_2 positive constants.

Further, we allow the kernel $K[S](\mathbf{v},\mathbf{v}',\mathbf{y})$ to depend explicitly on the output of the excitation–adaptation mechanism and on the chemoattractant concentration $S(t,\mathbf{x})$. The evolution of the latter is described by the reaction-diffusion equation

$$\partial_t S(t,\mathbf{x}) = \Delta S(t,\mathbf{x}) - \beta S(t,\mathbf{x}) + n(t,\mathbf{x}), \qquad S(0,\mathbf{x}) = S_0(\mathbf{x}), \tag{9.42}$$

[11]Following the terminology in literature, we shall say that equation (9.37) characterizes the *mesoscopic* scale of cell dispersal.

with $\beta > 0$ a constant quantifying the consumption of S. For other choices of S we refer to e.g., [61] (time-independent Gaussian kernel) and [26, 27] (linear, respectively piecewise linear function of the position).

We characterize the dependence of the turning kernel on the intracellular dynamics by[12]

$$K[S](\mathbf{v}, \mathbf{v}', \mathbf{y}) = \alpha_1(\mathbf{y})\bar{K}(\mathbf{v}, \mathbf{v}') + \alpha_2(\mathbf{y})\bar{K}(\mathbf{v}, \nabla S), \tag{9.43}$$

with $\alpha_1(\mathbf{y}) + \alpha_2(\mathbf{y}) = 1$, where $\alpha_i(\mathbf{y})$ $(i = 1, 2)$ are the probabilities of bacteria choosing the motion dictated by the reorientation kernel $\bar{K}(\mathbf{v}, \mathbf{v}')$ for the unbiased case, respectively, by $\bar{K}(\mathbf{v}, \nabla S)$ in presence of a chemoattractant gradient, where we can choose, for instance,

$$\bar{K}(\mathbf{v}, \mathbf{v}') = \frac{1}{2\pi\sqrt{\det \Sigma_K}} e^{-\frac{1}{2}(\mathbf{v}-\mathbf{v}')^T \boldsymbol{\Sigma}_K^{-1}(\mathbf{v}-\mathbf{v}')}, \quad \mathbf{v}, \mathbf{v}' \in \mathbb{R}^N, \tag{9.44}$$

with $\boldsymbol{\Sigma}_K \in \mathbb{R}^{N \times N}$ a given diffusion matrix for the corresponding stochastic velocity process.

Concerning the weights α_i we may choose e.g., $\alpha_i(\mathbf{y}) = \frac{\tilde{\alpha}_i(\mathbf{y})}{\tilde{\alpha}_1(\mathbf{y}) + \tilde{\alpha}_2(\mathbf{y})}$, with $\tilde{\alpha}_i(\mathbf{y}) = \exp(c_i |\mathbf{y}|^2)$, $i = 1, 2$, where the constants $c_i \in \mathbb{R}$ are such that $c_1 \cdot c_2 < 0$.

Alternatively, the influence of the subcellular dynamics can also be described by taking the kernel K to be the probability density function of a random variable, which is normally distributed with a given (constant) mean and a $\mathbf{y}$-dependent covariance matrix e.g., of the form $\mathrm{Var}(\mathbf{y}) = c_{1\sigma} \cdot \exp(c_{2\sigma}|\mathbf{y}|^2)$, with $c_{1\sigma} \geq 0$, $c_{2\sigma} \in \mathbb{R}$.

9.4.2 *Existence and Uniqueness Result for the Mesoscopic Model*

Explicitly solving equation (9.42) for S gives the solution [66]

$$S(t, \mathbf{x}) = e^{-\beta t} \int_{\mathbb{R}^N} S_0(\boldsymbol{\xi}) G(t, \mathbf{x}, \boldsymbol{\xi}) d\boldsymbol{\xi} + \int_0^t e^{-\beta(t-s)} \int_{\mathbb{R}^N} n(s, \boldsymbol{\xi}) G(t-s, \mathbf{x}, \boldsymbol{\xi}) d\boldsymbol{\xi} ds, \tag{9.45}$$

[12] Observe that this is a mixture of two simpler (Gaussian) kernels, whereby its weights may vary with the cell's inner dynamics. Thus, the cell motion experiences a higher bias in the direction of the chemoattractant gradient ∇S if the weight $\alpha_2(\mathbf{y})$ outperforms its unit conjugate $\alpha_1(\mathbf{y})$. In [38,39] the same type of turning kernel has been used in a multiscale model for tumor cell migration through tissue network. Unlikely most of the turning kernels proposed so far in the literature (see e.g., [28]) it explicitly involves the gradient of the chemoattractant.

with Green's function

$$G(t, \mathbf{x}, \boldsymbol{\xi}) := \frac{1}{2^N (\pi t)^{N/2}} \cdot e^{-\frac{|\mathbf{x}-\boldsymbol{\xi}|^2}{4t}}, \qquad N = 2, 3.$$

The characteristics of equation (9.38) are

$$\frac{d\tilde{\mathbf{x}}}{d\zeta} = \tilde{\mathbf{v}}(\zeta), \qquad \frac{d\tilde{\mathbf{v}}}{d\zeta} = 0, \qquad \frac{d\tilde{\mathbf{y}}}{d\zeta} = \mathbf{F}(S(\zeta, \tilde{\mathbf{x}}(\zeta)), \tilde{\mathbf{y}}(\zeta)) \tag{9.46}$$

and the back-in-time characteristics starting in $(t, \mathbf{x}, \mathbf{v}, \mathbf{y})$ are

$$\tilde{\mathbf{x}}(\zeta; t, \mathbf{x}, \mathbf{v}, \mathbf{y}) = \mathbf{x} - \mathbf{v}(t - \zeta) \tag{9.47}$$

$$\tilde{\mathbf{y}}(\zeta; t, \mathbf{x}, \mathbf{v}, \mathbf{y}, S) = \mathbf{y} - \int_{\zeta}^{t} \mathbf{F}(S(\xi, \mathbf{x} - \mathbf{v}(t - \xi)), \tilde{\mathbf{y}}(\xi)) d\xi. \tag{9.48}$$

We also denote $\tilde{\mathbf{y}}(0) =: \tilde{\mathbf{y}}_0(t, \mathbf{x}, \mathbf{v}, \mathbf{y}, S)$.

Then relying on the method of characteristics we deduce the integral form

$$\begin{aligned} f(t, \mathbf{x}, \mathbf{v}, \mathbf{y}) = {} & f_0(\mathbf{x} - \mathbf{v}t, \mathbf{v}, \tilde{\mathbf{y}}_0(t, \mathbf{x}, \mathbf{v}, \mathbf{y}, S)) \\ & \times \exp\Big(- \int_0^t \alpha(\zeta, \mathbf{x} - \mathbf{v}(t - \zeta), \tilde{\mathbf{y}}(\zeta; t, \mathbf{x}, \mathbf{v}, \mathbf{y}, S)) d\zeta \Big) \\ & + \int_0^t H(s, \mathbf{x} - \mathbf{v}(t - s), \mathbf{v}, \tilde{\mathbf{y}}(s; t, \mathbf{x}, \mathbf{v}, \mathbf{y}, S)) \\ & \times \exp\Big(- \int_0^{t-s} \alpha(\zeta, \mathbf{x} - \mathbf{v}(t - \zeta), \tilde{\mathbf{y}}(\zeta; t, \mathbf{x}, \mathbf{v}, \mathbf{y}, S)) d\zeta \Big) ds, \end{aligned} \tag{9.49}$$

where we denoted

$$\alpha(\zeta, \tilde{\mathbf{x}}(\zeta), \tilde{\mathbf{y}}(\zeta)) := \lambda(\tilde{\mathbf{y}}(\zeta)) + \nabla_{\mathbf{y}} \cdot \mathbf{F}(S(\zeta, \tilde{\mathbf{x}}(\zeta)), \tilde{\mathbf{y}}(\zeta)), \tag{9.50}$$

$$H(t, \mathbf{x}, \mathbf{v}, \mathbf{y}, S) := \lambda(\mathbf{y}) \int_V K[S](\mathbf{v}, \mathbf{v}', \mathbf{y}) f(t, \mathbf{x}, \mathbf{v}', \mathbf{y}) d\mathbf{v}'. \tag{9.51}$$

Using the notation $\mathbf{u} = (f, S)^T$, let us also denote the right hand side of our integral system (9.45), (9.49) by $\mathscr{A}\mathbf{u}$. As for the initial condition observe that we have

$$\mathbf{u}(0, \mathbf{x}, \mathbf{v}, \mathbf{y}) = (f_0(\mathbf{x}, \mathbf{v}, \mathbf{y}), S_0(\mathbf{x}))^T.$$

Definition 9.1. The function $\mathbf{u} \in C(0,T;L^1(\mathbb{R}^N \times V \times Y) \times L^1(\mathbb{R}^N))$ whose components satisfy the integral system (9.45), (9.49) is called a mild solution to our PDE system for the cell density and the chemotactic signal.

Further we make the following

Assumption 4.

(A_1) *There exist positive constants m_λ and M_λ such that $0 < m_\lambda \leq \lambda(\mathbf{y}) \leq M_\lambda$, for all $\mathbf{y} \in Y$.*

(A_2) *The sets V and Y are bounded. Moreover, we assume Y to be a domain in $\mathbb{R}^d_+$ such that there exist some linear forms $\mu_1, \dots \mu_r$ on $\mathbb{R}^d$ with*

$$Y = \{\mathbf{y} \in \mathbb{R}^d \ : \ \mu_1(\mathbf{y}) > 0, \dots, \mu_r(\mathbf{y}) > 0\}.$$

For all $\mathbf{y}^* \in Y$ and $j = 1,\dots,r$ such that $\mu_j(\mathbf{y}^*) = 0$, the inequality $\mu_j(\mathbf{F}(S,\mathbf{y}^*)) \geq 0$ is satisfied on $\mathbb{R}_+ \times \mathbb{R}^N$.

(A_3) *$\bar{K}$ satisfies the conservation condition $\int_V \bar{K}(\mathbf{v},\mathbf{w})d\mathbf{v} = 1$ and for all $\mathbf{v},\mathbf{w} \in \mathbb{R}^N$ and a generic constant $C > 0$*

$$\bar{K}(\cdot,\mathbf{v}) \leq C|\phi(\mathbf{v})| \quad on \quad \mathbb{R}^N \tag{9.52}$$

$$|\bar{K}(\cdot,\mathbf{v}) - \bar{K}(\cdot,\mathbf{w})| \leq C|\phi(\mathbf{v}) - \phi(\mathbf{w})| \quad on \quad \mathbb{R}^N \tag{9.53}$$

with $\phi : \mathbb{R}^N \to V$ smooth enough e.g., $\phi \in C^1(\mathbb{R}^N)$ and bounded. [13]

(A_4) *$\nabla_y \cdot \mathbf{F}(S(\cdot,\tilde{\mathbf{x}}(\cdot)),\tilde{\mathbf{y}}(\cdot))$ is integrable on $[0,T]$.*

The following result ensures the existence of a unique mild solution to our multiscale system.

Theorem 9.1. *Let $f_0 \in L^1(\mathbb{R}^N \times V \times Y) \cap L^\infty(\mathbb{R}^N \times V \times Y)$ and $S_0 \in L^1(\mathbb{R}^N) \cap L^\infty(\mathbb{R}^N)$. Then under Assumptions 4 the system (9.37)–(9.42) has a unique mild solution $(f,S) \in C(0,T;X)$, for any $T > 0$.*

Proof. For the detailed statements and proofs of this and the needed auxiliary results we refer to [76].

Let $\delta \in (0,1)$ to be specified later and $X = X_1 \times X_2$, with $X_1 = L^1(\mathbb{R}^N \times V \times Y) \cap L^\infty(\mathbb{R}^N \times V \times Y)$ and $X_2 = L^1(\mathbb{R}^N) \cap L^\infty(\mathbb{R}^N)$. We prove that the application $\mathscr{M}_{0,\delta} \ni \mathbf{u} \mapsto \mathscr{A}\mathbf{u} \in \mathscr{M}_{0,\delta}$ is a contraction, where

[13] For instance, it could be defined as in [39] by $\phi(\xi) := \xi$ for $s_1 \leq |\xi| \leq s_2$ and $\phi(\xi) := s_2\frac{\xi}{|\xi|}$ for $|\xi| > s_2$, respectively $\phi(\xi) := s_1\frac{\xi}{|\xi|}$ for $|\xi| < s_1$, whereby the set V of velocities is assumed to be symmetric, of the form $V = [s_1,s_2] \times \mathbb{S}^{n-1}$.

$$\mathscr{M}_{0,\delta} = \Big\{ \mathbf{u} \in C(0,\delta;X) \ : \ \|S(t) - e^{-\beta t} \int_{\mathbb{R}^N} S_0(\boldsymbol{\xi}) G(t,\mathbf{x},\boldsymbol{\xi}) d\boldsymbol{\xi} \|_{L^1(\mathbb{R}^N)}$$
$$+ \|f(t) - f_0(\mathbf{x} - \mathbf{v}t, \mathbf{v}, \tilde{\mathbf{y}}_0(t;S))$$
$$\times \exp\Big(- \int_0^t \alpha(\zeta, \mathbf{x} - \mathbf{v}(t-\zeta), \tilde{\mathbf{y}}(\zeta;S)) d\zeta \Big) \|_{L^1(\mathbb{R}^N \times V \times Y)} \leq 1, \ t \in [0,\delta) \Big\}.$$

The rest of the local existence and uniqueness proof follows by applying Banach's fixed-point theorem.

Upon iterating this reasoning the existence of the mild solution on each interval $(0, t_{\delta_j})$ follows, with $t_{\delta_j} \leq t_{\delta_{j-1}} + t_\delta$, which proves the global existence. Hence, a bootstrap argument is used: the time interval on which the solution exists (locally) is successively extended, eventually yielding existence on an interval $[0, T)$ for any $T > 0$. □

The multiscale model introduced and analyzed above accommodates the relevant levels of dynamics, however, its high dimensionality ($2N+2$ independent variables) renders its direct numerical simulation unfeasible, a problem which is further impaired by the necessity of handling different time scales for the (fast) intracellular dynamics, the kinetic motion of cells and the (slow) diffusion of chemotactic signal. One option would be to deduce some macroscopic (parabolic or hydrodynamic) limit for our system involving integro-differential equations and to perform the numerical simulations for the equations obtained in this way. This has been done for a similar system e.g., in [26] (in one dimension) and [27] for higher dimensions, however, under the assumption of a shallow chemoattractant gradient in order to ensure moment closure and by considering turning kernels that do not depend on the internal dynamics. Moreover, due to the complexity of the problem all those deductions of macroscopic limits are merely heuristic. The few rigorous results (see e.g., [35, 63]) have been obtained for drastically simplified settings and required rather restrictive conditions on the involved turning kernels. For instance, such an assumption is

$$\int_V K(\mathbf{v}, \mathbf{v}') d\mathbf{v}' = 1, \tag{9.54}$$

which is not fulfilled by many reorientation kernels, among others the one based on the von Mises distribution proposed in [12] and which relies on experimental evidence. This is the case, too, for all mixture-based kernels like the one proposed above in (9.43) or below in Example 9.2.

Since to our knowledge, there are so far no reliable numerical methods for handling the genuine multiscale model, in order to illustrate numerically the behavior of the cell population when accounting for all relevant scales we will use a nonparametric technique, which was applied in [73, 74] in a related context. This method avoids the use of differential equations for the cell density; instead, independent bacteria trajectories are simulated on the interval of interest by directly starting from the primary description of the involved biological processes. The data sets obtained in this way are then used to estimate the cell population density at an arbitrary time moment t.

The method allows for a great modeling flexibility, since it relies on simulations performed upon employing the stochastic processes which characterize the motion along with its triggering factors (e.g., intracellular events and chemoattractant).

9.4.3 Numerical Treatment: A Nonparametric Technique

We present a new approach to the numerical handling of some classes of biomedical problems.[14] As mentioned above, the classical techniques are not appropriate for systems of such complexity. In [73, 74] we proposed the use of a nonparametric method which we shortly describe in the following and which can be seen as a numerical approach to a large class of PDEs deduced from a stochastic framework. It also works for more complicated models—like those presented in the rest of this section—which are set up without involving PDEs.

The kernel density estimation technique is widely used for estimating complex density functions due to its flexibility and the fact that its consistency for various rates of convergence is already well established [70, 71]. The method uses computing power to allow a very effective handling of complicated structures. Often it is desirable to simply *let the data speak for itself* i.e., to look for an estimator of the population density, unconstrained (or as loosely as possible) by an a priori (parametric) form. This is in fact the aim of nonparametric density estimators. In the following we apply this method for independent simulations, however, it also works under fairly general conditions for dependent data, see e.g., [64, 83].

Starting from the model for cell movement, we simulate U independent bacterial trajectories on the interval of interest $[0, T]$ and we use the data sets obtained in this way to estimate the cell population density at an arbitrary moment of time $t \in [0, T]$. More details on how these simulations are performed are provided below.

The nonparametric estimators for the cell population density n at some moment t are defined by (see [70]):

$$\widehat{n_{\mathbb{H}}}(t, \mathbf{x}) = \frac{1}{U \det \mathbb{H}} \sum_{i=1}^{U} \mathscr{K}(\mathbb{H}^{-1}(\mathbf{x} - \mathbf{X}_i)), \qquad \mathbf{x} \in \mathbb{R}^N, \tag{9.55}$$

or, in an analogous way, for other relevant densities like e.g., for f:

$$\widehat{f_{\tilde{\mathbb{H}}}}(t, \tilde{\mathbf{x}}) = \frac{1}{U \det \tilde{\mathbb{H}}} \sum_{i=1}^{U} \tilde{\mathscr{K}}(\tilde{\mathbb{H}}^{-1}(\tilde{\mathbf{x}} - \tilde{\mathbf{X}}_i)), \qquad \tilde{\mathbf{x}} \in \mathbb{R}^{2N}, \tag{9.56}$$

[14] Some parts of this subsection have been reproduced from [74] with permission.

where $\mathscr{K}$ and $\tilde{\mathscr{K}}$ denote general kernel functions, $\tilde{\mathbf{x}} = (\mathbf{x}, \mathbf{v})^T$, $\tilde{\mathbf{X}}_i = (\mathbf{X}_i, \mathbf{V}_i)^T$, $\mathbf{X}_i = (X_{i1}, \ldots, X_{iN})^T$, $\mathbf{V}_i = (V_{i1}, \ldots, V_{iN})^T$, $i = 1, \ldots, U$ are the position, respectively, the velocity at the moment t of the simulated ith trajectory, while $\mathbb{H}$ and $\tilde{\mathbb{H}}$ are the corresponding bandwidth matrices, which are usually taken to be diagonal and invertible. For the concrete numerical applications here they are taken to be of the form $\mathbb{H} = h\mathbb{I}$, with $\mathbb{I}$ denoting the identity matrix and $h > 0$ being the so-called bandwidth parameter.

One of the most frequently used kernels in the univariate case is the Gaussian one, defined by $K(u) = \frac{1}{\sqrt{2\pi}} \exp(-\frac{1}{2}u^2)$, $u \in \mathbb{R}$. Other classical choices are: Epanechnikov and its variants, triangular, rectangular etc [70]. In the multivariate case the easiest form to be chosen for the kernel $\mathscr{K}$ is the multiplicative one: $\mathscr{K}(\mathbf{x}) = \prod_{j=1}^{N} K(x_j)$. Analogously for $\tilde{\mathscr{K}}$.

Thereby, the choice of the bandwidth matrix is important, whereas the choice of the kernel function is not so crucial, since it is possible to rescale the kernel function such that the difference between two given density estimators using two different kernel functions is negligible [54].

The choice of the bandwidth matrix is one of the most difficult practical problems in connection with the above method. The bandwidths are chosen according to the available information about the density to be estimated. For example, if it is known when the latter is very close to a normal density, then the bandwidths can be optimally chosen with the so-called *rule-of-thumb* [70] giving an explicit expression for the bandwidth matrix. However, this is rarely the case, so a more adequate choice is to compute the bandwidth according to one of the data driven bandwidth selection criteria. One of the most popular ones is the *least-squares-cross-validation* (LSCV), which has the goal to estimate the integral squared error (ISE) defined by

$$ISE(\mathbb{H}) = \int [\widehat{n_{\mathbb{H}}}(t, \mathbf{x}) - n(t, \mathbf{x})]^2 d\mathbf{x}, \tag{9.57}$$

where $\widehat{n_{\mathbb{H}}}(t, \mathbf{x})$ is the estimated density and $n(t, \mathbf{x})$ is the true density being estimated. The usual method for estimating ISE is the *leave-one-out* cross validation. The minimization of the estimated ISE leads to an optimal choice of the bandwidth matrix for a given kernel density function $\mathscr{K}$. It is this bandwidth selection criterion which we use in the present work. Alternatively, there is a plethora of other bandwidth selectors in literature (see e.g., [19] and the references therein).

Several results on consistency of the kernel density estimators settling the theoretical foundations of the nonparametric method have been derived e.g., by Cacoullos [8], Deheuvels [17], and Devroye and Györfi [18]. The issue of convergence speed has been addressed among others by Devroye and Györfi [18] and we refer for further, more specific convergence results and error estimates to Holmström and Klemelä [36] and the references therein. Similar results for the case with dependent data can be found e.g., in [83].

The computational cost of this method is tightly connected to the so-called curse of dimensionality: the amount of data necessary for an accurate estimation grows exponentially with the spatial dimension of the phenomena of interest. However, for the concrete biological problems handled in this work the spatial dimension is maximum $N = 3$ when estimating the macroscopic cell density and thus the size of the required data sets is a very reasonable one, for which the method performs well.

9.4.4 Extensions of the Classical Modeling Framework

The current models for describing cell dispersal rely on some classes of random processes either giving a geometrical description of the motion or considering stochastic increments of the cell velocity. The former build the class of so-called *velocity jump (VJ)* models, while in the latter the particle velocity obeys a multivariate *Ornstein–Uhlenbeck (OU)* process.

9.4.4.1 VJ Type Models

In the models of this class the changes in the cell velocity are dictated by a turning kernel. Based on particle kinetics one can deduce a Boltzmann like partial integro-differential equation where the usual collision term is replaced by an integral operator characterizing the turning events. A PIDE of this type has been deduced e.g., by Othmer et al. [62] in the absence of cell–cell interactions and external stimuli and has the form

$$\partial_t f(t,\mathbf{x},\mathbf{v}) + \mathbf{v} \cdot \nabla f(t,\mathbf{x},\mathbf{v}) = -\lambda f(t,\mathbf{x},\mathbf{v}) + \lambda \int_V K(\mathbf{v},\mathbf{v}') f(t,\mathbf{x},\mathbf{v}') d\mathbf{v}'. \quad (9.58)$$

Here, $K(\mathbf{v},\mathbf{v}')$ denotes the turning kernel characterizing the likelihood of a cell changing its velocity regime from $\mathbf{v}'$ to $\mathbf{v}$. The reorientations are modelled with a Poisson process with intensity λ, so the mean running time is $\tau = 1/\lambda$. The kinetic equation presented in Sect. 9.4.1 above aligns to this framework too, however, that setting is more realistic, since it also accounts for the effects of the intracellular dynamics and the chemoattractant concentration. These influences can be seen as acting like conditionings on the turning kernel.

In the following we avoid the already mentioned numerical simulation problems arising from the high dimensionality and the complexity of the kinetic formulation by considering an approach in which the cell velocities are seen as stochastic processes with dynamics characterized by the turning kernels. Their evolution—together with the knowledge of time distribution of reorientations—will allow to reconstruct the individual cell trajectories. These in turn will provide the data for the nonparametric estimations of the densities of interest.[15]

[15]Observe that the method outlined in Sect. 9.4.3 enables to directly compute the macroscopic cell density $n(t,\mathbf{x})$, without needing to go through the intermediate step of assessing its mesoscopic

We give here two examples of applying this method, both of which cannot be handled numerically in the PDE framework. For models featuring further interesting cell motion characteristics we refer to [73, 74]. For all simulations and estimations in this subsection we take $N = 2$, since this situation is easier to visualize than the 3D case, however, this is no restriction to the method which works very well in 3D too.

Example 9.1 (Multiscale model with intracellular dynamics). Consider the turning kernel $K[S](\mathbf{v}, \mathbf{v}', \mathbf{y})$ proposed in (9.43) above, which is allowed to explicitly depend both on the intracellular dynamics and on the gradient of the chemoattractant. In order to reduce the simulation effort we assumed (as in [26, 27, 61]) a given form of the chemoattractant signal

$$S(\mathbf{x}) = \frac{1}{(2\pi)^{N/2}\sqrt{\det \boldsymbol{\Sigma}_S}} e^{-\frac{1}{2}(\mathbf{x}-\mathbf{m}_S)^T \boldsymbol{\Sigma}_S^{-1}(\mathbf{x}-\mathbf{m}_S)}, \quad \mathbf{x} \in \mathbb{R}^N, \tag{9.59}$$

where $\mathbf{m}_S$ denotes the position of the signal source. In our simulations we take $\mathbf{m}_S = (5.5, 5.5)^T$ and $\boldsymbol{\Sigma}_S = 1.8\mathbb{I}_N$. For the intracellular dynamics the equations (9.41) have been considered. The reaction-diffusion equation for S involving the macroscopic cell density n could be solved as well in each time step, this being only a matter of computation costs. For the same reason we choose a constant turning rate λ, however, there is no challenge in considering one having a known form which depends on the intracellular dynamics and/or (possibly implicitly) on the current position and the concentration of the chemoattractant.

The parameters used in the simulations are—where applicable—consistent with those in literature [5, 12, 62]. Further parameter choices (for the weights in the turning kernel) are $c_1 = -5.2$, $c_2 = 3.75$ and $\boldsymbol{\Sigma}_K = 1.2\mathbb{I}_2$, $\gamma_1 = 1$, $\gamma_2 = 0.2$, $t_e = 0.001$ for the excitation time and $t_a = 12$ for the adaptation time in (9.41).

The simulation and estimation procedure then goes through the following steps:

1. solve the ODE system for the intracellular dynamics having as input the given concentration of the chemoattractant;
2. simulate U cell trajectories upon making use of the turning kernel proposed in (9.43);
3. estimate the macroscopic cell density with the nonparametric method.

All estimations have been performed using $U = 10{,}000$ simulations, though this is far more than is actually needed in order to ensure good accuracy. For further details on this issue we refer to [70, 74].

(higher dimensional) counterpart $f(t, \mathbf{x}, \mathbf{v}, \mathbf{y})$ as e.g., in the PDE approach. Hence this implies a dimension reduction from $2N + 2$ to N.

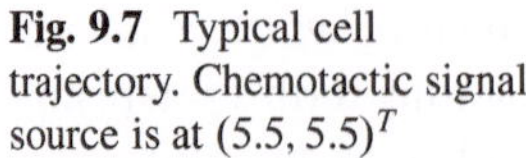

Fig. 9.7 Typical cell trajectory. Chemotactic signal source is at $(5.5, 5.5)^T$

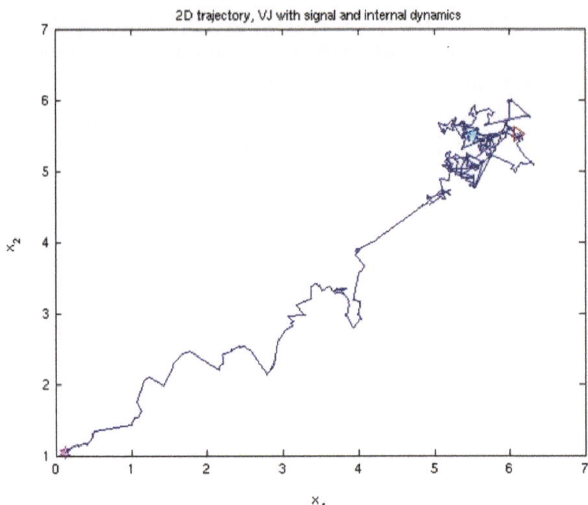

Figure 9.7 below has merely illustrative purposes[16] and shows a typical cell trajectory for this model.

Figure 9.8[17] collects a sequence of estimated macroscopic cell densities[18] at consecutive times.

Observe that when many enough cells reach the (closer) neighborhood of the signal source (at about $t = 10$) the population begins to split into a faster and a slower part. This can be clearly seen in Fig. 9.8d,e. Then the faster part seems to adapt (starting with $t = 12$) to the chemotactic signal, allowing the slower part to catch up before adapting itself. At about $t = 14$ the population has again the more compact form it had before splitting and moves on as such towards the chemotactic source, remaining for later times in its neighborhood and no longer splitting, which might be caused by all cells having adapted to the chemotactic signal.

Example 9.2 (Model for a bacterial population avoiding a hostile region). [19] Here we consider a VJ model for a cell population which has to avoid a hostile circular region of radius r centered at $\mathbf{x}_O \neq \mathbf{m}_0$, where $\mathbf{m}_0$ denotes the mean of the initial population density. A possible choice for the turning kernels would be to consider mixtures similar to the one in (9.43). However, since here we want to focus on featuring the avoidance of the unfavorable region the dependence on the chemotactic signal and the intracellular dynamics will be turned off.

[16] Of course, the influence of the subcellular dynamics cannot be seen on this level of a random individual path.

[17] Taken from [76].

[18] More precisely their respective projections on the $x_1 O x_2$ plane.

[19] See [73].

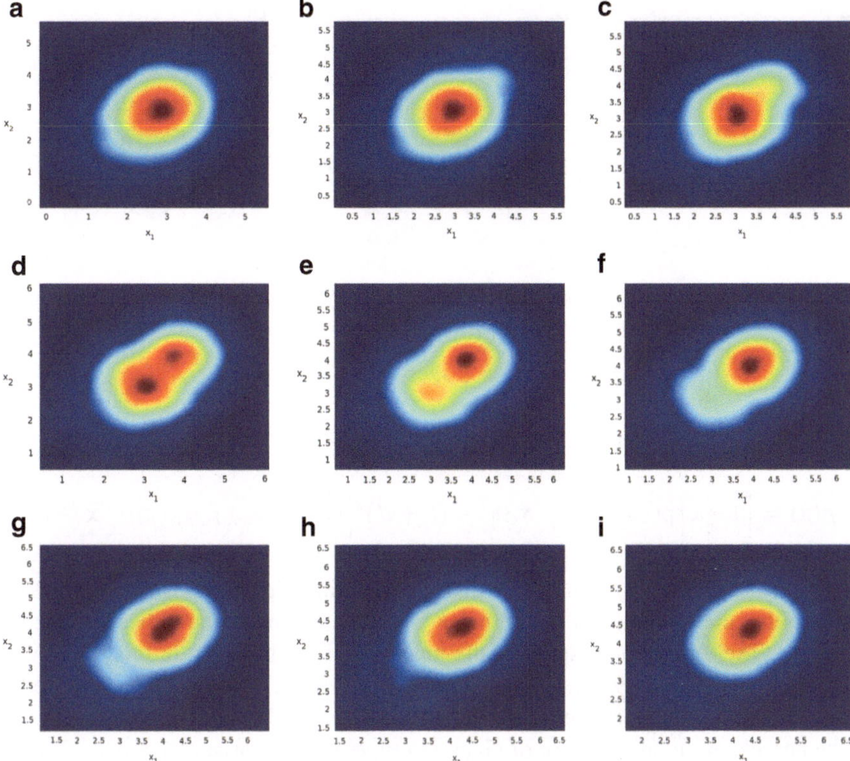

Fig. 9.8 Estimated macroscopic density at several time moments. Inner dynamics given by (9.41). **(a)** $t = 8$; **(b)** $t = 9$; **(c)** $t = 10$; **(d)** $t = 11$; **(e)** $t = 12$; **(f)** $t = 13$; **(g)** $t = 14$; **(h)** $t = 15$; **(i)** $t = 16$

The corresponding weights p, q in the mixture characterizing the turning kernel should also depend in this case on the cell position:

$$K(\mathbf{v}, \mathbf{v}', \mathbf{x}) = p(\mathbf{x}) K_1(\mathbf{v}, \mathbf{v}') + q(\mathbf{x}) K_2(\mathbf{v}, \mathbf{v}', \mathbf{x}), \tag{9.60}$$

where K_1 models a biased cell movement; e.g., it is a Gaussian kernel with mean $\frac{1}{2} v_M (\frac{\mathbf{v}'}{\|\mathbf{v}'\|} + \frac{\mathbf{x}_O - \mathbf{m}_0}{\|\mathbf{x}_O - \mathbf{m}_0\|})$, where v_M is a known speed for which a biologically relevant value is chosen (e.g., $10 - 20\,\mu$m/s, see [27]) and with a given covariance matrix Σ_{K_1}, while K_2 models the cell movement around a hostile region. Given the previous velocity $\mathbf{v}'$, K_2 is the conditional density of the random variable $\mathbf{V} = \mathrm{sign}(< \frac{\mathbf{x} - \mathbf{x}_O}{\|\mathbf{x} - \mathbf{x}_O\|}, \tilde{\mathbf{V}} >) \tilde{\mathbf{V}}$, where $\tilde{\mathbf{V}}$ is a random vector having any type of density which is suitable to describe a classical bacterial motion. For instance, it can be a Gaussian vector with mean $\frac{1}{2} v_M \left(\frac{\mathbf{x}_O - \mathbf{x}}{\|\mathbf{x}_O - \mathbf{x}\|} - \frac{\mathbf{v}'}{\|\mathbf{v}'\|} \right)$ and given covariance Σ_{K_2}. Thereby $< \cdot, \cdot >$ denotes the scalar product in $\mathbb{R}^N$.

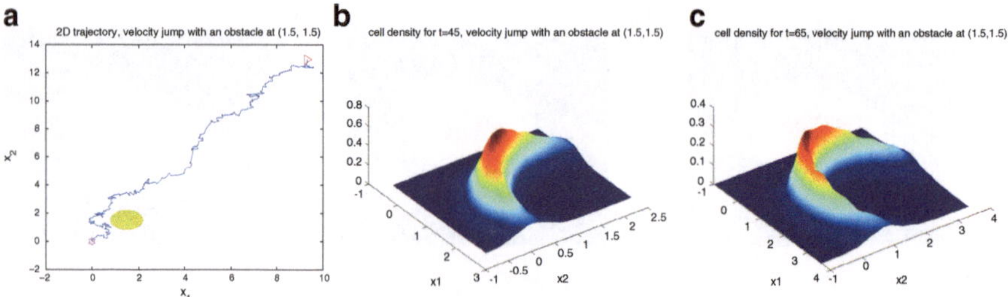

Fig. 9.9 Cell population avoiding a hostile region (**a**): a typical trajectory. Start (*magenta star*), end (*red right pointer*). Obstacle centered at (1.5, 1.5). (**b**) and (**c**): estimated macroscopic density at $t = 45$, respectively $t = 65$

Concerning the weights, we can choose for instance

$$p(\mathbf{x}) = \Big[1 - \exp(-\frac{1}{\sigma_O}\Big|\|\mathbf{x} - \mathbf{x}_O\|^2 - (r+d)^2\Big|)\Big] 1_{\mathbb{R}^n - C(\mathbf{x}_O,(r+d))}(\mathbf{x}), \quad \mathbf{x} \in \mathbb{R}^N,$$

and $q(\mathbf{x}) = 1 - p(\mathbf{x})$, where σ_O and d are some adequate positive parameters controlling the influence of the kernel K_2, while $1_{[\cdot]}$ denotes the indicator function of a set. The estimation results for the macroscopic cell density in this framework are illustrated in Figs. 9.9 and 9.10 below.[20] Concerning the parameters involved in the simulations, the obstacle is centered at $\mathbf{x}_0 = (1.5, 1.5)^T$ and has radius $r = 0.7$. The kernels in the mixture are of Gaussian type, with covariance matrices $\boldsymbol{\Sigma}_{K_1} = \boldsymbol{\Sigma}_{K_2} = 0.01\mathbb{I}_2$, the mean velocity is $v_M = 20\,\mu$m/s, and $\sigma_0 = 1$, $d = 0.001$.

For a model describing the evolution of a cell population in a heterogeneous medium with many (possibly randomly distributed) obstacles we refer to [74].

9.4.4.2 OU Type Models

The OU based models[21] can be written in the general form

$$d\tilde{\mathbf{x}}_t = \mathbf{b}(t, \tilde{\mathbf{x}}_t)dt + \boldsymbol{\sigma}(t, \tilde{\mathbf{x}}_t)d\mathbb{B}_t, \ t \geq 0 \tag{9.61}$$

where $\tilde{\mathbf{x}}_t \in \mathbb{R}^{2N}$ is the multivariate stochastic process with components $\mathbf{x}_t$ and $\mathbf{v}_t$, while $\mathbb{B}_t$ is a multivariate Brownian motion in $\mathbb{R}^{2N}$. In this stochastic differential equation the drift $\mathbf{b} \in \mathbb{R}^{2N}$ and the diffusion matrix $\boldsymbol{\sigma} \in \mathbb{R}^{2N\times 2N}$ are usually of the form

[20]Figures reproduced from [73].

[21]See [73, 74].

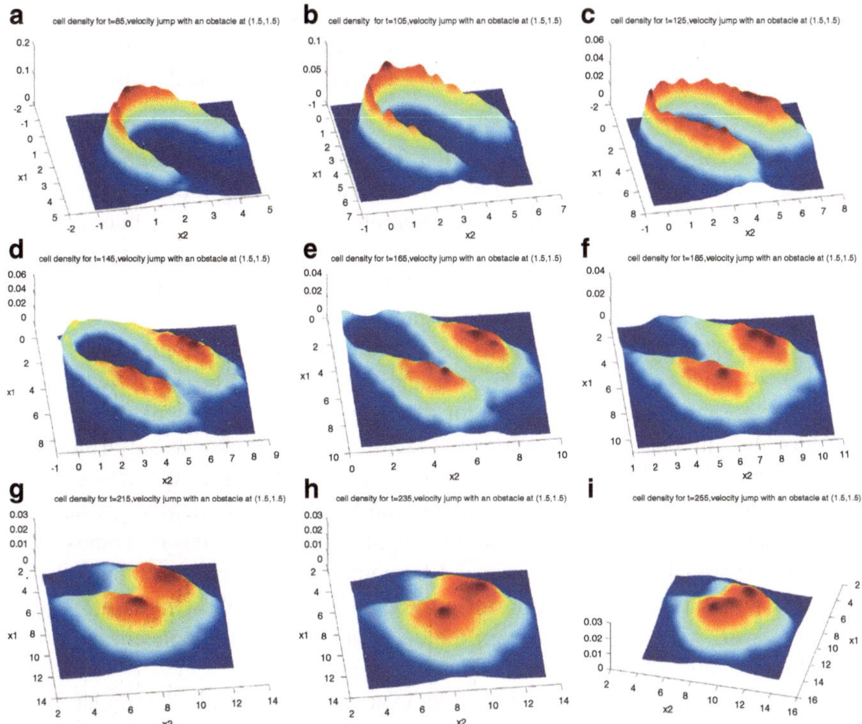

Fig. 9.10 Estimated macroscopic cell density at $t = 85$, $t = 105$, $t = 125$, $t = 145$, $t = 165$, $t = 185$, $t = 215$, $t = 235$ respectively $t = 255$

$$\mathbf{b} = \begin{pmatrix} \mathbf{v}_t \\ \dfrac{\chi}{\rho}(\mathbf{m} - \mathbf{v}_t) \end{pmatrix}, \quad \boldsymbol{\sigma} = \begin{pmatrix} \mathbf{0} & \mathbf{0} \\ \mathbf{0} & \dfrac{1}{\rho}\boldsymbol{\Sigma} \end{pmatrix}, \quad \text{with} \quad \boldsymbol{\Sigma} \in \mathbb{R}^{N \times N}, \quad \mathbf{m} \in \mathbb{R}^N, \tag{9.62}$$

where χ/ρ denotes the rate of mean reverting for the velocity process $\mathbf{v}$ and is related to the chemotactic sensitivity. $\mathbf{m}$ and $\boldsymbol{\Sigma}$ are constant matrices or some functions of stochastic processes. For $\boldsymbol{\Sigma}$ constant, the random variables are Gaussian distributed, which is often not realistic, see e.g., [13, 79, 82]. This problem, however, can be overcome upon allowing $\boldsymbol{\Sigma}$ to be a stochastic process of the OU type comprising several stochastic effects in the biological system. Details will be provided below.

When the coefficients in (9.61) are sufficiently smooth it is well known that the evolution of the density is described by the forward Kolmogorov equation (FKE)

$$\frac{\partial f}{\partial t} = -\sum_{i=1}^{2N} \frac{\partial}{\partial \tilde{x}_i}(b_i(t, \tilde{\mathbf{x}})f) + \frac{1}{2}\sum_{i,j,k=1}^{2N} \frac{\partial^2}{\partial \tilde{x}_i \partial \tilde{x}_j}(\sigma^{ik}(t, \tilde{\mathbf{x}})\sigma^{jk}(t, \tilde{\mathbf{x}})f), \tag{9.63}$$

where $\tilde{\mathbf{x}} = (\mathbf{x}, \mathbf{v})' \in \mathbb{R}^{2N}$, $\mathbf{b} = (b_i)_{1\leq i\leq 2N}$, $\boldsymbol{\sigma} = (\sigma^{ij})_{1\leq i,j\leq 2N}$ with b_i and σ^{ij} deterministic functions. For the model with nonsmooth coefficients to be presented below, however, it would be technically quite elaborate to deal with this PDE approach.

Example 9.3 (Model for a cell population inferring resting phases and switching between a biased and an unbiased motion regime). The OU model in (9.61) can be extended to allow for a random alternation between a motion biased in the direction of an external signal (e.g., chemoattractant) and an unbiased one. The latter is characterized by constant values of $\mathbf{m}$ and $\boldsymbol{\Sigma}$, while the random switching can be achieved by assuming them to be nonconstant.

To this purpose we choose $\mathbf{m}$ to be a stochastic process of the form $\mathbf{m}_t = \mathbf{g}(t, \Upsilon_t)$, with an appropriate function $\mathbf{g}$ and a stochastic process Υ_t, which can be seen e.g., as an individual indicator of the internal dynamics of the cell as a result of the influences of environmental factors like local abundance of nutrients or chemorepellents. For example, one can model Υ_t as another OU process with dynamics independent on the Brownian motion $\mathbf{W}_t$. In a more detailed description it could also be the logarithm of some weighted mean over the outcomes of an intracellular signaling pathway initiated by some input signal and modeled with SDEs. A possible parametrization for $\mathbf{g}$ to be used in the following is $\mathbf{g}(t, \Upsilon_t) = e^{-\gamma t}\mathbf{1}_{\{\Upsilon_t\leq 0\}}\mathbf{m}$, where $\gamma \geq 0$ is e.g., the decaying rate for the concentration of a stimulus (or some other environmental influence).[22]

The Gaussianity assumed for the random variables involved in the model for the case where the diffusion coefficient $\boldsymbol{\Sigma}$ is constant can be alleviated as well upon taking the volatility of the process $\mathbf{W}_t$ to be stochastic e.g., of the form $h(t, \Upsilon_t)\boldsymbol{\Sigma}$, with $\boldsymbol{\Sigma}$ being as before a constant matrix and h an appropriate real function. For instance, one could make the choice $h(t, \Upsilon_t) = ae^{-\gamma t} + (1 - ce^{-\gamma t})be^{\Upsilon_t}$ ($a, b \geq 0$, $a^2 + b^2 > 0, 0 \leq c \leq 1$), with $\gamma \geq 0$ and Υ_t having the same significance as above. Notice that the coefficient $h(t, \Upsilon_t)$ of the covariance matrix directly influences the velocity fluctuations: when the cell passes in a biased regime it reduces its velocity variance, whereas the latter is increased in the unbiased regime.

The new class of models obtained in this way takes the form

$$d\mathbf{x}_t = \mathbf{v}_t dt \tag{9.64}$$

$$\rho d\mathbf{v}_t = \chi(\mathbf{g}(t, \Upsilon_t) - \mathbf{v}_t)dt + h(t, \Upsilon_t)\boldsymbol{\Sigma} d\mathbf{W}_t, \qquad t \geq 0 \tag{9.65}$$

$$d\Upsilon_t = \alpha_\Upsilon(m_\Upsilon - \Upsilon_t)dt + \beta_\Upsilon dZ_t, \tag{9.66}$$

with $\mathbf{W}_t$ and Z_t independent Brownian motions, $\mathbf{g} : \mathbb{R}_+ \times \mathbb{R} \to \mathbb{R}^N$ and $h : \mathbb{R}_+ \times \mathbb{R} \to \mathbb{R}$ some given functions, and initial values $\mathbf{v}(0) = \mathbf{v}_0$ and $\Upsilon(0) = \Upsilon_0$, where $\mathbf{v}_0$, Υ_0 are random variables independent of $\mathbf{W}_t$ and Z_t. Here, ρ is a measure of persistence, $\alpha_\Upsilon, \beta_\Upsilon > 0$, $m_\Upsilon \in \mathbb{R}$.

[22]Observe that for $\gamma > 0$ the bacterial population asymptotically goes over into an unbiased regime.

The above choice of functions $\mathbf{g}$ and h is particularly suited to capture more features of the inter- and intracellular environments also in this framework,[23] where the velocity changes are continuous in time and prescribed by the OU system, unlikely the jumps dictated by a turning kernel of the type given in the previous paragraph.

For a model of the form (9.64) the application of the nonparametric technique is straightforward and includes the following steps:

1. solve the SDEs for the velocity process $\mathbf{v}_t$;
2. generate with the aid of $\mathbf{v}_t$ analogously to the velocity jump case the cell trajectories;
3. apply the nonparametric method to estimate the macroscopic cell density for the population of interest.

The models of the previous type can be easily adapted to allow for resting phases, a feature actually encountered in some bacteria species [23, 24, 30], but also in more evolved cells like e.g., fibroblasts [6].

In the PDE framework a transport equation model characterizing bacterial movement involving resting phase effects has been proposed in [62] under the assumption of the cells leaving the resting phase at random times governed by a Poisson process. A more recent transport based model with resting phases can be found in [33]. The results, however, were of rather theoretical nature and no numerical simulations have been performed yet to assess the behavior of cells predicted by any of those models.

Here we propose a more flexible way to model stationary phases upon using an OU driven stochastic switcher Γ_t, $t \geq 0$. This idea allows us to adapt all model classes presented above without PDEs (both for the cases OU and velocity jump) to account for resting phases.

In the OU case the pausing can be easily modelled in a similar way to the previous setting including environmental influences via stochastic processes:

$$d\mathbf{x}_t = \mathbf{v}_t dt \tag{9.67}$$

$$\rho d\tilde{\mathbf{v}}_t = \chi(\mathbf{g}(t, \Upsilon_t) - \tilde{\mathbf{v}}_t)dt + h(t, \Upsilon_t)\boldsymbol{\Sigma}\, d\mathbf{W}_t \tag{9.68}$$

$$d\Upsilon_t = \alpha_\Upsilon(m_\Upsilon - \Upsilon_t)dt + \beta_\Upsilon dZ_\Upsilon(t) \tag{9.69}$$

$$d\Gamma_t = \alpha_\Gamma(m_\Gamma - \Gamma_t)dt + \beta_\Gamma dZ_\Gamma(t) \tag{9.70}$$

$$\mathbf{v}_t = 1_{\{\Gamma_t \leq 0\}} \cdot \tilde{\mathbf{v}}_t. \tag{9.71}$$

$\mathbf{W}_t$, $Z_\Upsilon(t)$ and $Z_\Gamma(t)$ are independent Brownian motions, Υ_t has the same meaning as above and Γ_t is a classical OU process which is supposed to model the alternation between the moving and the resting phases regimes. The coefficients, the variables, and the functions $\mathbf{g}$, h involved have the same significance as in (9.64).

[23]Though not in an explicit way.

Thus, the cell will be in a resting phase as long as its corresponding switcher Γ_t takes positive values. Our choice of the switcher as an OU process was motivated by the good statistical properties of the latter and is more appropriate when time varying effects are to be included. Other choices are, however, possible as well. Moreover, we considered here the processes Γ_t and Υ_t to be uncorrelated, but this assumption can be easily dropped to allow for the extension to more general cases.

In the velocity jump case one can follow the same idea in order to account for resting phases: the new velocity after each reorientation is dictated by the turning kernel, while an OU process Γ_t allows switching between this new velocity and the zero one in the resting phase.

9.5 Conclusions

In this chapter we proposed some approaches based on SDEs or SDE-like processes to modeling biomedical problems which so far have been handled via ODEs or P(I)DEs. These new approaches are able to enhance the description of the relevant biological phenomena, while still keeping the corresponding settings manageable. Moreover, they allow statistical methods like the nonparametric density estimation for assessing the behavior of cell populations to be applied in a framework where the deterministic methods fail or are inefficient. We illustrated the new classes of models through three applications from biology and medicine: intracellular signaling pathways (here in particular the JAK–STAT signaling), evolution of tumor cells in response to radiotherapy, and cell dispersal. However, the problems studied here represent just a few paradigms and the model classes and methods for their mathematical handling presented in this work have a much wider applicability. Observe that some of these models can be set in the frame of nonautonomous (stochastic) dynamical systems, for which methods handling the deterministic case are provided in Chap. 1.

In Sect. 9.2 we considered some extensions of mathematical models for intracellular signaling. In doing so, we recalled the classical ODE and DDE settings by Timmer et al. [78], as well as some recent SDE approaches introduced in [75], and proposed some new models involving coefficients depending on stochastic processes and nonlocal SDEs. The latter allowed the uncertainties of the dynamics for the proteins interacting in the pathway to be captured and, at the same time, a more convenient handling of the time lag in this context. This has been achieved by treating the terms with delay in a deterministic way, while preserving the relevant stochasticity for those without delay. This also opens the possibility for parameter inference in this SDDE framework, an issue which is still out of reach in the more classical SDDE settings. The estimation of parameters for the new models is ongoing work.

The new model proposed in Sect. 9.3 for describing radiation therapy schedules enables randomness to be accommodated both in the treatment and in the evolution of the tumor. This also means that the commonly made (inadequate) assumption of a sufficiently large population of cells is no longer needed. Moreover, interesting features like the probability distribution of the persistence time for tumor cells under a certain treatment schedule (including the memory of treatment and repopulation effects during radiation pauses) have also been accounted for. The model used for describing the dynamics of the tumor cells under a specific treatment schedule involved a jump process which is a particular case of a Lévy process. Hence, the model (9.34)–(9.36) can be equivalently characterized by a system of SDEs driven by Lévy processes.

Eventually, in Sect. 9.4 we introduced a multiscale model for cell dispersal, in which the effects of subcellular dynamics on the behavior of the entire population were accounted for. Both the approaches via PIDEs and via SDEs have been considered. The former relies on the kinetic theory of (active) particles and leads to a Boltzmann-like transport equation for the cell density, coupled with a reaction-diffusion equation for the concentration of a chemoattractant and an ODE system for the excitation–adaptation mechanism of intracellular dynamics.[24] We provided a global existence and uniqueness result for the solution to this coupled system, however, due to the complexity and the high dimensionality of the problem, the issue of numerical simulations is still inaccessible. This motivated the alternative approach where we started from the underlying stochastic processes characterizing the cell movements in the same multiscale context and proposed some models for the evolution of their velocities, either driven by Brownian motions (the OU type models) or accounting for jumps (the VJ models). The latter too, can be seen as SDEs of the form (9.61), however, w.r.t. a pure jump Lévy process i.e. with $d\mathbb{L}_t$ instead of $d\mathbb{B}_t$ and of course with appropriate drift and diffusion coefficients. Then cell trajectories have been simulated with the aid of these OU and VJ models, which allowed for the estimation of the macroscopic cell density by using the nonparametric technique recalled in Sect. 9.4.3. In [74] we offered a new perspective to this method and interpreted it as a numerical procedure for solving PIDEs of the type presented in Sect. 9.4.1. Its applicability and advantages are addressed in that paper too. From a practical point of view one of its main assets is the fact that if real data become available (in the form of sufficiently numerous cell trajectories), then they can be directly used to assess the cell population density.

The high versatility of the method allows for handling a plethora of models featuring a complexity and a detail level which cannot be achieved in the P(I)DE framework. Section 9.4.4 provides merely a small selection; for further interesting models we refer to [73, 74].

Finally, in this chapter we assumed independent cell trajectories, however, the nonparametric technique can also be applied under fairly general conditions even in the case of correlated cell paths. We refer to e.g., [64] for the mathematical framework of nonparametric estimation with dependent observations. This opens the possibility of using this method also for self-organization models.

[24] Related multiscale models in the much more complex framework of tumor cell migration through a tissue network were addressed in [38, 39, 51].

Acknowledgement Christina Surulescu was partially supported by the Baden-Württemberg Foundation.

References

1. A. Bellen, M. Zenaro, *Numerical Methods for Delay Differential Equations* (Oxford University Press, Oxford, 2003)
2. N. Bellomo, A. Bellouquid, J. Nieto, J. Soler, Complexity and mathematical tools toward the modeling of multicellular growing systems. Math. Comput. Model. **51**, 441–451 (2010)
3. H.C. Berg, How bacteria swim. Sci. Am. **233**, 36–44 (1975)
4. H.C. Berg, D.A. Brown, Chemotaxis in Escherichia Coli analysed by three-dimensional tracking. Nature **232**, 500–504 (1972)
5. N. Bournaveas, V. Calvez, *Global existence for the kinetic chemotaxis model without pointwise memory effects, and including internal variables* (2008). arXiv:0802.2316v1 [math.AP]
6. D. Bray, *Cell Movements. From Molecules to Motility* (Garland, New York, 2001)
7. M.E. Burton, L.M. Shaw, J.J. Schentag, W.E. Evans, *Applied Pharmacokinetics and Pharmacodynamics. Principles of Therapeutic Drug Monitoring* (Lippincott Williams & Wilkins, Baltimore, 2006)
8. T. Cacoullos, Estimation of a multivariate density. Ann. Inst. Stat. Math. **18**, 179–189 (1966)
9. D. Carlson, R. Stuart, X. Lin et al., Comparison of in vitro and in vivo α/β ratios for prostate cancer. Phys. Med. Biol. **49**, 4477–4491 (2004)
10. C. Cercignani, *The Boltzmann Equation and Its Application* (Springer, Berlin, 1988)
11. K.N. Chadwick, H.P. Leenhouts. *The Molecular Theory of Radiation Biology* (Springer, Berlin, 1981)
12. E.A. Codling, N.A. Hill, Calculating spatial statistics for velocity jump processes with experimentally observed reorientation parameters. J. Math. Biol. **51**, 527–556 (2005)
13. A. Czirók, K. Schlett, E. Madarász, T. Vicsek, Exponential distribution of locomotion activity in cell cultures. Phys. Rev. Lett. **81**, 3038–3041 (1998)
14. S.B. Curtis, Lethal and potentially lethal lesions induced by radiation — A unified repair model. Radiat. Res. **106**, 252–271 (1986)
15. J.E. Darnell Jr., STATs and Gene Regulation. Science **277**, 1630–1635 (1997)
16. A. Dawson, T. Hillen, Derivation of the tumour control prabability (TCP) from a cell cycle model. Comput. Math. Method. Med. **7**, 121–142 (2006)
17. P. Deheuvels, Estimation non paramétrique de la densité par histogrames généralisés (II). Publications de l'Institut Statistique de l'Université de Paris **22**, 1–23 (1977)
18. L. Devroye, L. Györfi, *Nonparametric Density Estimation: The L_1 View* (Wiley, New York, 1985)
19. L. Devroye, Universal smoothing factor selection in density estimation: theory and practice. Test **6**, 223–320 (1997)
20. O. Diekmann, S. van Gils, S. Verduyn Lunel, H.-O. Walter, *Delay Equations, Functional-, Complex-, and Nonlinear Analysis* (Springer, New York, 1995)
21. S. Donnet, J.-L. Foulley, A. Samson, Bayesian analysis of growth curves using mixed models defined by stochastic differential equations. Biometrics **66**, 733–741 (2010)
22. J. Downward, The ins and outs of signalling. Nature **411**, 759–762 (2001)
23. M. Eisenbach, A. Wolf, M. Welch, S.R. Caplan, I.R. Lapidus, R.M. Macnab, H. Aloni, O. Asher, Pausing, switching and speed fluctuation of the bacterial flagellar motor and their relation to motility and chemotaxis. J. Mol. Biol. **211**, 551–563 (1990)
24. M. Eisenbach, J.W. Lengeler, M. Varon, D. Gutnick, R. Meili, R.A. Firtel, J.E. Segall, G.M. Omann, A. Tamada, F. Murakami, *Chemotaxis* (Imperial College Press, London, 2004)
25. L.E. Elsgol's, S.B. Norkin, *Introduction to the Theory and Application of Differential Equations with Deviating Arguments* (Academic, New York, 1973)

26. R. Erban, H.G. Othmer, From individual to collective behavior in bacterial chemotaxis. SIAM J. Appl. Math. **65**, 361–391 (2004)
27. R. Erban, H.G. Othmer, From signal transduction to spatial pattern formation in E. coli: A paradigm for multiscale modeling in biology. Multiscale Model. Simul. **3**, 362–394 (2005)
28. R. Erban, H.J. Hwang, Global existence results for complex hyperbolic models of bacterial chemotaxis. Discrete Continuous Dyn. Syst. Ser. B **6**, 1239–1260 (2006)
29. B. Firmani, L. Guerri, L. Preziosi Tumor immune system competition with medically induced activation disactivation. Math. Models Methods Appl. Sci. **9**, 491–512 (1999)
30. E. Greenberg, E. Canale-Parola, Chemotaxis in Spirocheta aurantia. J. Bacteriol. **130**, 485–494 (1977)
31. J.K. Hale, S. Verduyn Lunel, *Introduction to Functional Differential Equations* (Springer, New York, 1993)
32. L. Hanin, M. Zaider, Cell-survival probability at large doses: an alternative to the linear quadratic model. Phys. Med. Biol. **55**, 4687–4702 (2010)
33. T. Hillen, Transport equations with resting phases. Eur. J. Appl. Math. **14**, 613–636 (2003)
34. T. Hillen, G. De Vries, J. Gong, C. Finlay, From cell population models to tumor control probability: including cell cycle effects. Acta Oncologica **49**, 1315–1323 (2010)
35. T. Hillen, H.G. Othmer, The diffusion limit of transport equations derived from velocity-jump processes. SIAM J. Appl. Math. **61**, 751–775 (2000)
36. L. Holmström, J. Klemelä, Asymptotic bounds for the expected L^1 error of a multivariate kernel density estimator. J. Multivar. Anal. **42**, 245–266 (1992)
37. A. Jentzen, P.E. Kloeden, in *Taylor Approximations of Stochastic Partial Differential Equations*. CBMS Lecture Series (SIAM, Philadelphia, 2011)
38. J. Kelkel, C. Surulescu, On some models for cancer cell migration through tissue. J. Math. Biosci. Eng. **8**, 575–589 (2011)
39. J. Kelkel, C. Surulescu, A multiscale approach to cancer cell migration through network tissue. Math. Models Methods Appl. Sci. **22**, 1150017.1–1150017.25 (2012)
40. E.F. Keller, L.A. Segel, Model for chemotaxis. J. Theor. Biol. **30**, 225–234 (1971)
41. J.P. Kirkpatrick, J.J. Meyer, L.B. Marks, The linear-quadratic model is inappropriate to model high dose per fraction effects in radiosurgery. Semin. Radiat. Oncol. **18**, 240–243 (2008)
42. P.E. Kloeden, T. Lorenz, Stochastic differential equations with nonlocal sample dependence. Stoch. Anal. Appl. **28**, 937–945 (2010)
43. P.E. Kloeden, T. Lorenz, A Peano-like theorem for stochastic differential equations with nonlocal sample dependence. Stoch. Anal. Appl. **31**, 19–30 (2013)
44. P.E. Kloeden, E. Platen, *Numerical Solution of Stochastic Differential Equations* (Springer, Berlin, 1992)
45. F. Kozusko, Z. Bajzer, Combining Gompertzian growth and cell population dynamics. Math. Biosci. **185**, 153–167 (2003)
46. F. Kozusko, M. Bourdeau, A unified model for sigmoid tumor growth based on cell proliferation and quiescence. Cell Prolif. **40**, 824–834 (2007)
47. Y. Kuang, *Delay Differential Equations with Applications in Population Dynamics* (Academic, Boston, 1993)
48. R. Lapidus, R. Schiller, A mathematical model for bacterial chemotaxis. Biophys. J. **14**, 825–834 (1974)
49. D.F. Lea, *Actions of Radiations on Living Cells* (Cambridge University Press, New York, 1955)
50. T. Lorenz, in *Mutational Analysis: A Joint Framework for Cauchy Problems in and Beyond Vector Spaces*. Lecture Notes in Mathematics, vol. 1996 (Springer, New York, 2010)
51. T. Lorenz, C. Surulescu, *On a class of multiscale cancer cell migration models: Well-posedness in less regular function spaces*, pp. 1–67, TU Kaiserslautern. Preprint (2013) in reviewing at Math. Models Methods Appl. Sci. (submitted)
52. X. Mao, *Stochastic Differential Equations and Applications* (Harwood, Chichester, 1997)
53. G.I. Marchuk, *Mathematical Modeling of Immune Response in Infectious Diseases* (Kluwer, Dordrecht, 1997)

54. J.S. Marron, D. Nolan, *Canonical Kernels for Density Estimation*. Stat. Probab. Lett. **7**, 195–199 (1988)
55. O.V. Matvii, I.M. Cherevko, Approximation of systems with delay and their stability. Nonlinear Oscillations **7**, 207–215 (2004)
56. T.C. Meng, S. Somani, P. Dhar, Modelling and simulation of biological systems with stochasticity. Silico Biol. **4**, 293–309 (2004)
57. J. Murray, *Mathematical Biology I* (Springer, Berlin, 2002)
58. R.M. Nisbet, W.S.C. Gurney, *Modeling Fluctuating Populations* (Wiley, Chichester, 1982)
59. S.F.C. O' Rourke, H. McAneney, T. Hillen, Linear quadratic and tumour control probability modelling in external beam radiotherapy. J. Math. Biol. **58**, 799–817 (2009)
60. B. Øksendal, *Stochastic Differential Equations. An Introduction with Applications* (Springer, New York, 2003)
61. D. Ölz, C. Schmeiser, A. Soreff, Multistep navigation of leukocytes: a stochastic model with memory effects. Math. Med. Biol. **22**, 291–303 (2005)
62. H.G. Othmer, S.R. Dunbar, W. Alt, Models of dispersal in biological systems. J. Math. Biol. **26**, 263–298 (1988)
63. H.G. Othmer, T. Hillen, The diffusion limit of transport equations II: Chemotaxis equations. SIAM J. Appl. Math. **62**, 1222–1250 (2002)
64. A.R. Pagan, A. Ullah, *Nonparametric Econometrics* (Cambridge University Press, Cambridge, 1999)
65. S. Pellegrini, I. Dusanter-Fourt, The structure, regulation and function of the Janus kinase (JAK) and the signal transducers and activators of transcription (STATs). Eur. J. Biochem. **248**, 615–633 (1997)
66. A.D. Polyanin, *Handbook of Linear Partial Differential Equations for Engineers and Scientists* (Chapman & Hall/CRC, London/Boca Raton, 2001)
67. H. Qian, Nonlinear stochastic dynamics of mesoscopic homogeneous biochemical reaction systems — An analytical theory. Nonlinearity **24**, R19–R49 (2011)
68. Yu.M. Repin, On the approximation of systems with delay by ordinary differential equations. Prikl. Mat. Mekh. **29**(2), 226–245 (1965)
69. A.M. Reuther, T.R. Willoughby, P. Kupelian, Toxicity after hypofractionated external beam radiotherapy (70 Gy at 2.5 Gy per fraction) versus standard fractionation radiotherapy (78 Gy at 2 Gy per fraction) for localized prostate cancer. Int. J. Radiat. Oncol. Biol. Phys. **54**, suppl. 1, 187–188 (2002)
70. D.W. Scott, *Multivariate Density Estimation: Theory, Practice and Visualization* (Wiley, New York, 1992)
71. B.W. Silverman, *Density Estimation for Statistics and Data Analysis* (Chapman & Hall, London, 1986)
72. S.K. Srinivasan, R. Vasudevan, *Introduction to Random Differential Equations and Their Applications* (Elsevier, Amsterdam, 1971)
73. C. Surulescu, N. Surulescu, A nonparametric approach to cell dispersal. Int. J. Biomath. Biostat. **1**, 109–128 (2010)
74. C. Surulescu, N. Surulescu, Modeling and simulation of some cell dispersion problems by a nonparametric method. Math. Biosci. Eng. **8**, 263–277 (2011)
75. C. Surulescu, N. Surulescu, On some stochastic differential equation models with applications to biological problems. ECMTB **14**, 106–117 (2011)
76. C. Surulescu, N. Surulescu, *On two approaches to a multiscale model for bacterial chemotaxis*. Preprint, Institute for Numerical and Applied Mathematics, University of Münster (2011). http://wwwmath.uni-muenster.de/num/publications/2011/SS11a/
77. N. Surulescu, *On Some Continuous Time Series Models and Their Use in Financial Economics*, Ph.D. thesis, University of Heidelberg, 2010
78. I. Swameye, T.G. Müller, J. Timmer, O. Sandra, U. Klingmüller, Identification of nucleocytoplasmic cycling as a remote sensor in cellular signaling by databased modeling. PNAS **100**(3), 1028–1033 (2003)

79. H. Takagi, M.J. Sato, T. Yanagida, M. Ueda, Functional analysis of spontaneous cell movement under different physiological conditions. PLoS One **3**(7), e2648 (2008)
80. C.A. Tobias, E.A. Blakeley, F.Q.H. Ngo, T.C.H. Yang, The repair-misrepair model of cell survival, in *Radiation Biology and Cancer Research*, ed. by R.E. Meyn, H.R. Withers (Raven Press, New York, 1980), pp. 195–230
81. J. Touboul, G. Hermann, O. Faugeras, *Noise-induced behaviors in neural mean field dynamics* (2011) [ArXiv: 1104.5425v1]
82. A. Upadhyaya, J.-P. Rieu, J.A. Glazier, Y. Sawada, Anomalous diffusion and non-Gaussian velocity distribution of Hydra cells in cellular aggregates. Phys. A Stat. Mech. Appl. **293**, 549–558 (2001)
83. P. Vieu, Quadratic errors for nonparametric estimates under dependence. J. Multivar. Anal. **39**, 324–347 (1991)
84. M. Zaider, G.N. Minerbo, Tumour control probability: A formulation applicable to any temporal protocol of dose delivery. Phys. Med. Biol. **45**, 279–293 (2000)

Index

P.E. Kloeden and C. Pötzsche (eds.), *Nonautonomous Dynamical Systems in the Life Sciences*, Lecture Notes in Mathematics 2102, DOI 10.1007/978-3-319-03080-7,

LECTURE NOTES IN MATHEMATICS

Edited by J.-M. Morel, B. Teissier; P.K. Maini

Editorial Policy (for Multi-Author Publications: Summer Schools / Intensive Courses)

1. Lecture Notes aim to report new developments in all areas of mathematics and their applications - quickly, informally and at a high level. Mathematical texts analysing new developments in modelling and numerical simulation are welcome. Manuscripts should be reasonably selfcontained and rounded off. Thus they may, and often will, present not only results of the author but also related work by other people. They should provide sufficient motivation, examples and applications. There should also be an introduction making the text comprehensible to a wider audience. This clearly distinguishes Lecture Notes from journal articles or technical reports which normally are very concise. Articles intended for a journal but too long to be accepted by most journals, usually do not have this "lecture notes" character.

2. In general SUMMER SCHOOLS and other similar INTENSIVE COURSES are held to present mathematical topics that are close to the frontiers of recent research to an audience at the beginning or intermediate graduate level, who may want to continue with this area of work, for a thesis or later. This makes demands on the didactic aspects of the presentation. Because the subjects of such schools are advanced, there often exists no textbook, and so ideally, the publication resulting from such a school could be a first approximation to such a textbook. Usually several authors are involved in the writing, so it is not always simple to obtain a unified approach to the presentation.
   For prospective publication in LNM, the resulting manuscript should not be just a collection of course notes, each of which has been developed by an individual author with little or no coordination with the others, and with little or no common concept. The subject matter should dictate the structure of the book, and the authorship of each part or chapter should take secondary importance. Of course the choice of authors is crucial to the quality of the material at the school and in the book, and the intention here is not to belittle their impact, but simply to say that the book should be planned to be written by these authors jointly, and not just assembled as a result of what these authors happen to submit.
   This represents considerable preparatory work (as it is imperative to ensure that the authors know these criteria before they invest work on a manuscript), and also considerable editing work afterwards, to get the book into final shape. Still it is the form that holds the most promise of a successful book that will be used by its intended audience, rather than yet another volume of proceedings for the library shelf.

3. Manuscripts should be submitted either online at www.editorialmanager.com/lnm/ to Springer's mathematics editorial, or to one of the series editors. Volume editors are expected to arrange for the refereeing, to the usual scientific standards, of the individual contributions. If the resulting reports can be forwarded to us (series editors or Springer) this is very helpful. If no reports are forwarded or if other questions remain unclear in respect of homogeneity etc, the series editors may wish to consult external referees for an overall evaluation of the volume. A final decision to publish can be made only on the basis of the complete manuscript; however a preliminary decision can be based on a pre-final or incomplete manuscript. The strict minimum amount of material that will be considered should include a detailed outline describing the planned contents of each chapter.
   Volume editors and authors should be aware that incomplete or insufficiently close to final manuscripts almost always result in longer evaluation times. They should also be aware that parallel submission of their manuscript to another publisher while under consideration for LNM will in general lead to immediate rejection.

4. Manuscripts should in general be submitted in English. Final manuscripts should contain at least 100 pages of mathematical text and should always include
   - a general table of contents;
   - an informative introduction, with adequate motivation and perhaps some historical remarks: it should be accessible to a reader not intimately familiar with the topic treated;
   - a global subject index: as a rule this is genuinely helpful for the reader.

   Lecture Notes volumes are, as a rule, printed digitally from the authors' files. We strongly recommend that all contributions in a volume be written in the same LaTeX version, preferably LaTeX2e. To ensure best results, authors are asked to use the LaTeX2e style files available from Springer's web-server at
   ftp://ftp.springer.de/pub/tex/latex/svmonot1/ (for monographs) and
   ftp://ftp.springer.de/pub/tex/latex/svmultt1/ (for summer schools/tutorials).
   Additional technical instructions, if necessary, are available on request from:
   lnm@springer.com.

5. Careful preparation of the manuscripts will help keep production time short besides ensuring satisfactory appearance of the finished book in print and online. After acceptance of the manuscript authors will be asked to prepare the final LaTeX source files and also the corresponding dvi-, pdf- or zipped ps-file. The LaTeX source files are essential for producing the full-text online version of the book. For the existing online volumes of LNM see:
   http://www.springerlink.com/openurl.asp?genre=journal&issn=0075-8434.
   The actual production of a Lecture Notes volume takes approximately 12 weeks.

6. Volume editors receive a total of 50 free copies of their volume to be shared with the authors, but no royalties. They and the authors are entitled to a discount of 33.3 % on the price of Springer books purchased for their personal use, if ordering directly from Springer.

7. Commitment to publish is made by letter of intent rather than by signing a formal contract. Springer-Verlag secures the copyright for each volume. Authors are free to reuse material contained in their LNM volumes in later publications: a brief written (or e-mail) request for formal permission is sufficient.

Addresses:

Professor J.-M. Morel, CMLA,
École Normale Supérieure de Cachan,
61 Avenue du Président Wilson, 94235 Cachan Cedex, France
E-mail: morel@cmla.ens-cachan.fr

Professor B. Teissier, Institut Mathématique de Jussieu,
UMR 7586 du CNRS, Équipe "Géométrie et Dynamique",
175 rue du Chevaleret,
75013 Paris, France
E-mail: teissier@math.jussieu.fr

For the "Mathematical Biosciences Subseries" of LNM:

Professor P. K. Maini, Center for Mathematical Biology,
Mathematical Institute, 24-29 St Giles,
Oxford OX1 3LP, UK
E-mail: maini@maths.ox.ac.uk

Springer, Mathematics Editorial I,
Tiergartenstr. 17,
69121 Heidelberg, Germany,
Tel.: +49 (6221) 4876-8259
Fax: +49 (6221) 4876-8259
E-mail: lnm@springer.com